Seeing the Forest for the Trees

For centuries, people have understood that forests, and our utilization of them, influence the climate. With modern environmental concerns, there is now scientific, governmental, and popular interest in planting trees for climate protection. This book examines the historical origins of the idea that forests influence climate, the bitter controversy that ended the science, and its modern rebirth. Spanning the 1500s to the present, it provides a broad perspective across the physical and biological sciences, as well as the humanities, to explain the many ways forests influence climate. It describes their use in climate-smart forestry and as a natural climate solution and demonstrates that in the forest–climate question, human and sylvan fates are linked. Accessibly written with minimal mathematics, it is ideal for students in environmental and related sciences, as well as for anyone with an interest in understanding the environmental workings of forests and their interactions with climate.

Gordon Bonan is senior scientist at the National Center for Atmospheric Research. He is the author of *Ecological Climatology* (Cambridge University Press, 2015), *Climate Change and Terrestrial Ecosystem Modeling* (Cambridge University Press, 2019), and numerous publications on terrestrial ecosystems and climate. He is a Fellow of the American Geophysical Union, the American Meteorological Society, and the Ecological Society of America.

'Gordon Bonan is one of the world's leading experts on the carbon, water and energy dynamics of forests, and their influence on the Earth system. In Seeing the Forests for the Trees, he combines that scientific expertise with a deep understanding of how forests have influenced art and literature, as well as patterns of human settlement and land use. Bonan's understanding of the history of the forest-climate controversy (do forests affect climate, and for good or ill?) is encyclopedic, and in the first part of this book he tells the story in wonderful detail. This is followed by a clear and engaging description of how that controversy has been resolved through modern research, and an accessible telling of how forests actually function, from microclimates to the global carbon cycle. Case studies of climate sensitive regions and the potential for climate-smart forests bring the knowledge presented throughout the book to bear on important questions we face about conserving and managing these magnificent ecosystems.'

John Aber, University Professor emeritus at the University of New Hampshire and author of Less Heat More Light: A Guided Tour of Weather, Climate and Climate Change

'This is the third book in Gordon Bonan's exceptional series focussing on ecology and terrestrial ecosystems. It is the most accessible for a broader audience and will excite and intrigue readers from earth systems, ecology, environmental science and elsewhere. Even the expert will find a depth of history, and explanations of how our science fragmented and was renewed to become part of the solution to climate change.'

Andy Pitman, University of New South Wales

'Another must have book by Professor Gordan Bonan! Bonan takes a thoughtful, detailed and novel approach from both a historical and interdisciplinary scientific lens to examine how forests influence climate. This book will appeal to a range of audiences from detailed practitioners within the field to an interested undergrad!'

Christiane Runyan, Johns Hopkins University

Seeing the Forest for the Trees

Forests, Climate Change, and Our Future

Gordon Bonan

National Center for Atmospheric Research (NCAR)

CAMBRIDGE
UNIVERSITY PRESS

Shaftesbury Road, Cambridge CB2 8EA, United Kingdom

One Liberty Plaza, 20th Floor, New York, NY 10006, USA

477 Williamstown Road, Port Melbourne, VIC 3207, Australia

314–321, 3rd Floor, Plot 3, Splendor Forum, Jasola District Centre,
New Delhi – 110025, India

103 Penang Road, #05–06/07, Visioncrest Commercial, Singapore 238467

Cambridge University Press is part of Cambridge University Press & Assessment,
a department of the University of Cambridge.

We share the University's mission to contribute to society through the pursuit of
education, learning and research at the highest international levels of excellence.

www.cambridge.org
Information on this title: www.cambridge.org/9781108487528

DOI: 10.1017/9781108601559

First published 2023

A catalogue record for this publication is available from the British Library.

ISBN 978-1-108-48752-8 Hardback
ISBN 978-1-108-46784-1 Paperback

To old friends not forgotten and new friends found

Contents

Preface

The origins of this book lie in two separate events; little did I know the profound influence they would have on my scientific journey. In 1997, I published a study of the effects of deforestation on the climate of the United States. Many other studies had examined the climate effects of tropical deforestation, but temperate forests, many of which had been cleared over the past several centuries, were largely unstudied. That paper opened up to me a vast literature of historical writings on forests and climate change. What I had thought was a novel analysis of the climate influences of forests was instead but another paper in a centuries-long topic of study. I am indebted to the many historians of science who have written on this subject. Their works provided access to the rich primary literature of a controversy forgotten by modern-day climate scientists, but which profoundly shaped our study of nature. I am indebted, too, to the many organizations that digitized and made accessible the trove of books, papers, and documents from long ago, and also to the interlibrary loan department at the National Center for Atmospheric Research for assistance in finding the more obscure works.

At the same time, I had begun to consider how to convey the interdisciplinary aspect of the science – part ecology, part atmospheric science. The second pivotal event, then, was the publication of *Ecological Climatology: Concepts and Applications* (1st ed., 2002). *Ecological Climatology* describes why terrestrial ecosystems are an essential component of climate science; explains the physical, chemical, and biological processes by which ecosystems affect climate; and discusses the commonality between climate science and ecology. A subsequent book, *Climate Change and Terrestrial Ecosystem Modeling* (2019), describes how to mathematically represent terrestrial ecosystems in numerical models of climate. Both books are rich in complexity and nuances and both delve deeply into the details of the science and the models.

Seeing the Forest for the Trees: Forests, Climate Change, and Our Future is a companion to those two works. Matt Lloyd at Cambridge University Press, who has supported my books over the years and shepherded them into production, and I had been discussing for some time the need for a primer to complement the other books. *Seeing the Forest for the Trees* is that primer, told in the context of forests and climate. In this, I am indebted also to Sarah Lambert of Cambridge University Press for her patience with the project. *Seeing the Forest for the Trees* combines the historical controversy over forests and climate change, spanning several centuries of debate, with our modern understanding of climate science. There is growing scientific and public interest in the use of forests to mitigate anthropogenic climate change – a nature-based solution to solve a human-made problem. The notion of preserving forests or planting trees to improve climate is not new. It is an age-old idea, reborn with new concern about climate change. The subject brings together concepts of climate science, ecology, human impacts on climate, and more broadly our relationship with forests. There is also a need to consider forests beyond the lens of science; to consider them not just as a public utility that provides climate services but also from a humanities vantage point. The rationalism of science must be balanced with the romanticism of forests. This book tells the story of forests and climate change from multiple points of view, interwoven with central tenets of forest ecology and climate science and with an understanding of past controversies. Complexity is reduced and the science is distilled down to its essence so that the larger message of forest influences on climate emerges while also broadening our perception of forests and the natural world. The book allows us to see the forest for the trees.

Part I

Historical Perspective

Nature ties, by secret links, the destiny of mortals to that of forests

Alexandre Moreau de Jonnès (1825)

1　The Forest–Climate Question

November 10, 1888, was a fine autumn day before the dreary span of winter in Washington, DC. The afternoon was warm, one of several days with pleasant temperatures, and there was only a trace amount of rain.[1] The agreeable weather likely lifted the spirits of Washingtonians, and as night fell and the sky darkened and the temperature dropped, there was an air of excitement and anticipation among leading intellectuals in the nation's capital. Nearly 300 years of European settlement had greatly altered the landscape of the United States.[2] The Virginia and New England countryside found by the first English colonists contained extensive old-growth forest. Thereafter, the landscape was cleared of forests, felled for their wood products and replaced by homesteads, towns, and farmland. The vast virgin forests of New England – towering white pines, giants of the forest, famed for the masts they had provided the navy; mixtures of beech, maple, and birch along with hemlock and spruce in northern states; oaks and hickories in southern New England – had long since been felled.[3] Similar widespread clearing had altered the forests of Virginia and other states along the eastern seaboard, and the same pattern of forest clearing was repeated in the Midwest and Great Lakes regions with the westward expansion of the population.[4] A popular belief at the time was that deforestation was decreasing rainfall. Conservationists advocated this viewpoint in their desire to protect forests from further destruction. Land developers picked up the theme and promoted planting trees in the arid western portions of the country to increase rainfall, and the US Congress enacted just such legislation. The emerging science of meteorology provided a counternarrative. As weather observations became more widespread throughout the country and extended over longer periods of time, meteorologists could find no evidence for a changing climate. In this backdrop, fifty scholars, politicians, and concerned citizens – members of the Philosophical Society of Washington and their guests, the largest assembly in recent weeks – gathered at the Cosmos Club in northwest Washington to hear a lively debate on the subject.[5]

The Society was founded in 1871 as a forum to discuss relevant scientific topics of the day. At the previous meeting two weeks earlier, Bernhard Fernow, society member and chief of forestry for the federal Department of Agriculture, had lectured on forest–rainfall influences, arguing that forests do indeed increase rainfall.[6] Fernow was an ardent forest conservationist. He believed that "no other people of the earth have consumed virgin forests as lavishly as have the people of the United States."[7] The challenge of the forester was to balance conflicting uses of forests, both private and communal interests, to provide forest products while preserving "the forest conditions which are favorable to climate and waterflow." Fernow's presentation had struck a nerve among society members, but there was little time for in-depth discussion. The Society organized a special symposium for its 324th meeting on November 10 to consider the question, "Do forests influence rainfall?"

Fernow returned and faced a withering rebuttal from fellow society member Henry Gannett, chief geographer for the US Geological Survey. That same year, Gannett had published a study that refuted the premise that forests affect rainfall.[8] Now, before society members and Fernow, he denounced Fernow's reasoning as "a case of theory run riot."[6] Although Gannett's previous article was "sufficiently clear for the average reader," Fernow had "misunderstood" the analysis. Gannett was obliged to "state it once more" and "to state this plainly" to ensure no further confusion: "Forests exercise no influence whatever upon rainfall." Gannett closed his lecture by asking, "Is it worthwhile to go on planting trees for their climatic effects?" His own reply affirmed the "uselessness of it," and he concluded with an admonishment not to cultivate trees in the western lands in an effort to increase rainfall. Rather, because forests evaporate considerable quantities of rainwater that otherwise might be harnessed for irrigation, Gannett advised that the trees be cleared "in order that as much of the precipitation as possible may be collected in the streams."

The proceedings of the meeting were duly reported in the scholarly journal *Science*, where the editors commented on the unseemliness of the forest–rainfall topic with the view that "much of the discussion of it, unfortunately, has not been of a purely scientific character."[9] The premise that forests increase rainfall, as expressed in the editorial, was not being studied for the betterment of society but rather was being advocated for financial or personal gain at the expense of truth and public good. None of this satisfied the appetite of society members for a learned resolution, and the arguing carried over for a third meeting. Fernow and Gannett were joined by society members and meteorologists Cleveland Abbe and Henry Hazen from the federal Weather Bureau.[10] Abbe was the first chief meteorologist for the government and laid the foundations for modern weather prediction, for which he is recognized today by the American Meteorological Society in its distinguished service award.[11] In his lecture, Abbe cautioned the audience on ascribing undue influences to forests given the difficulty of accurately measuring rainfall.[12] He subsequently published a sharp critique of the forest–rainfall theory, which was not supported by "rational climatology."[13]

The notion that forests influence the climate of a region had captured the imagination of scholars and the public alike since Christopher Columbus encountered the new lands. It swelled as the first European settlers reached the shores of America and found a landscape of boundless forests with a climate markedly different from that of their homelands. The debate coursed throughout the first 300 years of the nation, first with a confident belief that clearing forests was improving climate, only to be followed with concern for a reduction in rainfall upon loss of forests, and then an optimistic understanding that planting trees would increase rainfall. With time, the rhetoric became even more personal and still more vitriolic than expressed in the autumn of 1888. European states, too, looked inward at the loss of forests across their own lands brought on by centuries of inhabitation and likewise asked if the climate was changing. Similar concern reached across the world to India, Australia, New Zealand, and Africa with the spread of European empires. But on those November evenings in Washington, DC, the die was cast: the study of climate was the domain of physical meteorologists. There was no room for foresters, and there was without doubt no influence of forests on climate.

Except this was wrong. The science was not settled. It was only in hiatus, suppressed during the birth of meteorology as a scientific endeavor by rigid opposition from those adamantly seeking a geophysical explanation for climate. Modern climate science developed over the last few decades shows that the views expressed by Gannett, Abbe, and others were mistaken. Climate is not solely the study of atmospheric physics and fluid dynamics. It requires knowledge of the ecology, physiology, and biogeochemistry of terrestrial ecosystems and the many physical, chemical, and biological processes by which not just forests but also grasslands, croplands, and other ecosystems affect climate. In an era in which humans are now the dominant climate-altering factor on the planet, the way in which we treat forests – felling or planting trees, managing their growth – is one way in which we change climate. This time, however, it was atmospheric scientists voicing the alarm at the end of the twentieth century.

The rising concentration of carbon dioxide (CO_2) in the atmosphere had sparked concern about climate change. Climate science emerged as a facet of atmospheric science, and new theories and mathematical models of atmospheric circulation and global climate were needed to understand how human activities alter climate. Some visionary scientists, in building their models, rediscovered that forests influence climate. These scientists found that they had to include mathematical equations for plants and their roots, leaves, and stomata in their models. They discovered that the characteristics of forest canopies – the amount of solar radiation absorbed and reflected by leaves; the height of the canopy; the openness of stomata; the depth of roots – affect the accuracy of the models. Atmospheric scientists, in developing their theories and applying their models, rediscovered that forests mattered, and a few prescient ones, seeing the destruction of the Amazonian rain forest, asked once more if deforestation could be changing climate.

And so, the debate about forests and climate continues today, but now in the context of human-caused climate change. Trees remove CO_2 from the atmosphere during their growth and store the carbon molecules in biomass. As leaves, branches, and other organic debris fall to the ground and decay, carbon accumulates in the soil. For this reason, ecologists advocate forest protection and tree planting as a nature-based solution to mitigate planetary warming by removing CO_2 from the atmosphere. Indeed, the socioeconomic pathways devised by economists and policymakers to limit planetary warming below critical thresholds require sizable storage of carbon in forests to counter anthropogenic emissions. Because of their large carbon storage potential, stopping the clearing and degradation of tropical rain forests, preserving and restoring forests in other regions, and planting new forests rank high in public imagination as solutions to fight climate change.[14]

Yet carbon storage is but one way in which forests influence climate. As the developers of climate models discovered, forests affect temperature, precipitation, and atmospheric circulation through a variety of meteorological processes, sometimes in ways that counter the carbon benefits of forests. Climate scientists speak of the absorption, reflection, and emission of solar radiation; evaporation of water from soil and transpiration by plants; the turbulent exchange of heat with the atmosphere; momentum absorption by plant canopies; and the production of biogenic aerosols – climate processes that differ among forests, grasslands, and other types of vegetation. Climate scientists use their mathematical models of Earth's climate to study these processes, the climate consequences of deforestation, and the potential for forests to lessen planetary warming. Some scientists use satellites orbiting the planet to ascertain the temperature difference between forest and non-forest land, and still others contrast meteorological measurements made at forests with those at open fields to discern climate differences. Hydrologists, too, study deforestation and forest management, but in the context of water supply to towns and cities. That forests determine climate at regional, continental, and global scales is a profound change in our understanding of climate.

Two images taken from space illustrate starkly different perspectives of the science of our planet (Figure 1.1).[15] The blue marble view of Earth – the famed full-color photograph taken on December 7, 1972, by Apollo 17 astronauts looking homeward while speeding to the moon and the first to show Earth as an entire disk – emphasizes white clouds swirling over blue oceans. The image shows the African continent with the Atlantic Ocean to the west and the Indian Ocean to the east. Antarctica is a white mass at the bottom of the photograph. North Africa and the Arabian Peninsula are visible at the top of the image as

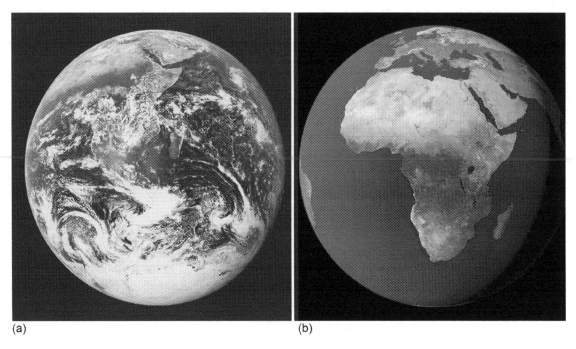

(a)   (b)

Figure 1.1 Two spaceborne images of Earth as a "blue marble" (a) and as an "emerald planet" (b).
Panel a is the iconic "blue marble" photo taken from Apollo 17 and represents the geophysical view of the atmosphere in motion. The original is a full-color image of blue oceans, white clouds, and tannish-brown land. Panel b is a "Blue Marble: Next Generation" image showing Africa and portions of Europe. This "emerald planet" view of Earth emphasizes the biosphere. The original NASA image, courtesy of Reto Stöckli and Robert Simmon, is colorized to show blue oceans, green forest vegetation, and tannish-brown grasslands and open lands.

tannish-brown land devoid of vegetation. Clouds obscure much of the lush, green tropical forests of central Africa. The blue marble embodies the geophysical perspective in which climate is understood as fluid air and water in motion. Newer satellite technologies recreate the imagery, but peer beneath the clouds to see the forests. These satellite images depict extensive green continents plush with vegetation. In the right panel of Figure 1.1, the tropical forests of central Africa emerge as a wide swath of dark gray extending across the continent. The full-color image presents Earth not as a blue marble but as an emerald planet. It is a biological view of Earth – of life, and the trees and plants that inhabit the world.

Both perspectives are correct in their own context, but a complete understanding of our planet requires merging the geophysical understanding of climate with the biological understanding of ecosystems. This blended viewpoint recognizes the deep interconnections between atmospheric science and ecology. It is a perspective that has gained prominence in the last few decades as scientists seeking to understand how the world is changing have adopted a holistic view of Earth as a system and the interconnectedness of the physical and living realms. It is a view that some earlier scholars, before the specialization that character-izes science today, embraced in their studies of forests and climate, and it is this interdisciplinary view of climate that meteorologists so resoundingly rejected at the end of the nineteenth century. How has the notion that forests influence climate and that management of forests can improve climate, which was so readily advocated and then so vehemently dismissed as fanciful, reemerged in the twenty-first century?

And what does the narrative of the forest–climate controversy tell us as we embark on the new paradigm of Earth system science and planetary stewardship? Like forests themselves, there are many ways to see the story line.

The first central element is the telling of the narrative itself. Historians of science have rebuked the men who argued for an influence of forests on climate, both in the belief that climate was changing and for linking the supposed change to the destruction of forests.[16] The intellectual debate about forests and climate played out on a broader stage of European colonialism and the accompanying settling of newfound islands and continental lands. In the narrative of historians, deforestation was first seen as a means to improve the different, and therefore unhealthy, climates of faraway lands compared with the more favorable climate of Europe. When deforestation was recognized for its harmful effects, reforestation and forest conservation became the new cry. In this sense, the intellectual, governmental, and popular interest in forests and climate was an effort to promote colonization and settlement by improving the climate while also endorsing the superiority of Europeans, not the indigenous populace, in their knowledge of how to best manage the lands.[17] Concern over forests and climate change must also be understood in terms of conflict over forest regulation and state versus individual rights to manage forests. Individuals advocating a national policy to manage forests for climate benefits clashed with equally spirited individuals asserting private ownership of forests and denying any influence of forests on climate.[18] The emerging science of meteorology, too, looms over the historical narrative.[19] The early arguments for a changing climate were largely anecdotal, lacking in observation. Meteorologists, as they measured temperature and precipitation and advanced their understanding of climate, provided the missing quantitative and predictive rigor. In this story line, it was the critical skepticism of meteorologists that righted the duplicity of forest advocates.

The story of forests and climate, thus, is mostly told in a historical narrative as being scientifically wrong; advanced for political, economic, or cultural reasons; or perhaps a necessary instruction as we once more confront the intertwined science and politics of human-caused climate change. One writer called the reasoning "occasionally ingenious and often devious."[20] The heated debate during the years 1877–1912 over forest influences on rainfall has been described as a "period of propaganda," a view that seeped into subsequent historiographies.[21] The origins of modern-day environmentalism can be found in the forest–climate debate, but the actors in that history have been described as "propagandists" or as "propagandizing."[22] A review of the controversy in America noted "its false assumptions," in which warnings of a changing climate "formed effective propaganda" and were "secure from speedy refutation."[23] To a forest historian, foresters chased "this will-o'-the-wisp" to boost their standing as a profession.[24] A writer of historical ecology described the subject as "the realm of armchair speculation" in which proponents "managed to construct grandiose theories on the basis of a few known facts."[25] These criticisms predate our current understanding of forest influences on climate, but recent scholars still cite the period as "a checkered history that should elicit caution."[26] It was "a few bold scientists" who countered widespread claims about the effects of deforestation in the Hapsburg Empire during the late nineteenth century.[27] A review of the forest–rainfall controversy in nineteenth-century France, told through the perspective of a hydrologist, characterized the debate in terms of romantic and emotional foresters on one side and scientifically rigorous engineers on the other.[28]

What has been called the forest–climate question or debate or controversy has been told as a cautionary tale, without doubt a politicization of science that used exaggerated threats to influence governmental policy and even a xenophobic expression of European supremacy, but it has not been told from a modern scientific perspective. When interpreted in that light, the proponents of forest influences on climate, while not always correct, were also not necessarily wrong. Many of the scientific questions posed in today's research on forests and climate change have their origins in the forest–climate question. In focusing on the rhetorical excesses of the controversy, historians have been blinded to the emerging new science of what was then being called forest meteorology.

The second element of the story line, therefore, is to see the narrative not as an unscientific misstep, but rather as the foundation for the interdisciplinary study of Earth as a system. The modern paradigm of Earth system science presents the world as an interconnected system, not only the atmosphere and land with its ecosystems, watersheds, and soils, but also the oceans, sea ice, glaciers, and the people that inhabit the planet. Earth system science has been portrayed as new, emerging only in the last few decades, and as providing the scientific basis to responsibly manage the planet in the era of the Anthropocene.[29] The multi-century controversy over deforestation and climate change confounds that view. It presents an alternative narrative in which inadvertent, or even purposeful, modification of climate is longstanding, but by felling or planting trees.[30] Further to this, the central tenet of Earth system science – the interconnection between the biosphere and atmosphere – is, in fact, a centuries-old idea, first conceived in the long-held belief that forests influence climate and doomed to fail by the disciplinary specialization of the sciences. Narrow-mindedness prevented a vision of the world as an interconnected system.

The legacy of that failing lives on today, seen in the different scientific tools and research methods by which not only ecologists and climate scientists, but also hydrologists study forests and climate change. Atmospheric scientists, at the close of the nineteenth century, narrowly defined climate science to the exclusion of ecology, and now, in partnership with oceanographers and other physical geoscientists building the new generation of Earth system models of climate, are once again defining the relevance of ecology for the Earth system. If the rebirth of Earth system science is to be successful and not collapse under the burden of interdisciplinarity, we must understand the forest–climate controversy. *Seeing the Forest for the Trees* is the necessary story of the history and science of forest–climate influences and how the disciplinary geophysical perspective of climate became the interdisciplinary study of Earth as a system.

Thirdly, as the forest–climate controversy so distinctly evinces, the narrative of forests and climate must be presented through the lens of a robust interdisciplinary viewpoint, one deeper than can be attained through the standpoint of a single field of study. Many academic fields are engaged in the science, but the methods used, the questions asked, and even the language of science vary across disciplines. The various scientific communities too often talk across academic disciplines rather than exchange knowledge among disciplines. When faced with a complex, all-consuming problem, it is too easy to fall back on the narrowness of academic specialization and the comfort found in knowledge and scholarship in a limited, disciplinary context. Yet narrowness often brings hubris. The science of forests and climate change requires broad interdisciplinarity; not the self-assurance of narrow expertise, but rather the humility fostered by multiple, and often contrasting, viewpoints. If not, if proffered through traditional disciplinary study, we are doomed to repeat the hubris and disciplinary chauvinism voiced by Gannett, Abbe, and others. *Seeing the Forest for the Trees* takes a broad approach to forests and climate by integrating perspectives across the physical and biological sciences and between the humanities and sciences. We have still not broken through the disciplinary barriers that doomed the first birth of the science or that have shaped its historical telling. We need improved communication across disciplines if we are to better advocate for forests in a changing climate.

The fourth element of the narrative is hubris. Just as in the past, today's advocacy of forests as a nature-based solution to the perils of climate change – forests for climate protection – can be overheated. The science, as before, is uncertain and the evidence conflicting. There is no one common voice with which to speak of forests and climate. One climate scientist in the historical debate, presenting a review of the differing accounts on forest–climate influences at the end of the nineteenth century, drew on the tale of Ariadne from Greek mythology, in which Ariadne gave her lover Theseus a clew (a ball of thread) to guide him out of the Minotaur's labyrinth.[31] Today's scientists are still seeking the clew to solve the forest–climate puzzle. The influence of forests on climate has been argued for more than five centuries without resolution. Some readers may be disappointed to find that this book, too, provides no unqualified verdict. Others, however, may celebrate the mysteries and complexities of nature and reflect upon humankind's connection to forests, marveling in the question posed by Alexandre Moreau de Jonnès in 1825: does nature, with its intricate interdependencies, tie the fate of humanity to that of forests?[32]

The fifth element to the forest–climate controversy, then, is the story of our relationship with nature, our stewardship of the planet, and our hubris in engineering nature for a better climate. It is a belief in human superiority over nature and that we control climate, first by deforesting the land and then by replanting it. In rejecting the science at the end of the nineteenth century, humankind's ability to change climate was also rejected and a decades-long dissonance between human actions and climate ensued throughout the twentieth century.[33] Now, we once again talk of purposely geoengineering climate. Yet deep in the background, hidden behind the anthropocentric view that forests are a means to control climate, is the majesty, mystery, and wonder of the natural world. Nature continues to amaze us, and there are many things about the functioning of planet Earth that we still do not understand, either scientifically or morally. Yet for all the secrets still to be discovered and the morality to be reasoned, one truth is evident: Peoples across the ages and in all regions of the world have voiced a close connection between forests and climate and a knowledge that trees influence the climate around homes, in towns, and wherever there are forests. This is a recurring theme throughout history and speaks to our sense of place in the world. The narrative of deforestation traces across the history of humankind, and the controversy about forests and climate is one chapter in that narrative. It is an expression of the indelible way in which forests have shaped humanity. The forest–climate controversy is the history of an idea: that our past has been intertwined with forests, and our future is too.

The sixth element to the narrative is that if we see forests only in the context of climate services, we miss the larger meaning of trees in our lives. Humankind has had a long and complex relationship with forests. Forests have inspired artists and writers who have captured their wonder in paintings and in prose. Their preservation has been advocated by conservationists who have touted the environmental, recreational, and aesthetic benefits of forests; by medical experts for physical and mental health; and forests have garnered the attention of scientists seeking to understand their inner workings. Forest timber has fueled the economic growth of countries throughout the world, and their mysteries and wilderness have fed the dark imaginations of people. Societal views of forests have evolved from foul and sinister – to be conquered – and a source of timber – to be exploited for wealth – to advocacy of the many societal benefits, including climate stabilization; but forests must be seen as more than a public utility. The climate benefit of forests is an anthropocentric reading of nature – that nature exists to serve us. It is utilitarian in its rationale, in which society values the goods and services provided by forests. But what if the climate benefit is not so great after all? What if some forests provide benefits while others do not? And what if the science is not exact enough to provide an answer with a high level of confidence? These very questions span the 400-year history of the forest–climate question from 1500 to 1900, and they are still relevant today. There is no one simple story to tell. The science and advocacy of forests for their climate services, like before, is contentious. The science is complex and messy, and our relationship with forests is confounded. We need to effectively cut the Gordian knot presented by that messiness to voice a coherent meaning; we have to see the forest for the trees.

Seeing the Forest for the Trees presents a broad interdisciplinary perspective of forests, climate change, and our future. Chapters 2–6 examine the forest–climate question throughout history and the conflict between meteorologists and forest conservationists in the worldwide debate over forest influences on temperature, rainfall, and streamflow. Chapter 2 focuses on temperature and the belief that deforestation was improving the cold winters found in America. Chapter 3 shifts the focus to water – the rain that falls from the sky and the water conveyed in streams – and the advocacy for planting trees to increase the supply of water. Chapter 4 traces the spread of planting trees for rain throughout the world. Chapters 5 and 6 examine the convergence of science and public policy as the science of forest meteorology was formalized (Chapter 5) and as meteorologists forcibly pushed back (Chapter 6). As the debate reached its zenith, the naysayers prevailed and forest influences on climate faded from consideration. Chapter 7 looks at the history of forests in environmental thought and how trees and forests are portrayed in the humanities. Forests, after all, are much more than climate regulators. Can romanticism blend with science to answer the forest–climate question?

The remaining chapters present the science: How do we know in what way forests affect climate, and what is the scientific basis for forest–climate influences? Chapter 8 offers a primer on global physical climatology. This is the traditional geophysical view of climate. In addition, there is the biological understanding of climate in which forests and other ecosystems influence climate at local, regional, and global scales. Chapter 9 on forest biometeorology presents the processes by which forests interact with the atmosphere. Chapter 10 considers the scientific tools used to study forest biometeorology. Then, Chapters 11–15 show the influences of forests on climate through various physical, chemical, and biological processes. Chapter 11 examines forest influences at the local (micro-)scale of forest microclimates. Subsequent chapters look at the larger (macro-)scale. Chapter 12 elaborates on the hydrologic cycle to understand how forests affect water availability. Chapter 13 delves into the carbon cycle and the role of forests to remove CO_2 from the atmosphere. Chapter 14 examines forest influences at the macroscale in various regions throughout the world, and Chapter 15 provides specific case studies. The final chapters consider climate-smart forests and how forests of the future can be managed for their climate benefits. Chapter 16 defines what is meant by a climate-smart forest. Chapter 17 examines the processes controlling forest biogeography and the stresses forests face in the coming years. Chapter 18 is a concluding essay that ties together the various themes and provides perspective.

The language of today's science was not available to scholars of the nineteenth century and before, and so the forest–climate history can be foreign to modern sensibilities. Before the advent of academic specialization, indeed before the designation "scientist" was crafted in the early nineteenth century, scholars of nature studied natural history, which was defined as "a description of any of the natural products of the earth, water or air."[34] From this general usage emerged the broad classification of science into the modern branches of physics, chemistry, and biology. Meteorology was developing as a science in the nineteenth century, and climatology was used specifically to mean the study of climate. Actors in the debate can be identified as meteorologists or climatologists. With the close connection between physics and meteorology, many were physicists. Today's academia embraces the broader terminology of atmospheric science, and climate science is preferred to climatology.

Botany was established as a field of study, and forestry was establishing its scientific basis. "Ecology" was not coined as a word until 1866, when Ernst Haeckel defined it as the science of the relations of organisms to the surrounding environment, both physical and biological.[35] Instead of ecologists as they exist today, there were botanists, foresters, and horticulturalists. "Ecosystem" became a concept only in 1935, and it took many years to define what an ecosystem is and how to study it.[36] The general usage today embodies the interrelationship between the physical and biological environments. A terrestrial ecosystem combines living organisms and their physical environment into a functional system linked through biological, chemical, and physical processes. Closely associated with ecosystem is the term biosphere, which in fact precedes ecosystem in the lexicon of ecology. Eduard Suess introduced the term in 1875 to designate the living component of Earth, one of the enveloping spheres, along with the atmosphere, hydrosphere, and lithosphere, and Vladimir Vernadsky, in 1926, provided the modern conceptualization of the biosphere in terms of energetics and biogeochemical cycles.[37] Today, biosphere is used interchangeably with ecosystem, or to also mean specifically the sum of all ecosystems on the planet. None of this semantics, however, detracts from a general understanding of the forest–climate controversy as a deep splintering in the science of the environment that can be seen in today's disciplinary study of atmospheric science and ecology.

2   Tempering the Climate, c. 1600–1840

By the time that Bernhard Fernow, Henry Gannett, Cleveland Abbe, and others debated forest influences on rainfall in November 1888, there was a 2,000-year-old tradition that forests affect climate. The Greek philosopher Theophrastus (c. fourth century BCE) wrote of the effects of marsh draining and deforestation in *De causis plantarum*.[1] He described how large bodies of standing water moderate climate; draining marshes to create farmland had made the countryside colder. Elsewhere, clearing forested land had warmed the climate because opening the woods allows sunlight to heat the ground. Later, the Roman architect and engineer Vitruvius, in *De architectura libri decem* (c. first century BCE), observed that the best spring water is found in mountainous regions, where thick forests prevent drying by the sun.[2] The shade of trees, furthermore, protects snow from melting rapidly. Similar spring water, Vitruvius explained, is not found in flatlands without tree cover. The Roman scholar Pliny the Elder (23–79 CE), too, wrote of the effects of deforestation on springs and streamflow.[3]

These works are part of a recognition by Greek and Roman intellects that humans not only exist within nature, but also modify nature – and specifically, through deforestation, climate.[4] It is a theme that continued through subsequent centuries. In *De vegetabilibus*, part of his writing on the transformation of unhealthy land into healthy fields, the thirteenth-century scholar and theologian Albertus Magnus again told of the changes in climate brought by cutting down forests.[5] Forests, he similarly wrote in *De natura locorum*, are unwelcoming. The air inside a forest is suffocating – thick and moist and foggy – and with whirlwinds, and some trees give off noxious odors that poison the air, but this is rectified once the trees are removed.[6] The belief that deforestation changes climate arose again in the sixteenth century during European settlement of the Americas and continued over the subsequent centuries with the destruction of the native forests, to be replaced with worldwide calls in the nineteenth century to preserve forests and plants trees for climate benefits.

The Quandary of America

The cold winter climate of North America puzzled European explorers and early colonists. The prevailing knowledge of climate was largely based on latitude, and with a latitude comparable to that of southern regions of Europe, a similar climate was expected by French and British settlers of northern America.[7] One obvious difference was the vast forests of America, which were lacking in Europe after many centuries of human inhabitation, and explorers, colonists, and scholars saw in the forests the cause of the coldness. By blocking warm sunlight from heating the ground, forests were thought to decrease temperature. It was a continuation of ideas expressed in classical literature and revived by medieval scholars, and it grew in scope as Europeans arrived on the shores of America. The widespread conversion of forests to fields throughout Europe was believed to have warmed climate, causing less harsh winters, and the less severe climate was thought to have advanced the arts, sciences, and culture.[8] The colonists expected that cutting trees and clearing land would similarly make the climate of America more habitable and elevate the American experience.

It was a task that British colonists undertook with pride.[9] As one government officer would later say when the global wave of deforestation reached the shores of New Zealand, "in the matter of forests, the Anglo-Saxon is the last man in the world that ought to be let alone."[10] European visitors commented on the excessive felling of forests during the settlement of America (Figure 2.1).[11] The Swedish naturalist Pehr Kalm thought the colonists to be lavish, even hostile toward forests, because "their eyes are fixed upon the present gain, and they are blind to the future."[12] Isaac Weld, an Irish visitor, remarked that Americans "have an unconquerable aversion to trees."[13] Weld found beauty in the untouched woodlands he encountered in his travels, but in making settlements "they cut away all before them without mercy; not one is spared; all share the same fate." With the rapid and widespread felling of trees, America, in the view of one forest historian, presented "a gigantic laboratory" within which to study deforestation and climate change.[14]

Figure 2.1 Sketches of land clearing in western New York.
Proceeding clockwise: In panel a, the initial forest clearing is small, only to fell trees for the small log house and to raise a few livestock. Panel b shows the settler has cleared a few acres of land. Ten years have passed in panel c. Thirty to forty acres of land have been cleared, and neighbors have cleared their land. Panel d depicts the land 45 years following the initial clearing. Reproduced from ref. (11).

Climate improvement by deforestation was expressed early in the colonization of America.[15] The Frenchman Marc Lescarbot traveled to the New France settlement at Port Royal, located in present-day Nova Scotia and, in a 1609 account of his expedition, could not explain the cold "unless we say that the thickness of the woods and the greatness of the forests prevent the sun from heating the earth."[16] French Jesuit missionaries complained of the long, cold winters, thought to be caused by forests.[17] The endless forest blocks sunlight from warming the soil and prevents the snow from melting, but this is rectified by clearing the forests. The priest Pierre Biard provided just this reasoning in his 1616 account of the country: "Thus, from these lands nothing can arise except cold, gloomy, and mouldy vapors."[18] The missionary Paul Le Jeune later gave the same reason for the cold: "Experience teaches us that the woods engender cold and frosts."[19] The colonist Nicolas Denys, likewise, attributed the cold to the snow that accumulates in the forests, but he informed his countrymen back in France that cutting down the trees is shortening the winters. "The new France can produce everything as well as the old" if many people worked to clear the forests.[20] It evidently worked, or at least was thought to, because the botanist, agronomist, and erudite student of science Henri-Louis Duhamel du Monceau reported in 1746 that "the people of Canada claim that the winters are not as harsh as they were formerly, which they attribute to the large amount of land that has been cleared."[21]

Similar sentiments appeared in English narratives. Richard Whitbourne helped colonize Newfoundland and wrote a 1620 account of the area to promote its settlement. "The land is overgrown with woods and bushes," he described, but if the trees were cleared "so as the hot beams of the sun might pierce into the earth," the climate would warm and make the land fit for farming.[22] Faced with the harsh northern climate, later settlers of Canada continued to assert the benefits of land clearing. In his travels through the region, Pehr Kalm found that the people of Montreal and Quebec believed that felling trees made the climate milder, opening the land to sunlight and speeding the ripening of corn.[23] A 1750 pamphlet to encourage British settlement of Nova Scotia acknowledged the exacting climate but optimistically asserted that "it would certainly grow better and better every day, in proportion as the woods are cut down, and the country cleared and improved."[24]

This hopeful view – that climate could be improved by cutting down forests – extended into the colonies of the future United States. An early history of New England printed in 1637 compared that region to northern Spain and southern France, at a similar latitude, and proclaimed it is "as pleasant, as temperate and as fertile as either, if managed by industrious hands."[25] A 1654 description of Massachusetts, in relating the harsh winter, claimed that the winter cold had moderated "very much, which some impute to the cutting down the woods."[26] The Virginia of 1672 was said to be "blest with a sweet and wholesome air, and the clime of late very agreeable to the English, since the clearing of woods."[27] These views made their way across the Atlantic, where the chemist and physicist Robert Boyle, remembered today for his law relating the pressure and volume of a gas, agreed that the climate of America had become warmer.[28] Cotton Mather, too, opined on climate change. Mather was a prominent Massachusetts religious leader; he also was a scholar and a fellow of the Royal Society of London. In *The Christian Philosopher* (1721), he wrote of the known sciences of the day and sought a harmony between scientific and religious thought. He described the causes of rain, snow, hail, thunder, lightning, and rainbows; the properties of air; and the circulation of winds. In his entry on cold climates, he wrote that "our cold is much moderated since the opening and clearing of our woods."[29]

Many people writing about America in the latter half of the eighteenth century agreed. Hugh Williamson, an intellect and later a surgeon in the Revolutionary War, signer of the US Constitution, and politician from the state of North Carolina, expressed the ardor of the day in a 1770 paper read before the American Philosophical Society. A common wisdom was that "within the last forty or fifty years there has been a very observable change of climate, that our winters are not so intensely cold, nor our summers so disagreeably warm as they have been."[30] With continued clearing of interior lands "we shall seldom be visited by frosts or snows, but may enjoy such a temperature in the midst of winter, as shall hardly destroy the most tender plants."[31] Williamson's enthusiasm for climate betterment remained unabated throughout his life, when in his later years he authored a patriotic retort to the perceived inferiorities of the American climate.[32] One contemporary of Williamson pronounced the climate of the New England as having been "vastly improved since the country has been cleared of wood and brought into cultivation. The cold in winter is less intense, the air in summer purer, and the country in general much more wholesome."[33] After clearing away the woods "both the heat and cold are now far more moderate, and the constitution of the air in all respects far better, than our people found it at the first settlement," another writer asserted.[34]

Not all succumbed to the enthusiasm for climate change, though. An early skepticism appeared in 1676 in the Royal Society of London's *Philosophical Transactions*.[35] The letter writer disputed the common belief that the climate of America was becoming more temperate as a result of deforestation. John Mitchell, who produced an authoritative map of eastern North America, complained about the poor climate in a 1767 account of the colonies. The century and a half of British experience in North America showed that "neither the soil nor climate will admit of any such improvements, and there is nothing to be done against nature."[36] Johann Schöpf, a physician with Hessian troops during the American Revolution, expressed a dislike for the climate of America. Experience showed him that the winters had not become milder, yet "the credulous Americans have long flattered themselves that, by the great progress of cultivation and by the destruction of the forests of the country, their climate has for some years been rendered much milder, and the severity of their winters been moderated."[37]

Some intellects spoke against intervening in nature. The German philosopher Johann Gottfried Herder questioned the wisdom of deforesting the land to change climate and instead invoked a respect for the goodness in nature.[38] Europe presented a dilemma: once "a dank forest," but now brimming with cultivated fields "exposed to the rays of the Sun."[39] Yet Herder cautioned against clearing forests to transform foreign lands into another Europe.[40] Nature must be viewed in its entirety, and there is a vital interdependence among forests, the abundance of wildlife, the waters in lakes and rivers and springs, and humanity. Strength arises from the woods, and clearing forests to cultivate the land deprives people of necessary shade and moisture. Rather than elevating the peoples of a land, deforestation causes them to decline, and "their souls are left behind in the woods." Herder was aware of Williamson's study, which had been translated in German, but was critical of the intended climate improvement. Nature is "not mastered by force," and he was skeptical of humanity's conquest of climate: "Future ages may decide, what benefit, or injury, our genius has conferred on other climates."

Forest–Climate Processes

Colonial observers of nature lacked specific measurements to show that climate was changing, but they were not short of explanations for why it should be changing. Proof that deforestation moderates the harshness of winter could be found in the warmer European winters compared with ancient times, or so many academics contended. In an elaborate theory of climate determinism of human culture, French scholar Jean-Baptiste Dubos asserted that the climate of modern Rome was warmer than previously.[41] Other scholars advanced the premise of climate warming in their histories of Europe, and from this emerged the concept that clearing the forests had warmed the climate, thereby allowing the enlightenment of eighteenth-century society. Simon Pelloutier put forth just this explanation in his 1740 history of European peoples. Europe was formerly covered with forests, but "since these forests have been cut down and uprooted," sunlight now reaches the ground "and provides us therefore a bigger degree of heat."[42] The Scottish philosopher David Hume seconded this finding in a 1752 essay on population.[43] Historical writings confirmed climate has warmed, and Hume saw an explanation in deforestation. The warming, Hume reasoned, has occurred because "the land is at present much better cultivated, and [. . .] the woods are cleared, which formerly threw a shade upon the earth, and kept the rays of the sun from penetrating to it."[44] The same could be seen across the Atlantic: "Our northern colonies in America become more temperate, in proportion as the woods are felled." Hume's countryman and fellow philosopher James Dunbar put the matter more directly: "By opening the soil, by clearing the forests, by cutting out passages for the stagnant waters, the new hemisphere becomes auspicious, like the old."[45] From this idea, colonial American intellects rationalized still more causes of climate change.

The American politician, statesman, and student of nature Benjamin Franklin recognized, as had French Jesuit missionaries more than a century earlier, that the shade cast by forests delays the melting of snow, thereby preventing warming of the land. Sunlight, Franklin explained in a 1763 letter, "melts great snows sooner than they could be melted if they were shaded by the trees. And when the snows are gone, the air moving over the earth is not so much chilled."[46] However, Franklin admitted that "whether enough of the country is yet cleared to produce any sensible effect, may yet be a question," and he was skeptical that temperature measurements were of sufficient accuracy for confirmation. Franklin knew of another consequence of deforestation, "by clearing America of woods, and so making this side of our globe reflect a brighter light" as seen from space.[47] That open fields reflect more sunlight than do forests is a central means by which deforestation changes climate, but in the faulty reasoning of the day, the reflected sunlight was thought to warm the air.[48] Today, this is known to cool the surface climate.

Scholars of the day identified an additional process by which forests reduce temperature. They knew that evaporation, by which liquid water changes to vapor in the air, cools temperature, and they knew that forests "perspire" considerable quantities of water. This, some learned men recognized, was a means by which forests cool the atmosphere. William Robertson, principal of the University of Edinburgh, advocated this view in a 1777 history of America. Robertson believed in the degeneracy of the American climate compared with that of Europe, and forests were the culprit.[49] Relating the work of his colleague John Robison, a physicist and mathematician at the university, Robertson reiterated the diminished solar heating of the ground under thick forest canopies.[50] Furthermore, "it is a known fact, that the vegetative power of a plant occasions a perspiration from the leaves in proportion to the heat to which they are exposed; and, from the nature of evaporation, this perspiration produces a cold in the leaf proportional to the perspiration. Thus, the effect of the leaf in heating the air in contact with it, is prodigiously diminished." Today, this process is known as transpiration and is a primary mechanism by which forests cool the surface climate.

Hugh Williamson's belief in climate betterment was based on an understanding of atmospheric dynamics. Scholars knew that differences in temperature create atmospheric circulations. The cold land during winter normally drives surface winds toward the warmer ocean along the eastern seaboard. The warmer air over cleared land reduces the land–sea temperature contrast, Williamson reasoned, and the cold winds from the northwest weaken. Later writers on climate change in America would latch on to this idea.

Deforestation does not similarly increase the heat of summer, but rather increases summer winds that mix cool air from aloft with warm air near the surface. Williamson understood that "tall timber greatly impedes the circulation of the air, for it retards the motion of that part which is near the surface."[51] Proof is seen in that it is often "extremely sultry and warm in a small field, surrounded by tall woods," but not so on a larger cleared field. Williamson further described how a mosaic of "vast tracts of clear land, intersected here and there by great ridges of uncultivated mountains" generates freshening surface breezes because the warm air over the cleared land rises and is replaced with cool air rushing down from the mountains.[52] In this, Williamson was advancing a concept of mesoscale circulations generated by heterogeneity in the landscape.

The French naturalist Georges-Louis Leclerc, Comte de Buffon, loomed large in the discourse on forests and climate change. His multivolume *Histoire naturelle, générale et particulière*, begun in 1749 and continuing over several decades, was a widely read study of the natural sciences and the history of the world, and it profoundly influenced subsequent scholarship in natural history, ecology, biogeography, and evolution.[53] Buffon espoused a broad theory in which climate shapes both animal life and humanity, even molding national character, social customs, the arts, and intellect, but the progress of humankind could improve upon nature and overcome its limitations.[54] The stark contrast between Europe and America was proof of the theory. With its thick forests, marshes, and poor climate, America lacked large animals and its inhabitants were inferior to Europeans.

Williamson's essay had been reprinted in France, and it influenced Buffon's thinking.[55] In *Des époques de la nature*, published in 1778 as an installment of *Histoire naturelle*, Buffon outlined how humankind was improving climate by clearing forests and cultivating the land.[56] Williamson had provided the proof, and further evidence could be seen in comparison of Paris and Quebec. Though situated at similar latitude and elevation, the climate of Paris was much more hospitable than that of Quebec because the French countryside, as well as bordering countries, had been cleared of forests. Indeed, the climate of France was improved from centuries of deforestation, which had reversed a long-term trend of climate cooling. So, too, had a century of clearing thick tropical forests, "where the sun can barely penetrate," warmed the climate of Cayenne in French Guiana.[57] Buffon also introduced the notion of a feedback with rainfall that further enhances the cooling produced by forests: "While the trees are standing, they bring cold, they decrease the heat of the Sun by their shade, they produce moist vapors which form clouds and fall back in rain," and the rain brings still more cold as it falls to the ground.[58] By managing the land, man can modify the climate and "set so to speak the temperature at the point that suits him."[59] Trees, Buffon enthusiastically asserted, should be cleared in lands where heat is needed, but planted in deserts to lessen the heat. In this, Buffon was implying that trees are more useful in some climes than in others.

Thomas Jefferson – politician, statesman, naturalist, and polymath – was well-read in climate writings, and he expressed an understanding of forests and climate change. His *Notes on the State of Virginia* (1787) offered a strident retort to Buffon's theory of the poorer state of climate and nature in America. Jefferson provided a table of the weights of various animals to prove his point, and he even sent to Buffon the remains of a moose as further evidence of the prodigious size of American wildlife.[60] Jefferson also defended the climate of America. Though Virginia receives considerably more rainfall annually than in Europe, Jefferson asserted that "we have a much greater proportion of sunshine here than there."[61] He further remarked that "a change in our climate however is taking place very sensibly. Both heats and colds are become much more moderate within the memory even of the middle-aged."[62] Jefferson found that breezes often extended farther inland from the seacoast than they had previously, and "as the lands become more cleared, it is probable they will extend still further westward."[63] Jefferson was aware of Williamson's paper,[64] and he proposed a similar concept of mesoscale circulation to explain the changing winds in Virginia.[65] Clearing vast tracts of Virginian woodlands created a warmer landscape because the air over forests is cooler than that of cultivated lands. As central Virginia was cleared of forests, cool air from the ocean to the east and virgin forest to the west flowed in to replace the warm, ascending air over the cultivated land.

The French botanist Bernardin de Saint-Pierre pointed to another agent in the climate influences of forests.[66] The conifers found in northern lands – firs, larches, pines – have small needles arranged in various orientations. Bernardin likened them to the hair on animals and claimed that they produced a similar effect. To him, forests in northern lands are not a source of cold but rather their needles protect against the cold. In the tropics, to the contrary, palms, bananas, and other trees have large, mostly horizontal leaves that cover the ground in shade and provide shelter from the heat. In Bernardin's reasoning, tropical and northern forests differ in their climate influences – cooling the tropics, warming northern latitudes. While not exactly as hypothesized, the size and orientation of foliage do, in fact, affect the absorption of solar radiation and exchange of heat with the atmosphere. Furthermore, today's climate science recognizes the contrasting influence of tropical and boreal forests to cool and warm the climate, respectively.

The discovery that plants produce a purified air, first known as dephlogisticated air and later identified as oxygen, produced a new explanation for the winter cold. Edward Holyoke was a member of the American Philosophical Society and a founding member of the American Academy of Arts and Sciences. In a paper presented to the Academy in 1788 and later published in 1793, Holyoke linked the cold American winters with the evergreen canopies of northern forests.[67] He hypothesized that the dephlogisticated, or purified, air emitted by trees becomes bitterly cold. Production of the cold air stops with the shedding of deciduous leaves in autumn but continues unabated by the pine, fir, spruce, hemlock, and other evergreen trees that retain their foliage in winter. These trees "do in fact yield, during the winter season, such a pure air."[68] There is little doubt, he later affirmed, that "our pine woods are a source of cold."[69] Holyoke's ideas were not considered outlandish. Parker Cleaveland, a professor of mathematics and natural philosophy at Bowdoin College in Maine and surrounded by evergreen trees, planned to further study the idea.[70] Timothy Dwight, president of Yale, agreed with the premise of evergreen trees as a source of coldness, but he did not think the trees were sufficiently widespread to create the cold New England climate.[71] He further observed that southerly regions of the country experienced mild temperatures despite an abundance of pine trees.

The Debate Heightens

As the eighteenth century came to a close, the notion of climate improvement through deforestation, after nearly two centuries of thought, reached a crossroad. Samuel Williams, a professor of mathematics and natural philosophy at Harvard, reflected on the tremendous changes that had occurred over the preceding years.[72] Previously, "this country was in a state of nature. A vast uncultivated wilderness every where presented itself to view. The country in every place was covered with tall and thick trees." Instead, Williams now saw a landscape in which "the forests near the sea coasts have been cut down. The swamps have been drained. The face of the earth has been laid open to the influence of the wind and sun. And the wilderness has been changed into meadows, pastures, orchards and fields of grain." Williams viewed these changes not as a destruction of nature, but as a mark of progress. Felling trees, opening forests, and cultivating the land was diminishing the coldness, and "this will continue to be the case so long as diligence, industry and agriculture shall mark the conduct of mankind."

Williams further expressed his opinion on the changing climate in his *Natural and Civil History of Vermont* (1794).[73] The change "is so rapid and constant, that it is the subject of common observation and experience." As the area of land under settlement increases, "the cold decreases, the earth and air become more warm; and the whole temperature of the climate, becomes more equal, uniform and moderate." The most obvious change is diminution of the winter cold, which "has been observed in all the settled parts of North America." Williams relied on earlier written descriptions of the region to ascertain the extent of the warming, and he, like others before him, used classical writings to deduce that similar change had accompanied land clearing throughout Europe.[74] There were other indicators of change, too.[75] The sunlight on open lands evaporated water and dried the surface. The declining snow cover further confirmed that climate was warming. Winds were changing, coming more often inland from the seacoast. The seasons were adjusting: longer summers; a shift in autumn to later months; a shorter winter; the weather much more variable and uncertain.

Williams brought a new line of inquiry to the forest–climate problem. Whereas much of the thought to date had been qualitative reasoning, Williams provided quantitative measurements – a contrast in temperature between a forested site and an open field – as proof that deforestation warms the climate.[76] He thought the warming should be evident in the soil, and, indeed, his measurements showed that the field soil was several degrees warmer than the forest soil. The additional heat, Williams reasoned, is "communicated to the lower parts of the atmosphere. Thus the earth and the air, in the cultivated parts of the country, are heated in consequence of their cultivation, ten or eleven degrees more, than they were in their uncultivated state." Williams was partly correct in his reasoning – the warmer temperature of open land compared with forest is widely observed in today's science – but the warmth he observed occurred during summer months. There was little temperature difference during autumn, and when Williams compared temperatures on a January day, he found the soil in the open land was frozen whereas the forest soil, covered by deep snow, was warmer.[77] Williams, furthermore, incorrectly extrapolated the soil warming to air temperature. These flaws in reasoning did not escape the attention of Noah Webster.

Webster, best known today as the famed lexicographer, would have none of these thoughts. Webster was not inexperienced in meteorological research. He had previously studied the formation of dew, and he understood that a green leaf is cooler than dry soil during summer.[78] While criticizing the premise of widespread climate change in America, he thought that Holyoke's ideas on evergreen trees had merit.[79] But Webster is remembered by historians of science for a 1799 paper read before the Connecticut Academy of Arts and Sciences in which he objected to the notion that deforestation had moderated the summer heat and winter cold during the almost 200 years of settlement.[80] The arguments of Hume, Buffon, Williams, and others relied on biblical and classical writings, historical accounts, and personal anecdotes. Webster dismissed the reasoning as flawed, the past writings and remembrances as inaccurate, and the analysis prone to overgeneralization. Webster was especially harsh in his critique of Williams, who was "unfortunate in his facts" and "still more so in his reasonings and deductions."[81] As for Thomas Jefferson, he "seems to have no authority for his opinions but the observations of elderly and middle-aged people."[82] Webster was not, however, unopen to patriotism in the face of European critics. In an earlier essay, he had declared "the climate of America is as salubrious, as that of any country in the same state of cultivation."[83]

Webster did not dismiss the idea that cutting down forests changes the climate, but he did present an alternative view. Previously, Webster described how the climate was more variable because "by levelling the forests, we lay open the earth to the sun, and it becomes more impressible with heat and cold."[84] Forests moderate the heat of summer and cold of winter; cutting them down has the reverse effect. In his 1799 address to the learned men of Connecticut, Webster further explained his thinking. The temperature measurements of Williams proved so, and this is opposite to the assertion that deforestation makes winter milder. "So far is it from truth," he declared, "that the clearing and cultivation of our country, has moderated the rigor of our cold weather."[85] Rather, "while a country is covered with trees, the face of the earth is never swept by violent winds; the temperature of the air is more uniform, than in an open country; the earth is never froze in winter, nor scorched with heat in summer."[86] However, "the clearing of lands opens them to the sun, their moisture is exhaled, they are more heated in summer, but more cold in winter near the surface; the temperature becomes unsteady, and the seasons irregular."[87] In supplementary remarks presented in 1806, Webster again concluded that "the warm weather of autumn extends farther into the winter months, and the cold weather of winter and spring encroaches upon the summer," which he attributed to "the greater quantity of heat accumulated in the earth in summer, since the ground has been cleared of wood, and exposed to the rays of the sun; and to the greater depth of frost in the earth in winter, by the exposure of its uncovered surface to the cold atmosphere."[88] William Dunbar, a naturalist and explorer of the American West for Jefferson, stated a similar position.[89]

The question of whether deforestation was changing the climate continued to stir passions at the turn of the century and as the nineteenth century progressed. Many writers still endorsed the premise. One account of America said that "opening the country has greatly tended already to lessen the cold" and that cultivation "must tend to moderate the climate."[90] Constantin-François Volney, a French traveler throughout the country, commented that "for some years it has been a general remark in the United States, that very perceptible partial changes in the climate took place, which displayed themselves in proportion as the land was cleared."[91] He relied on the works of Williams and Jefferson, but Volney heard similar testimonies himself during his travels.[92] The changing climate "is an incontestable fact," and any "doubts vanish before the multitude of witnesses and positive facts."[93] Even the climate of Canada in the region of the Great Lakes would be improved "in proportion as the forests are cut down" by allowing warm southern air to extend farther north.[94] David Ramsay described the oppressive summer heat, marshes, and stagnant waters of South Carolina, but he thought that the climate would be improved by clearing the forests and cultivating the land, just as it had in Britain since Roman times.[95] Late in life, Jefferson continued to advocate for climate surveys "to show the effect of clearing and culture towards changes of climate."[96] Jefferson expressed a sincere respect for trees. In a conversation on the unbridled cutting of trees throughout the nation's capital to obtain firewood, he declared to his dinner companions that "the unnecessary felling of a tree, perhaps the growth of centuries seems to me a crime little short of murder."[97] The wanton loss of the towering, majestic trees pained Jefferson "to an unspeakable degree."

Academic publications gave credence to forest influences on climate. In its entry explaining climate, a scientific dictionary repeated assertions about climate improvement in the United States and tropical America.[98] The pioneering geologist Charles Lyell thought it "unquestionable" that deforestation had made the American winters less cold and the summers less hot.[99] The geographer Conrad Malte-Brun, too, believed that forest clearing and cultivation of the land had altered climate.[100] In his textbook, he asserted that deforestation and cultivation are but some ways in which "man exercises a slow but powerful influence upon the temperature of the air."[101] Without changes to the land, "few climates would be salubrious and agreeable." By opening the woods and sowing crops, a "vanquished nature yields its empire to man, who thus creates a country for himself." Yet Malte-Brun accepted limits to human industry. Too much deforestation "becomes a scourge which may desolate whole regions."[102]

Others, like Webster, were skeptical of the evidence that climate had changed; after two centuries of faith, proof was still lacking. Benjamin Smith Barton, a professor at the University of Pennsylvania, noted the lack of observations, but suggested useful climate data might be found in the annual growth of tree rings.[103] The Yale president Timothy Dwight complained that "the observation of this subject has been so loose, and the records are so few and imperfect, as to leave our real knowledge of it very limited."[104] He did agree that forests, by shading the ground, cause snow to persist longer than in open land, but this only creates a perception of a colder winter in the presence of forests. In describing climate change in America, the *Edinburgh Encyclopedia* asserted more forcibly that "the theory now sketched, is, we fear, to be regarded rather as the birth of a lively fancy, than the offspring of accurate science."[105] John Leslie, a Scottish mathematician and physicist at the University of Edinburgh known for his investigations on heat, brought the rigor of physics to bear upon the study of climate.[106] In a lecture read before the Royal Society of London in 1793, he articulated an understanding of climate in which the atmosphere, set in motion by solar heating of the planet, redistributes heat between the tropics and the poles. As would later meteorologists, Leslie dismissed the influence of forests on temperature.

Divergent Science

Resolution to the forest–climate question required a theory to explain forest influences on temperature as well as observations to support the theory. The acclaimed naturalist and geographer Alexander von Humboldt sought both. Humboldt established in Europe an experimental approach to quantitatively compare forested and open lands. It was a science grounded in observations taken at the local scale and extrapolated to the larger macroscale encompassed by geographic regions or continents. Other scholars, however, started at the macroscale and examined the pattern of temperature across large domains that differed in forest cover, looking for a signature of forests in their temperature maps. From these two different scientific approaches, a sharp contrast of opinions arose.

Humboldt was one of the foremost scholars of the natural world in his day.[107] His expeditions through tropical America and Central Asia informed his views of the interrelationship between vegetation and climate and laid the foundations for modern-day physical geography, climatology, and ecology. Humboldt traveled extensively throughout Spanish America during 1799–1804. In his expeditions across the vast savannas, through the lush tropical forests, and up mountainsides from the steamy lowland jungles to the cold, snow-covered summits, he saw vegetation change in relation to the climate.[108] Humboldt also witnessed the destructive outcome of deforestation in the tropics (Figure 2.2).[109] It was the effects of forest clearing on the supply of water to Lake Valencia in Venezuela that sharply focused this thinking on deforestation, and it was there that he began to formulate his ideas about forests and climate. Trees, Humboldt noticed, "surround themselves with an atmosphere constantly cool and misty."[110] It was a thought that Humboldt returned to throughout his lifetime.

One topic of study was the geographic variation in temperature across the planet, latitudinally from the hot equatorial tropics to the frigid poles and longitudinally across continents. In a foundational analysis published in 1817, with its famous map of temperature depicted by isothermal lines (delineations of equal

(a)   (b)

Figure 2.2 Illustrations of tropical forests in Brazil from *Voyage pittoresque dans le Brésil* by Johann Moritz Rugendas (1827). Panel a shows a virgin forest (Chasse dans une forêt vierge). Panel b depicts clearing a forest (Defrichement d'une forêt). Reproduced from ref. (109).

temperature), Humboldt identified an overall macroscale pattern dictated by latitude, continental geography, and oceans along with local and regional variations created by, among other factors, soil and forests.[111] Humboldt continued to refine his thinking on climate over the ensuing years. As had his earlier exploration of tropical America, Humboldt's travels across Russia in 1829 through sweeping tall grass steppes and unfamiliar Siberian wilderness shaped his understanding of forests and climate. Upon arriving back at St. Petersburg at the end of his expedition, Humboldt called for further climate studies to assess the impact of deforestation. The goal of science, Humboldt said in an address before the Russian Imperial Academy of Sciences, is to increase knowledge and to shape intellectual reasoning.[112] Scientific bodies must "regularly observe, measure, monitor" the natural world.[113] The scientific study of climate is an area of scholarship, not just of theoretical interest but also important for "the material needs of life,"[114] and the effect of deforestation on climate is a central topic that requires more investigation.[115]

The insight gained firsthand and from years of scholarship is evident in the resulting work, an 1831-issued two-volume description of the geography and climate of Russian Asia entitled *Fragmens de géologie et de climatologie asiatiques*. Humboldt included in the book a lengthy update to his previous research on isotherms.[116] Although other factors are primary, Humboldt knew that the vegetated state of the land, with its sandy deserts, grassy savannas, and forests, is a necessary piece of a "mathematical theory of climate."[117] Deforestation is one of the causes of warm climates, while vast forests create cold climates.[118] Humboldt understood the unique climate of the extensive savannas in South America, but it was the lush tropical forests that drew his interest.[119] He described how forests decrease temperature by shading the ground and by transpiring water.[120] The latter process is readily evident in "the vapor trails that can be seen in broad daylight in the virgin forests between the treetops."[121] He explained a third process of cooling by radiative exchange, enhanced by the many layers of leaves in a forest canopy.[122] These three influences – shade, transpiration, radiative exchange – are of such importance that knowing the extent of forests compared to bare ground or grassland "is one of the most interesting numerical elements of the climatology of a country."[123] In revising his thoughts some years later, Humboldt changed the wording to read "the most interesting and the most neglected."[124]

Humboldt returned to these ideas in two other works. *Ansichten der Natur*, or *Views of Nature* in an English translation, was Humboldt's first masterpiece. Published in 1808 and again with expanded editions in 1826 and 1849, Humboldt, through his revisions and amended notes, reveals a progression in his thinking on forests and climate. Humboldt included in the 1808 edition a passage on the interrelatedness of plants and desert dryness,[125] but he did not mention the climate influences of forests until the 1849 edition.[126] *Cosmos*, the first volume of which was published in German in 1845, was Humboldt's unfinished lifetime triumph and an internationally acclaimed best seller. In his description of the physical geography of the world, Humboldt conveyed the interconnectedness between forests and climate (Figure 2.3).[127] Climate may be thought of in terms of the temperature, humidity, wind, and other properties of the atmosphere, but it requires knowledge of the reciprocal interactions of the atmosphere with the land, "whether bare or clothed with forests, or with grasses or other low growing plants."[128] As he had previously, Humboldt explained the importance of forest shade, transpiration, and radiative exchange, by which the presence of forests cools temperature and their absence raises temperature.[129]

Other French researchers added to Humboldt's theories. One scholar inspired by Humboldt was Jean-Baptiste Boussingault. A chemist remembered today for his study of nitrogen and plant growth,

Figure 2.3 *Heart of the Andes* (Frederic Church, 1859).
The accurate portrayal of vegetation and climate epitomizes the union of art and natural science described by Alexander von Humboldt in *Cosmos*. Image provided courtesy of the Metropolitan Museum of Art (New York).

Boussingault traveled to South America and, like Humboldt, reflected on forests and climate. Upon analyzing temperature measurements in forested and non-forested locations in tropical America, Boussingault concluded that "the abundance of forests and the resulting humidity tend to cool the climate of a country" while drought and aridity "on the contrary tend to increase its heat."[130]

A different scientific framework was emerging in the United States, where meteorological observing networks extending over wide geographic domains provided a new means to study climate; indeed, their advent was a pivotal moment in the development of meteorology as a science.[131] They also brought forth an alternative interpretation to the forest–climate question. While Humboldt advanced a theory in which forests are an essential component of climate science, contemporary American meteorologists analyzed temperature observations over large regions and concluded forests have only marginal impact on climate. The United States, which was undergoing rapid changes in its land, was a prime country to obtain data, and an 1826 register of meteorological observations at military posts scattered throughout the country was prepared to address whether climate was changing.[132] Joseph Lovell, surgeon general of the army and author of the register, believed the temperature measurements provided "an opportunity of bringing the question to the test of experiment and observation."

Samuel Forry, a surgeon in the army, followed through with Lovell's intent and published in 1842 an authoritative climate analysis based on observations taken at 31 military posts.[133] "As a test of the truth of theories, statistical investigations are of vast importance," Forry wrote.[134] He acknowledged an influence of forests, even allowing that changes in the characteristics of the land must be accounted for when comparing temperature measurements over a lengthy period of time, but he thought this to be "extremely subordinate" compared with other determinants of climate such as continental geography, proximity to oceans, and the presence of mountains.[135] Forry was suspicious of claims that temperature had changed over the ages because of deforestation and cultivation of the land, and he was especially critical of the climate ideas of Jefferson, Williams, and Volney.[136] Their arguments were, he thought, more fanciful than factual, and he advised that "it is a good rule in philosophy to ascertain the truth of a fact before attempting its explanation."[137] Few were exempt from Forry's reproach. He lashed into the notion of climate improvement expressed earlier by Malte-Brun.[138] In a later work, the geologist Charles Lyell, too, incurred Forry's wrath; his views on deforestation were "unsustained by any well observed facts."[139] Forry recognized that forests "doubtless tend considerably to diminish the temperature of summer" by evaporating water and shading the ground,[140] but his climate science had a limited role for forests – at best a change in the seasons similar to that described by Webster: "that in countries covered with dense forests, the winters are longer and more uniform than in dry, cultivated regions, and that in summer, the mean temperature of the latter is higher."[141]

Humboldt's understanding of forests and climate was groundbreaking. It was a vision of the interdependences of the natural world that would not be realized until a century and a half later with modern mathematical models of climate. Later researchers such as Antoine-César Becquerel in France and Ernst Ebermayer in Germany extended Humboldt's observational framework with paired meteorological observatories in forested and open locations. Their forest observatories are the foundations from which modern meteorological measurements of forest–atmosphere interactions arose, and their writings, along with Boussingault's, would be used by conservationists throughout the far reaches of the world in calls to protect forests. Yet Forry's explanation of a more minor influence of forests became the prevailing paradigm. The *Encyclopedia Britannica* espoused a similar understanding in its contemporaneous entry on physical geography.[142] Forests cool the air in hot climates and prevent the loss of heat from the ground in cold climates. "The clearing of forests may thus affect the temperature," but, the author continued, "the mere effect of cultivation can never be very considerable in changing a climate." Humboldt's own thinking on climate was influenced by Forry, and in the third edition of *Views of Nature*, Humboldt weighed in on deforestation and climate change in the United States.[143] As to whether deforestation had made the climate more moderate, with milder winters and cooler summers as had been claimed for so long, Humboldt thought such statements "are now generally discredited."

■

Though the notion of creating a more temperate climate proved to be false, a more advanced understanding of forest influences on climate – one now seen in modern climate science – emerged during the more than 200 years of thought. The climate influences of snow cover and transpiration have withstood the test of time, though the modern science reveals additional complexities not envisioned by eighteenth-century scholars. The mesoscale circulations described by Williamson and Jefferson can, in fact, be created by the differential heating of forested and open lands arranged in a patchwork mosaic across the landscape. That drought and aridity increase temperature, as Humboldt and Boussingault proposed, is a staple of modern climate science. The contrast between the climate influences of tropical and boreal forests proposed by Bernardin de Saint-Pierre, while not exactly as he envisaged, is a key but poorly understood piece of present-day plans to plant trees for climate benefits.

Yet the fundamental dilemma of spatial scale – the difference between the small-scale microclimates of forests and clearings and the large-scale influence of forests on the macroclimate of a region – remains today. Scientists such as Williams, Humboldt, and Boussingault studying forests at the local scale drew different conclusions than did Forry and others analyzing larger-scale climate patterns. This divergence in the science would only grow throughout the nineteenth century as Becquerel, Ebermayer, and others established forest meteorological observatories throughout Europe while meteorologists, especially in America, sought a large-scale dynamical framework to understand climate. As meteorology advanced as a science and explanations of climate were put forth, many meteorologists, while accepting the microclimate influences of forests, dismissed the premise that forests influence the large-scale, or macroscale, climate of a region. One leading cause for concern was the popular belief that forests enhance rainfall.

3   Destroying the Rains, c. 1500–1830

As Europeans explored and settled the lands that would become Spanish and British America, men returning home told of the damp, rainy climate with thick forests, and they described notable changes in rainfall since the first colonies were established. With the arrival of the Spanish in the New World, Europeans associated the plentiful rainfall and moistness of the Americas with the lush forests of the region. The moistness was thought to cause ill health, but deforestation would make the climate healthier. Yet concern about a decrease in rainfall, particularly in island territories, soon appeared. Advocates for forest conservation joined the debate, reframing the forest–climate problem to encompass rainfall and from one of deforesting the land to make a more temperate climate to one of preserving forests and planting trees to protect the supply of water. In the rainfall benefits of forests, conservationists found a way to convey the environmental destruction wrought by deforestation.

Forests and Rainfall

Spanish explorers commented on the changing climate that followed their arrival in the New World. Gonzalo Fernández de Oviedo y Valdés, who was active in the settlement of Hispaniola in the years shortly after Columbus, provided one of the earliest accounts of deforestation and climate change in the Americas. His 1548 history of the region tells of a change in the weather after the forests surrounding the city of Santo Domingo were felled and the land was cleared and opened: "I have seen these lands much changed in those provinces that I have traveled to, and they continue to change every day, particularly as it concerns periods of cold and hot weather."[1] The changing weather was "a sign of how this region and its processes have been tamed and calmed." Learned men knew so, "and it is a very natural, reasonable, and evident thing that it would be thus, because as this land is very humid and deeply forested and impenetrable, having never been trodden upon nor opened up." One report that proved enduring was that forests enhance rain and that deforestation, consequently, decreases rainfall. Christopher Columbus gave an early expression to this thought. In a biography published in 1571, his son Ferdinand wrote that Columbus attributed the plentiful rainfall found in Jamaica to "the great forests of that land; he knew from experience that formerly this also occurred in the Canary, Madeira, and Azore Islands, but since the removal of forests that once covered those islands, they do not have so much mist and rain as before."[2] Deforestation was not yet seen as a threat, however. Spaniards entering southern Mexico in the late 1500s associated the forests they encountered with rain, the dampness of the region, and unhealthfulness; clearing the forests would dry the countryside.[3]

The association between trees and rainfall grew as men ventured across the oceans and reported on their findings. An often-repeated tale was of a tree on El Hierro, one of the Canary Islands, from which sailors would obtain fresh water. A description is found in Ferdinand Magellan's circumnavigation of the world (1519–22). Antonio Pigafetta, who accompanied the expedition and later recounted the voyage, told of a tree that provides water when a cloud envelops it, "the leaves and branches of which distil a quantity of water."[4] The travelogue of another voyager, homeward bound in 1639 from India, said of the tree: "its branches are covered with a cloud, which is never dispelled, but resolved into a moisture, which causes to fall from its leaves, a very clear water, and that in such abundance."[5] The tale of the magical tree would be repeated many times over the next two centuries. It likely influenced Columbus in his consideration of deforestation and rain, as well as subsequent actors in the forest–rainfall debate.[6]

The English physician Henry Stubbe, arriving in the Caribbean, was in agreement on the usefulness of trees, "there being certain trees which attract the rain."[7] As a result, "if you destroy the woods, you abate or destroy the rains." Proof was evident at Barbados and Jamaica, where "they have diminished the rains, as they extended their plantations." John Evelyn included Stubbe's account in the third edition of his celebrated *Sylva* (1679) and added a warning that deforestation endangered the settlers of the Caribbean islands "so that if their woods were once destroyed, they might perish for want of rains."[8] John Clayton described in 1688 the violent thunderstorms found in tidewater Virginia, which he attributed to the forests prior to their clearing. "When the country was not so open, the thunder was more fierce," he reported.[9] This view would carry weight over the ensuing centuries. More than 250 years after Columbus's thoughts were first voiced, Alexander von Humboldt, in describing the physical geography of Haiti, said that the deforestation of the island had decreased rainfall.[10] His source was Ferdinand's biography of Columbus.

The belief that forests contribute to rainfall was not without scientific basis. It reflected knowledge of plant physiology and the hydrologic cycle that preceded our modern science by three centuries. Scholars of the era were aware of pores on leaves, known now as stomata, and reasoned that they allow materials to enter or exit plants. In a 1682 treatise on plant anatomy, Nehemiah Grew illustrated the pores (Figure 3.1) and likened leaves to the skins of animals, which "are made with certain open pores or orifices, either for the reception, or the elimination of something for the benefit of the body: so likewise the skins, of at least many plants, are formed with several orifices or passports, either for the better avolation of superfluous sap, or the admission of air."[11] Marcello Malpighi of Bologna also observed the pores on leaves and deduced a similar purpose to expel moisture from plants.[12] Subsequently, the Englishman John Woodward demonstrated in 1699 that the water consumed from the soil by plants "passes through the pores of them, and exhales up into the atmosphere."[13] This process is today called transpiration. It is distinct from evaporation, which is the direct conversion of liquid water to vapor in the air, most obviously from oceans, lakes, and other large water bodies, but also from the soil. The rate at which water evaporates is determined by meteorological conditions such as the radiant energy, wind speed, and the humidity of the air. Transpiration is the evaporation of water from plants and additionally involves the physiology of stomata and transport of water from the soil through the stem and out the leaves to the atmosphere. The two processes together are referred to as evapotranspiration.

A linkage between transpiration and rainfall was readily established. Edmond Halley had explained the circulation of the trade winds in the tropics.[14] In a subsequent paper that described the cycling of water between land and the oceans, Halley related that the water vapor transpired by plants contributes to rainfall.[15] In his 1699 study, Woodward also reasoned that the water passing through leaves to the atmosphere is why forested lands "should be very obnoxious to damps, great

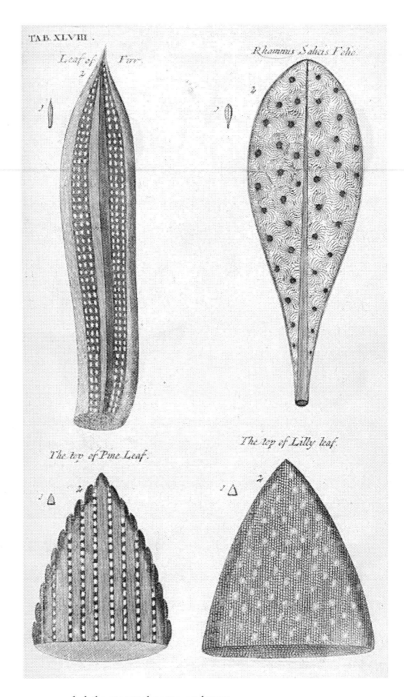

Figure 3.1 Microscopes revealed the stomatal pores on leaves.
Shown is an illustration published by Nehemiah Grew in 1682 of leaves with stomata depicted as regularly spaced openings in the surface of the leaf. Reproduced from ref. (11).

humidity in the air, and more frequent rains, than others that are more open and free."[16] Woodward further remarked that the humid air "was a mighty inconvenience and annoyance" to the first settlers of America, when the country was wooded, "but as these were burnt and destroyed, to make way for habitation and culture of the earth, the air mended and cleared up apace: changing into a temper much more dry and serene than before." Evelyn, too, while concerned about diminishing rainfall in the Caribbean, expressed the more common mores of that era when describing the effects of deforestation in America. Influenced by Woodward, Evelyn's fourth edition to his book, published in 1706, ascribed ill effects to forests.[17] Thick forests, he wrote, "render those countries and places, more subject to rain and mists, and consequently unwholesome." The climate of America was "so much improved by felling and clearing these spacious shades, and letting in the air and sun, and making the earth fit for tillage, and pasture, that those gloomy tracts are now become healthy and habitable."

Physiological experiments published in 1727 by Stephen Hales in his *Vegetable Staticks* heightened interest in the influence of forests on rainfall. Hales provided a clearer understanding of the physiological controls of transpiration, and he is remembered today for uncovering the mechanisms by which transpiration draws water from soil through roots to leaves in a process called cohesion-tension theory (Figure 3.2).[18] A further key finding was to measure the surface area of leaves from which water is transpired. The leaves of a one-meter tall sunflower, Hales discovered, comprise nearly four square meters in area.[19] Humboldt would later use this fact to speculate on the enormous quantity of water transpired by the Amazon rain forest, with its many layers of leaves.[20] Contemporary French scholars seized upon Hales's work as the physiological basis for rainfall over forests.[21] So common was the knowledge that the entry on rain in a French encyclopedia (1744) reported that forests are "one of the causes of rain; because trees sweat a large amount of vapors."[22] In contrast, deserts, without trees, are dry. Deforestation decreases rainfall, as could be seen in the Caribbean, because "the forests are always filled with a humid air, thick, charged with the exhalations of trees, which form clouds by their rise in the atmosphere." Likewise, when writing of Jamaica, a French novelist related that when the forests are cleared "the clouds and consequently the rains become more rare and less thick."[23]

Students of the American climate advanced similar ideas. In his history of Jamaica (1774), Edward Long wrote of the local climate and the changes it had undergone since the first settlement of the island. Being well-read in the studies of Halley, Hales, and others, Long showed an understanding of the fundamentals of tropical island meteorology – temperature, precipitation, the trade winds, land and sea breezes, thunderstorms, and hurricanes; the varied climates found on the island; evaporation and streamflow; and he wrote of a decrease in rainfall since the woods were cleared.[24] Hugh Williamson fretted over "a copious perspiration through the leaves of trees" while the country was forested, whereby "the air was constantly charged with a gross putrescent fluid."[25] Samuel Williams understood that the extensive canopy of leaves in a forest transpires vast quantities of water to the atmosphere; his own experiments proved so.[26] Williams knew, too, that the emitted water vapor condenses to form clouds and rain.[27] Perhaps this is why, though he had no measurements to prove so, he could not account for the greater rainfall in America than in Europe "without supposing that the immense forests of America, supply a larger quantity of water for the formation of clouds, than the more cultivated countries of Europe."[28] In his study of the climate of America, Constantin-François Volney wrote that cutting down forests "diminishes the quantity of rain, and the springs resulting from it, by preventing the clouds from stopping and discharging their waters on the forests."[29]

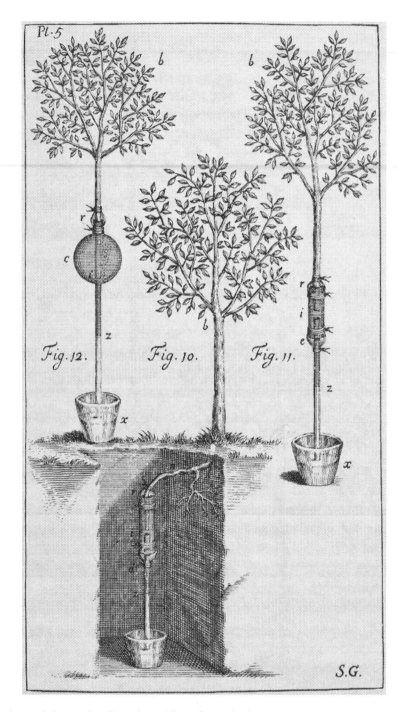

Figure 3.2 Experiments led to a scientific understanding of transpiration.
Shown are experiments by Stephen Hales published in 1727 to study sap flow in the root of a pear tree (Fig. 10) and in the branches of two apple trees (Fig. 12 and Fig. 11). Reproduced from ref. (18).

Discovery of Oxygen

Studies on stomata and transpiration were one part of an explosion of knowledge at the time in plant physiology, and they were paralleled by advancements in the chemical emissions of plants. Evelyn, as did Woodward, believed that the water evaporating from plants bore odors. Despite centuries of tree clearing, Evelyn claimed that many English estates were "unhealthful, by reason of some grove, or hedge-rows of antiquated dotard trees ... filling the air with musty and noxious exhalations."[30] Scientific studies throughout the eighteenth century further established the chemical influence of plants on air. Hales also studied air, and his experiments with plants grown in glass jars led him to speculate on an additional functional role of leaves, which he likened to the lungs of animals so that plants obtain "some part of their nourishment from the air."[31]

The subsequent discovery of photosynthesis, and of oxygen as a product of photosynthesis, overthrew the prevailing eighteenth-century understanding of chemistry and presented forests in a beneficial viewpoint. The broad conceptual basis for chemistry during the mid-eighteenth century was known as phlogiston theory.[32] The theory postulated that combustible bodies contain a substance called "phlogiston" that is released to the air during combustion. Combustion ceases in an enclosed space because the air is saturated with phlogiston, at which point it is said to be completely "phlogisticated." That the flame of a candle enclosed by a jar eventually extinguishes was taken as proof of the production of phlogisticated air. Respiration by animals, too, removes phlogiston from the body and produces phlogisticated air, which cannot support life, evidenced by placing a mouse in an enclosed jar.

In his experiments with plants, Joseph Priestley, a multifaceted English minister, philosopher, and chemist, effectively discovered oxygen, which spelled the end of the phlogiston theory and led to a new oxygen theory of combustion. Priestley used an experimental setup similar to that developed by Hales, in which he introduced a plant into a glass jar, together with a burning candle or a live mouse, and placed upside down in a bowl of water (Figure 3.3).[33] In a 1772 publication of his findings, Priestley described how the candle flame did not go out; nor did the mouse suffer: "I have accidentally hit upon a method of restoring air which has been injured by the burning of candles, and that I have discovered at least one of the restoratives which nature employs for this purpose. It is vegetation."[34] Plants, Priestley reasoned, "reverse the effects of breathing, and tend to keep the atmosphere sweet and wholesome."[35] Priestley understood the profound significance of his finding. Respiration by animals and the decay of organic material "injure" the atmosphere (producing CO_2), but plants "repair" the air (producing oxygen) so "that the remedy is adequate to the evil."[36] It would take several more years for the gas to be identified as oxygen (the French chemist Antoine Lavoisier coined the word; Priestley, himself, was an adherent to phlogiston theory and called it "dephlogisticated air"), but Priestley's experiments stirred other researchers to conduct their own investigations, and by the end of the century the process of photosynthesis, which consumes CO_2 and produces oxygen, was largely understood.[37]

It was the dephlogisticated air produced by evergreen trees while deciduous trees are dormant that Edward Holyoke thought caused the winter coldness,[38] but Priestley's discovery of dephlogisticated air opened a new facet to the forest problem in America. Upon hearing of Priestley's results, Benjamin Franklin exclaimed: "I hope this will give some check to the rage of destroying trees that grow near houses ... from an opinion of their being unwholesome."[39] Samuel Williams calculated that forests contribute "immense quantities of air," produced "in the purest state."[40] Benjamin Rush, a medical doctor

Figure 3.3 Experiments led to a scientific understanding of leaf gas exchange.
Shown is the inverted glass apparatus used by Stephen Hales to study the effects of candles and plants on air. Reproduced from ref. (33).

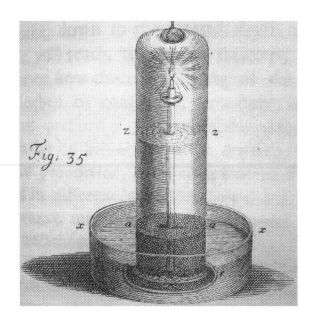

and professor of chemistry at the University of Pennsylvania, saw benefits in planting trees to prevent diseases arising from the stagnant water of millponds. Trees, he explained, act mechanically by providing shade and by obstructing winds, "but they act likewise *chemically*. It has been demonstrated that trees absorb unhealthy air, and discharge it in a highly purified state in the form of what is now called 'deflogisticated' air."[41] It was a concept that appealed to others, and the oxygen benefits of forests would become a recurring theme for many forest advocates. Concern that the destruction of forests, by reducing photosynthesis, will deplete oxygen in the atmosphere remains prevalent today.

A Precipitation–Evapotranspiration Feedback

Students of nature began to speak of forests and the atmosphere as a dynamic system. Forests, by transpiring copious amounts of water, increase rainfall and cool temperature, thereby lessening the tropical heat and, conversely, accentuating the desert heat where forests are lacking. Today, these feedbacks are recognized as essential to understand the planetary boundary layer, which is the layer of the atmosphere close to the ground. Georges-Louis Leclerc, Comte de Buffon, voiced this thinking in his theory of forest influences on climate. Buffon admired the *Vegetable Staticks* of Hales and thought much was to be learned from the work; indeed, he personally translated the book into French.[42] In *Des époques de la nature* (1778), Buffon described how forest clearing in French Guiana was not only increasing temperature, but the rains started later and stopped earlier and were less abundant and less continuous.[43] Conversely, planting trees in the desert would lessen the scorching heat, bring rain, and "all the mildness of a temperate climate."[44]

Humboldt likewise conceived a theory linking forests, evapotranspiration, and rainfall. An early expression of this thought is found in *Ansichten der Natur*, published in 1808. In describing the barren desert environment, Humboldt wrote of an interrelationship in which the lack of plants contributes to the desert dryness.[45] Loss of vegetation initiates feedback that prevents precipitation and further reduces the vegetation: "Thus do the lack of rain and the absence of plants share a mutual causality. It does not rain, because the naked and barren plain of sand grows hotter and radiates more heat. The desert does not become a steppe or a grassland, because no organic development is possible without water." Elsewhere, Humboldt wrote of a decrease in rainfall caused by deforestation.[46] It had been known since the time of Columbus, but, Humboldt lamented, "this warning has remained almost unheeded for three centuries and a half."[47] From Hales, Humboldt knew of the large amount of water moved by transpiration, and he saw this himself in that "torrents of vapor rise above an equinoctial land covered with forests."[48]

Williamson, so outspoken in his belief of climate betterment, expressed the same understanding in his 1811 book on the climate of America. He reaffirmed his premise of several decades earlier that the climate was improved through deforestation, but he also expressed a newer understanding of forest influences.[49] He described the cooling achieved by evapotranspiration, the enhancement of evapotranspiration by forests, and the contribution of forests to rainfall: "The vapours that arise from forests, are soon converted into rain, and that rain becomes the subject of future evaporation, by which the earth is further cooled. Hence it follows, that a country, in a state of nature, covered with trees, must be much colder than the same country when cleared." This process makes the heat of the tropics more tolerable because "a perpetual verdure and thick foliage, within the tropical regions, tend greatly to moderate the heat, by copious evaporations," and conversely makes the heat of the deserts more insufferable because similar such cooling does not occur where vegetation is lacking.

The Problem of Streamflow

The historical argument over forests and rainfall is rooted in conflicting ideas. That forests increase precipitation, and consequently deforestation decreases the amount of rain, was widely expressed since Columbus arrived on the shores of America. Subsequent naturalists further advanced these ideas in the Americas, island territories, and worldwide. The drying of springs and streams following deforestation was evidence that the rains cease after trees are cut down, or at least signaled a general desiccation of the land. Others, however, reported an increase in torrential floods after clearing the forests. Volney expressed these contradictions in his writings on the American climate.[50] The people of Kentucky "already complain of drought, which increases in proportion as the country is cleared of wood." Yet it is also a fact "well established in Kentucky, that many of the streams have become more abundant, since the woods in this neighbourhood have been cut down." Volney believed the former to be more general and more important because, in his view, forest clearing decreases rainfall. Proof is found in the number of small streams that "now fail every summer" or that "have totally disappeared." These themes – reduction of rainfall, drying of freshwater springs and streams, torrential floods – permeated the forest–rainfall controversy of the nineteenth century.

That forest soils are moist and that cutting down forests, therefore, dries the land has long been recognized. The Roman scholars Vitruvius and Pliny the Elder described the influence of forests on springs and streamflow more than 2,000 years ago.[51] In their writings of climate change, observers of the American climate, too, noticed that springs were drying up and the flow of water in streams and rivers was changing. An early account on British settlements in America reported that "as the country is cleared of wood and brush, small streams dry up."[52] Samuel Williams expressed the same thought in his history of Vermont:

"Another remarkable effect which makes part of the change of climate, and always attends the cultivation of the country, is an alteration in the moisture or wetness of the earth. As the surface of the earth becomes more warm, it becomes more dry and hard, and the stagnant waters disappear."[53] Williams saw that "many of the small streams and brooks are dried up," and he expressed concern that "mills, which at the first settlement of the country, were plentifully supplied with water from small rivers, have ceased to be useful." The reasoning of Williams was not without merit. He and other scholars knew that the shade cast by forests reduces the solar energy available to evaporate water and that the trunks, branches, and leaves of trees block drying winds. Williams had previously performed experiments to measure evaporation, and his further experiments demonstrated that the evaporation of standing bodies of water is less under a forest canopy than in an open site exposed to sun and wind.[54] The naturalist Gilbert White commented on the moisture benefits of forests in his famous natural history of Selborne (1789): "Trees perspire profusely, condense largely, and check evaporation so much, that woods are always moist; no wonder therefore that they contribute much to pools and streams."[55]

While evaporation was understood to dry out swamps and marshes, the more common observation was that deforestation increases the supply of water to streams and rivers. Noah Webster believed that trees and leaves help to retain water so that "the water runs off suddenly into the large streams" when the land is cleared of forests.[56] Volney similarly wrote of the retention of rainwater by leaves on the forest floor.[57] Timothy Dwight attributed the surging flow of the Connecticut River in upper New England, "often fuller than it probably ever was, before the country above was cleared of its forests," to melting snow in open land.[58]

French engineers developed an advanced knowledge of forest influences on streamflow. A 1797 study by Jean-Antoine Fabre of floods in the high mountains of France identified deforestation as the main cause.[59] Fabre's explanation is prescient of our modern understanding of forest hydrology. He wrote that the foliage and branches of trees intercept much of the falling rain before it reaches the ground. That which is not held in the forest canopy drips slowly to the earth and infiltrates into the soil. Decaying leaves and debris on the forest floor absorb much of the rainwater. Small shrubs and other groundcover hinder the movement of water that does flow over the land. The destruction of forests disrupts these processes so that rainwater runs off into streams and rivers. The solution to torrential flooding was to reforest the mountains and preserve what forests still stand. Later French engineers such as Alexandre Surell further advanced a scientific understanding of forest hydrology and advocated similar solutions.[60] Surell's recommendation to alleviate floods, like Fabre, was simple: "reforest the high regions of the mountains."[61] Still more reports on deforestation and floods in France followed.[62]

Humboldt's travels to Lake Valencia in Venezuela profoundly shaped his view of forests.[63] Widespread deforestation of the lands around the lake had reduced the supply of water. Humboldt understood that removing trees eliminates the shade of forest canopies and increases evaporation from the soil. As a result, the amount of water flowing in rivers decreases at some times of the year. Yet the forest undergrowth also impedes the flow of water over the ground. When trees are removed, the unprotected soil allows heavy rains to collect as torrential floods at other times. "The destruction of forests, the want of permanent springs, and the existence of torrents, are three phenomena closely connected together," he explained. Humboldt's report on Lake Valencia inspired other scholars to investigate the links between deforestation and water.

Jean-Baptiste Boussingault, who had earlier studied the effects of tropical forests on temperature, addressed the contradictions in the hydrologic regime after deforestation in an analysis published in 1837.[64] Whether deforestation changes the climate of a region is a question of great importance and much heated debate, Boussingault noted.[65] On the one hand, the observed decrease in streamflow is evidence that forest removal decreases rainfall.[66] Yet clearings also produce torrential floods during heavy rainstorms, which calls into question the expectation that cutting down forests decreases rainfall. Boussingault's own investigation of lakes in South America and elsewhere led him to conclude that extensive deforestation does, in fact, decrease streamflow, but whether this results from less rainfall, greater evaporation, or a combination of both was unclear to him.[67] However, Boussingault reported that even small clearings insufficient to alter rainfall can change the hydrologic regime. The torrential floods following clearing are evidence that forests regulate the flow of water independent of changes in precipitation. Although Boussingault could not definitively say that deforestation reduces rainfall in all lands, "meteorological facts" led him to conclude that extensive clearing of forests in tropical regions decreases the annual rainfall. Boussingault returned to the topic of deforestation, climate change, and streamflow in later works.[68]

Protecting Forests for Climate

The belief that forests provide rain and maintain streams was used to justify forest conservation, and even planting trees, in France's island possessions and at home during the latter half of the eighteenth century and into the early nineteenth century. One advocate for forest conservation was Pierre Poivre, who was influential in developing policies to curb deforestation on the Île de France (modern-day Mauritius) in the Indian Ocean. In a 1763 address to an agricultural society in Lyons, he faulted the islanders for cutting down the forests and drying the soils because the rains "follow the forests exactly, stop there and no longer fall on the cleared lands."[69] His 1766 appointment as a colonial administrator on the island further underscored governmental concern about rain.[70]

Poivre's exact motivations for remaking the island's forests are uncertain, but climate change was central to his plan.[71] Poivre addressed the islanders upon his arrival and described the radical changes required in forests.[72] The island had made gains during its settlement, but he asked the colonists to consider "the immense loss of its woods and of the deterioration which results from it."[73] He chided them for having left their successors "only arid lands abandoned by the rains."[74] What had been an island paradise was now ruined: "Nature has done everything for the Isle of France: men have destroyed everything there."[75] Previously, the "magnificent forests" acted upon the passing clouds, which "resolved into a fruitful rain."[76] But after years of converting forests to farmland, the land was less productive, the rivers ran dry, and the sky, by ceasing to provide rain, "seems to avenge the outrages done to nature and to reason."[77] The island was on a path to ruin and in a few short years "would no longer be habitable; it should be abandoned."[78] Poivre's policies did produce substantial land reforms – even the creation of forest preserves. When the British gained possession of the island in 1810, they were astonished to find good land covered with trees and promptly cut them down to grow crops, only to later lament a perceived decline in rainfall.[79] Deforestation and its effects on climate continued to raise fear for several decades after Britain's seizure of the island.[80]

François-Antoine Rauch also raised French awareness of deforestation and climate change.[81] In his call for a nurturing plan to improve food supplies, published in 1792, he wrote of the influence of forests on "the harmony of elements," among which are "the clouds that they attract, the springs that they fertilize, the streams, the rivers that they feed."[82] With *Harmonie hydro-végétale et météorologique* (1802), Rauch explained his concept of harmony in nature.[83] Central to this harmony are forests and their influence on climate through the cycling of water with the atmosphere. The tops of trees "command from afar the wandering waters of the atmosphere, to come and pour into their protective urns the waters that nourish the springs, make the streams flow, refresh the green meadows."[84] Forests absorb vast quantities of water, and "as nature does nothing in vain" the water is returned to the atmosphere "by the transpiration of plants, to form dews, mists and new clouds."[85] Forests are, in Rauch's mind, "the nurturing mothers of fountains, limpid streams, rivers full of fish, commercial and navigable rivers."[86] Deforestation breaks this harmony, with devastating effects on climate, but Rauch had a plan to restore the climate by reforesting the countryside (Figure 3.4).[87]

The British also devised regulations on islands in the Caribbean and elsewhere to preserve forests for their climate benefits. Widespread deforestation throughout the Caribbean with the introduction of plantations had prompted concern about a decrease in rainfall, and a common theme to the legislation was the belief that forests attract clouds. Fear of drought led to forest preserves by the mid-1700s on Tobago "for the purpose of attracting the clouds."[88] Forests were similarly protected at Barbados in 1765.[89] On the island of St. Vincent, legislation enacted in 1791 created the King's Hill Forest Reserve "for the purpose of attracting the clouds and rain."[90] Alexander Anderson, director of the island's botanical garden, advocated for the reserve because forests provide shade to lessen the tropical heat, "nor is there any reasonable doubt that trees have a very considerable effect in attracting rain."[91] But the effectiveness of the tree planting was later questioned, and still more and taller trees were needed to attract and detain the clouds that produce rain.[92] Likewise, an account on "raining trees" published in Britain in 1831 retold of the miraculous water-producing tree on El Hierro in the Canary Islands and advocated reforesting the islands of the Caribbean to produce rain.[93]

British officials governing St. Helena, a remote tropical island located in the southern Atlantic Ocean, encouraged the growth of forests in 1794 "because it is well known that trees have an attractive power on the clouds."[94] Problems with drought could have been avoided "had the growth of wood been properly attended to." Alexander Beatson, governor of the island from 1808 to 1813, further promoted forest preservation to ensure the rains.[95] Beatson thought this was likely because "trees have a power of attracting clouds, as well as moisture from the atmosphere," but he admitted that lack of measurements prevented confirmation.[96] A British naval surgeon offered one reason why the islanders associated forests with rainfall. He observed that the frequent fog and mist found on the island coalesces into water droplets on leaves so that "the trees drop perpetual moisture."[97] Tree planting was encouraged on Ascension Island as well, to increase rainfall as was thought to have occurred following reforestation at St. Helena.[98]

■

Alongside the belief that deforestation moderates the winter cold was a long-standing convention that links forests with rainfall. The plentiful rainfall in the Americas was associated with the thick forests of the region, and deforestation was believed to reduce the rains. A similar belief unfolded elsewhere in the world, and conservationists seized upon the idea to advocate forest preservation and replanting to maintain or enhance rainfall. The scientific basis for forest influences on rainfall was found in new knowledge of stomata, transpiration, and leaf gas exchanges – advancements that were closely tied to discoveries in photosynthesis. Eighteenth- and nineteenth-century naturalists sought further evidence for changes in precipitation in the flow of water in streams and rivers. Interception of rainwater by the leaves in forest canopies, infiltration into the soil, runoff over the ground, and evaporation from the soil were identified as key determinants of streamflow. From this emerged the science of forest hydrology.

Figure 3.4 François-Antoine Rauch's depiction of a reforested France in the frontispiece to *Harmonie hydro-végétale et météorologique* (1802).
Rauch called for the reforestation of France and "to recreate this beautiful nature on all her surface."

Many of the thoughts on forests and rainfall voiced at the time are seen in today's science. Stomata, transpiration, and photosynthesis are recognized as key physiological processes by which plants influence climate and are included as components of climate models. The thoughts first expressed more than two centuries earlier by Buffon, Humboldt, and Williamson on tropical forests, the intertwining of evapotranspiration, precipitation, and temperature in feedback loops, and the dynamic coupling of forests with the atmosphere are not uncommon to modern-day climate scientists studying tropical deforestation. As was supposed, tropical forests lessen the heat of the region and contribute to the rainfall so that deforestation warms and dries the land. However, the science, while supportive of forest influences on rainfall, is much more nuanced than the broad claim that forests everywhere increase precipitation. The distinction is that this does not apply to all forests in all regions of the world, and the influence on rain varies with the spatial extent of deforestation. While modern science does accept forest–rainfall influences, albeit in a narrow context, the premise was sharply questioned 200 years ago. The rebuke only grew as the nineteenth century lengthened and the advocacy of forests galvanized into a worldwide campaign for forest conservation and tree planting to enhance rainfall.

4   Planting Trees for Rain, c. 1840–1900

The period from the mid-nineteenth century to the turn of the century brought heightened interest in the harmful effects of forest clearing on rainfall. There was growing fear that deforestation was turning prosperous lands into deserts, accompanied by efforts to conserve remaining forests or replant denuded lands. In India, Australia, New Zealand, South Africa, Russia, and the United States, the cry was raised: governments must, for the benefit of humanity, intervene to protect and manage forests for rainwater. These efforts followed prior concern in France and on British island possessions; indeed, the earlier tales of deforestation and drought in those lands were often recycled as the narrative spread throughout the world. In this, the proponents of forest conservation advanced a need for government control of forests and used forest influences on rainfall as justification. Opponents of state interference in forested lands, in turn, attacked the premise of forest–rainfall influences and decried the lack of evidence for climate deterioration. The ensuing debate was a narrative of misunderstanding, misuse of data, and hyperbole. It is this aspect of the forest–climate problem – the oft-called "desiccation theory" and its misuse to achieve policy goals – that has largely formed the historiography of the controversy, but beneath the rhetoric is found a fledgling knowledge of forest influences on climate that can be seen in today's science.

India

Reports about forests and rainfall appeared in popular British literature in the middle of the nineteenth century. In the eighth edition of the *Encyclopedia Britannica* (1859), the astronomer and multifaceted British scientist John Herschel considered lack of vegetation to be one reason for sparse rain, especially in warm climates.[1] Among the many plants and seeds listed in a weekly British horticultural periodical that same year was a communique on the lack of rain in England. The author attributed the dryness to the destruction of forests.[2] The water transpired by trees is discharged "into the atmosphere, where it forms clouds, condenses, and comes back in the state of rain." This is "the prodigious hydraulic power of vegetation." The meteorological society of Scotland sponsored a contest in 1859 to consider whether rainfall in Scotland and western parts of Europe had decreased. To the contrary, the prizewinning essayist concluded that it had not, and, moreover, there was no evidence for, nor need to worry about, a decline in rainfall from deforestation.[3] Later publications, however, furthered the cause for forests. A history of the Bible linked the deforestation of ancient biblical lands to a decline in rainfall.[4] The weekly magazine *Chambers's Journal* printed an article in 1876 on the uses of forests; one was to increase rainfall.[5] It is not surprising, then, to find alarm in all reaches of the British Empire that deforestation was decreasing rainfall, as well as calls to conserve forests or plant trees to ensure the well-being of British colonies.

Foresters working in British India embraced the forest–rainfall question, and in doing so shaped public perception of forest conservation elsewhere in the world.[6] The tropical forests of India provided needed timber, and Britain, in the latter part of the nineteenth century, established an encompassing bureaucracy to administer the forests. By the late 1830s, reports linking deforestation to drought had begun to appear.[7] Medical officers serving British interests in India widely echoed the concerns over deforestation, rainfall, and water supply that had been expressed previously in other countries.[8] Donald Butter, a military surgeon, described the benefits of the "sylvan vesture" presented by the forest cover.[9] Cutting down the forests would warm the temperature, diminish the rains, and decrease the water in springs and rivers, he warned. Alexander Gibson, a surgeon and botanist, was a leading proponent for forest conservation in India.[10] The climate benefits of forests necessitated their protection, he argued in 1846.[11] Edward Balfour also extolled the virtues of forests.[12] While admitting the existing measurements were insufficient and further investigation was needed, it was evident to him that "the quantity of rain which falls alters as the trees are diminished or increased."[13] His writing invoked the perils of a positive feedback in which forests protect against drought and their removal initiates a cascading decline: "Nature when left to herself provides a compensatory influence in the dense leafy forests but if these are consigned to destruction every successive drought will prove more baneful than the preceding."[14] Balfour's solution to the climate destruction evident in India was to plant trees.

The medical officers were effective in making their case for the climate benefits of forests. In an 1851 report, the governor-general of India decried the destruction of forests. The shortage of timber was alarming, but so, too, was loss of the "salubrious and fertilizing effect of foliage."[15] Nicol Dalzell, another actor in the Indian forest service, also told of the detrimental consequences of deforestation.[16] While acknowledging some benefits, more commonly deforested lands "are condemned to the miseries of dearth, drought, and barrenness."[17] The lesson of history is to protect forests "so that those salutary effects on the climate and water-supplies . . . may continue permanent."[18] In his instruction for the practice of forestry, William Schlich included the climate influences of forests as an "indirect utility of forests."[19] A German-trained forester, Schlich attended to British forestry interests in India and developed a formal education program for British foresters that was tailored to tropical forests. His summary of forest–climate influences was based on European studies of forested and open lands. As to the influence of forests on rainfall in India, Schlich was more cautious.[20] John Nisbet, a British forester in India, penned a lengthy essay in the scientific journal *Nature* on the climate significance of forests, not just in India but elsewhere in the world.[21] He ended with a plea to plant trees "for the purpose of ameliorating the climatic conditions for man." In his 1900 survey of forestry in India, Berthold Ribbentrop, too, was sure of the climate benefits of forests. "The wholesale destruction of forests has had the most deteriorating effect on the climate of India," he asserted.[22] This "is certain."

Much of the advocacy of forest conservation for climate benefits was, as elsewhere, lacking in observations. In an 1851 report to the British Association for the Advancement of Science, Hugh Cleghorn tempered his conclusions with the comment that there was "a deficiency of *exact* or experimental information."[23] As rainfall observations became more widespread, meteorologists developed knowledge on the geography and dynamics of Indian rainfall. The focus was to understand the seasonal monsoons of the region, but the data presented an opportunity to infer the influence of forests. A series of papers published in the 1880s shows both the scientific interest and also the counternarrative of misunderstanding and misuse of observations that was expressed in other countries. Russian meteorologist Alexander Voeikov discerned forest influences through comparative analysis of regions that differed in forest cover, and answers to the forest–rainfall question could be found in the forests of India.[24] He observed differences in rainfall in sections of India that he attributed to the amount of forest cover.[25] The belief that forests influence rainfall was, however, met with skepticism at the British Royal Meteorological Society.[26] In a discussion of Voeikov's findings, George Symons, a past president of the society, said that "he had been unable to find any reliable statistical evidence in support of the general belief." The president at the time, Robert Scott, likewise claimed that "no precise numerical values existed by which the truth of the belief could be tested." Nonetheless, a report of Voeikov's work in *Nature* affirmed that "man by afforestation and dis-afforestation can modify the climate around him."[27]

Henry Blanford, the chief of the meteorology department in India, took a different approach.[28] He analyzed precipitation at fourteen meteorological stations in a large region extending over nearly 160,000 square kilometers that had been cleared of trees. The forest was subsequently allowed to regrow and covered 80 percent of the land. The rainfall was larger for a ten-year period after reforestation than for the years immediately prior to regrowth. More importantly, rainfall increased progressively as the forest regrew. "The evidence, thus afforded, in favour of the influence of forests on rainfall appears to me to be of considerable weight and importance," Blanford concluded.[29] The Austrian meteorologist Julius Hann concurred.[30] Blanford's analysis led Dietrich Brandis, a German-born forester who directed forest management for the British in India at the time, to enthusiastically assert that better care of forests would increase precipitation and even if only a small amount, "the benefits for the country and its people would be incalculable."[31] The reliability of the measurements Blanford used was questioned, but his findings were highlighted in US governmental reports on forests and climate.[32]

Australia and New Zealand

Concern about deforestation and drought spread to other British lands, some with far fewer trees than in India (Figure 4.1).[33] Stories that linked deforestation to a decline in rain were common in the Australian and New Zealand press beginning in the mid-nineteenth century and continuing into the twentieth century, and governments in both lands wrestled over the need to preserve forests to protect the climate.[34] An 1845 British translation of Boussingault's *Économie rurale* encouraged settlers to plant "a thick screen of leafy trees" around Sydney, on "every knoll within a hundred miles," to help conserve rainwater.[35] A later writing by William Branwhite Clarke, an Anglican clergyman, geologist, and accomplished natural scientist in New South Wales, recounted the dire effects of deforestation told by others.[36] Clarke had previously considered the matter in an 1835 publication.[37] Forty-one years later and near the end of his life, Clarke revisited the topic in the context of the droughts common to Australia. Nature provides solutions to drought that "defy the wisdom of man to parallel," Clarke related.[38] Science reveals "the mysteries of the visible creation," and, with knowledge in hand, it is humankind's duty to employ "the friendly forces of Nature to our advantage." One of those friendly forces was the use of trees to protect water supplies, and the trees of Australia presented advantages not found in other forests of the world. Ferdinand von Mueller, a government botanist in Melbourne, had found in the drought-resistant acacia and eucalyptus "the means of obliterating the rainless zones of the globe, to spread at last woods over our deserts, and thereby to mitigate the distressing drought."[39] Clarke, in describing the many benefits of forests, agreed.[40] The climate benefits of forests expansion in Australia were reported in *Nature* as being "well worthy of attention."[41]

Figure 4.1 Bush clearing near Oeo (Thomas Good, 1893).
Depiction of forest clearing on the North Island of New Zealand. Mount Taranaki is shown in the distance. Reproduced from ref. (33) with permission of the Alexander Turnbull Library (Wellington, New Zealand).

Fear that deforestation was changing the climate of New Zealand arose in the 1840s and reached New Zealand's Parliament in 1868 when the government debated legislation on forest conservation.[42] One rationale for the legislation was the desirability of forests from "a climatic point of view."[43] William Travers, a politician, naturalist, and a founder of the scholarly New Zealand Institute, later characterized the changing climate brought about by deforestation as a "violation of natural laws."[44] That changes were occurring in many parts of New Zealand "there can be no doubt." Parliament again debated the merits of forest conservation for climate benefits in 1873 and 1874.[45] The sponsor of the 1873 legislation encouraged his fellow members of Parliament to pass the bill "so that history might not be able to relate that they received a fertile country, but, by a criminal want of foresight, transmitted to posterity a desert."[46] It was a claim echoed by others. One contributor to the intellectual argumentation claimed that "to strip a semi-tropical country of its forests is to convert it into an arid desert."[47]

The 1874 legislation again highlighted the climate effects of deforestation.[48] The legislation was necessary because "it is expedient to make provision for preserving the soil and climate by tree planting." Julius Vogel, who as premier headed the government, sponsored the bill and presented a lengthy oration on the perils of deforestation and the many efforts in other countries to conserve forests. He urged passage of the bill because the forest question "is of something for New Zealand to cling to for generations; to shape its future; to decide its climate, its adaptability for settlement, its commercial value, its beauty, its healthful-ness, and its pleasure-bestowing qualities." New Zealand subsequently appointed a conservator of forests, who argued that the climate influences of forests needed to be considered in managing New Zealand's forests.[49] Other calls to replant and conserve forests followed, to prevent consequences "of the most disastrous nature to the ensuing generation" and "to maintain the protection given by nature against the disturbance of the climatic equilibrium, the occurrence of droughts, the disastrous effects of flood-waters, etc., etc."[50] Reports on the climate effects of deforestation and calls for action continued to appear late in the century, but the debate played out after a few decades and there was less interest in deforestation and climate change after the turn of the century.[51]

South Africa

Concern about drought, a drying of the land, and a decline in water supply shaped forest policy in South Africa. John Fox Wilson authored a report in 1865 asserting that the desiccation was due to the clearing of shrubs and trees.[52] He repeated the many narratives of others as proof of the effects of deforestation, and he described a positive feedback in which drought kills trees and shrubs for lack of moisture, resulting in "the effect of drought in this instance becoming in its turn an auxiliary cause of drought."[53] His understanding of the process by which loss of vegetation increases aridity in drylands is prescient of the modern science of land–atmosphere feedbacks in drylands:

> The more denuded of trees and brushwood, and the more arid the land becomes, the smaller the supply of water from the atmosphere. The greater the extent of heated surface over which the partially exhausted clouds have to pass, the more rarefied the vapour contained in them necessarily becomes, and the higher the position which the clouds themselves assume in the atmosphere under the influence of the radiating caloric; consequently the smaller the chance of the descent of any rain on the thirsty soil beneath.

John Croumbie Brown shared Wilson's views and promoted the use of forests to combat drought.[54] A Scottish minister and missionary, he became a prolific writer on forests and climate. He relied on the tried-and-true narratives of climate change told so many times by others, but, in contrast to other avid conservationists at the time, Brown also brought a scientific focus to his writings. His 1863 report as the government botanist for the Cape Colony revealed his passion for forest conservation, even for expanding the area of forests, and his dismay at the current state of the landscape. As mankind has colonized new lands, "he has found the country a wilderness; he has cut down trees, and he has left it a desert."[55] This passion – an almost reverence for forests – colored his many subsequent writings, but Brown's interest in science was already apparent. He described the physics of evaporation and believed that increased soil evaporation upon forest clearing contributed to the desiccation of the region.[56]

Brown's main focus was on the aridity of South Africa, and he expanded his thinking in a series of books. His first work on the hydrology of South Africa described the general desiccation of the region.[57] One cause of dryness was the destruction of forests, which allowed for greater evaporation from soil exposed to heat and drying winds. As to the influence of forests to increase rain, Brown was more reserved: "This is a matter which must be determined by the testimony of the rain-gauge, and not by *a priori* reasoning."[58] Nonetheless, he thought it "not improbable" that there is greater rainfall over forests in a recycling of rainwater that is evaporated and precipitated back to the soil and evaporated again.[59] Brown's solution to the aridity of South Africa was to construct dams and plant trees. The atmosphere, rivers, reservoirs, and aquifers could provide the water needed to irrigate the land and grow trees.[60] Brown's views on deforestation and water were greatly influenced by French research. His *Reboisement in France* (1876) provided translations of the extensive French studies on deforestation and mountain floods, including the works of Jean-Antoine Fabre and Alexandre Surell.[61] Brown saw in France, with its science of forest hydrology, its laws to protect forests, and its reforestation, the solution to deforestation in South Africa. A later publication on the famous French forest ordinance of 1669 and the history of forests in France continued this theme.[62]

Brown's 1877 book *Forests and Moisture* put forth his overall concept of forest influences on climate and water.[63] He examined the influence of forests on the humidity of air, the formation of clouds and rain, and the supply of water on land. Other forest advocates employed alarming anecdotes and repeated oft-told tales of desiccation in their calls for forest conservation. Brown, as well, succumbed to the biases and rhetorical excesses of the day – too accepting of unreliable accounts and too dismissive of incongruous meteorological observations.[64] Nonetheless, in *Forests and Moisture* he brought a scientific focus that was wanting in the writings of others. While still lacking in specific observations, Brown effectively outlined a theory for forest–rainfall influences that was interdisciplinary in breadth and combined aspects of plant physiology, the physics of evaporation, the meteorology of clouds and rain, and forest hydrology. Brown considered the workings of trees and their interaction with the atmosphere at multiple spatial scales: from the meteorological effect of a single leaf, imperceptible at a quick glance; to that of a tree, more evident but still small; to "the effects produced by a forest with its countless trees, boughs, and leaves."[65] Brown was "reasoning from the lesser to the greater," but also vice versa "from the greater to the less," to obtain "clearer and more comprehensive views of the truth." To understand transpiration, Brown described the workings of stomata, the rise of sap in trees, the physics of water movement, and the absorption of water by roots. To Brown, the amount of water transpired by a forest is determined by "the number of stomata, or stomates, on a leaf, multiply this by the number of leaves on a branch, the product by the number of such branches on a tree, and the product of this by the total number of trees in the clump, or the total number of trees in the forest."[66] Modern-day scientists still struggle with the nuances and complexity of this scaling.

Russia

Russia, too, was swept into the forest–climate debate. As early as 1828, a report on the forest of Białowieża, located on the border between present-day Poland and Belarus, related the cold climate and abundant flow of water in streams to the vast forest of the region.[67] Portions of primeval forest remain today and the ancient woodland is designated a United Nations World Heritage site, but already then logging had reduced the forest. Clearing the trees was believed to have ameliorated the harshness, but to have also dried up the streams. Upon returning from his expedition across Russian Asia in 1829, Alexander von Humboldt advanced the view that deforestation produced a drying of the land and called on Russian academics to study the effects of deforestation on climate.[68] Thereafter, reports linking deforestation to a more severe climate – extreme aridity, scorching summer heat, acute winter cold – appeared in Russian forestry publications and elsewhere in the 1830s and 1840s.[69] An 1845 book on the geology of Russia furthered the claim that logging was "a prime cause" of drought in the country.[70]

As in other countries, the forest–climate debate grew louder in the middle of the century before crashing at the end of the century, and, as elsewhere, it centered on the belief that deforestation was desiccating the land – in this case the steppes of Russia. Replanting the forests would increase rainfall and restore the productivity of the land. The forest question grabbed the imagination of Russians and can be seen in the arts of the day.[71] In one telling example, a character in Anton Chekhov's play "Uncle Vanya" – Mikhail Astrov, a medical doctor and a strident forest enthusiast – bemoans the deterioration of climate and the desiccation of rivers that follows deforestation, but he asserts his empowerment over climate by planting trees and by saving forests from destruction: "When I walk by the peasant forests that I've saved from the axe, or when I hear the rustling of my young forest, planted with my own hands, I realize that I have a little power over the climate too, and if a thousand years from now people are happy, then it's partially my doing, too. When I plant a small birch tree and then watch it turn green and sway in the wind, my heart fills with pride."[72] And as in the similar debate raging across the United States, familiar arguments and counterarguments were voiced by proponents and skeptics of planting trees for climate benefits: tales of climate deterioration found in other lands and throughout history, personal experience of climate change over a lifetime, but memories of better climates in the past are not infallible; meteorological observations prove that climate is stable, or if it is changing, it is for reasons other than deforestation; large-scale features of the atmosphere control rainfall; precipitation determines the occurrence of forests and not vice versa.[73]

The climatologist Alexander Voeikov, a leader in the Russian meteorological community, was a key voice advocating for forest influences.[74] He understood that people modify the climate near the ground – temperature, humidity, rainfall, wind speed – through their actions on vegetation, and especially the forests.[75] Voeikov favored geographic analyses to discern the influence of forests on climate. In looking at climatological maps, Voeikov found notable temperature anomalies that he attributed to the presence of forests.[76] Tropical forests of South America and India, for example, cool temperature because of high evapotranspiration and greater cloud cover over the forests. A reduction in temperature due to forests could be found in parts of Europe during the summer. Voeikov also thought that forests can increase rainfall in certain conditions, as seen, for example, in India. Malaysia presented another example where forests increase rainfall.[77] These studies formed the basis for a chapter on the effects of forests on climate in Voeikov's influential book on climatology published in 1884 and translated to German in 1887.[78] Voeikov's writings particularly influenced British and American views on forests and climate.[79]

The United States

Conserving forests for their beneficial influence on climate and reforesting previously cleared lands, even foresting the prairie, became the focus of horticulturalists, botanists, and foresters in the United States in the mid-1800s. A survey on the trees of Massachusetts in 1846 touted the climate benefits of forests in protecting against the extremes of winter cold and summer heat, storing rainwater, and blocking strong winds.[80] In an 1844 account of his travels in the western arid lands, Josiah Gregg described the plains as "dry and lifeless." Cultivating the soil might increase the rain, and so, too, "shady groves, as they advance upon the prairies, may have some effect upon the seasons." Evidence could already be found in that "the droughts are becoming less oppressive."[81] A governmental report on agriculture for the year 1849 included an article on "agricultural meteorology," which, relying on the authority of Humboldt, Boussingault, and others, told of the harmful effects of deforestation.[82] Richard Upton Piper, in his *Trees of America* (1855), extolled the benefits of forests for climate and water supply.[83] Other reports to agricultural societies followed, and state legislatures requested information on the status of forests and agriculture in their states.[84] Many of these reports spoke of the harmful effects of deforestation and the climate benefits of forests. A report commissioned by the Wisconsin legislature in 1867 to investigate "the injurious effects of clearing the land of forests upon the climate" had a particularly alarming title: "Report on the disastrous effects of the destruction of forest trees."[85]

The reason for this interest was George Perkins Marsh. The 1864 publication of *Man and Nature* brought deforestation and the forest–climate question to the forefront of scientific, political, and public debate in the United States.[86] Hailed as a milestone in the birth of the forest conservation movement, Marsh broadly described the deleterious effects of humanity on the environment and warned of the collapse of civilization through deforestation and environmental degradation. In a lengthy chapter on forests, Marsh touched upon deforestation and climate change. He expressively summarized the consequences of forest clearing in the following passage:

> With the disappearance of the forest, all is changed. At one season, the earth parts with its warmth by radiation to an open sky – receives, at another, an immoderate heat from the unobstructed rays of the sun. Hence the climate becomes excessive, and the soil is alternately parched by the fervors of summer, and seared by the rigors of winter. Bleak winds sweep unresisted over its surface, drift away the snow that sheltered it from the frost, and dry up its scanty moisture.[87]

As to rainfall, Marsh wrote that forest influences are "so exceedingly complex and difficult" and the evidence "conflicting in tendency, and sometimes equivocal in interpretation."[88] Nonetheless, he felt that many foresters and physicists were like-minded in the view that deforestation decreases rainfall. The bareness and aridity of deserts presents a powerful image of the dynamic coupling between vegetation and rainfall, and Marsh, like Humboldt and others before him, seized the thought. "It has long been a popularly settled belief," Marsh wrote, that vegetation and rainfall "are reciprocally necessary to each other." As proof of the commonness of this opinion, Marsh quoted the following poem:

> Afric's barren sand,
> Where nought can grow, because it raineth not,
> And where no rain can fall to bless the land,
> Because nought grows there.

With climate change and other calamities from deforestation, "thus the earth is rendered no longer fit for the habitation of man."[89]

The timing of Marsh's publication coincided with the settlement of arid western lands and led to advocacy of tree planting to increase rainfall.[90] The federal government provided credibility to the claims. Joseph Henry, secretary of the Smithsonian Institution, wrote in 1855 that the climate effect of deforestation "deserves more attention than it has usually received."[91] To his thinking, "it may be profitable to allow forests of considerable extent to remain in their pristine condition." His Smithsonian Institution later translated the findings of the noted French physicist Antoine-César Becquerel for American readers.[92] Protecting the supply of timber motivated an 1865 call to action for Congress to enact a national policy on forests, but so, too, did the benefits forests provide for climate.[93] The central interior region of the country lacked the moderating influence of oceans found along the coastal regions, but foresting the lands west of the Mississippi River would provide a similar effect. Why, pleaded the author, "shall we leave the lands to neglect and comparative barrenness, when, by adding forests as great modifiers and controllers of temperature and precipitation, they may probably become as desirable as any lands we possess?"[94]

Joseph Wilson, commissioner of the General Land Office (US Department of Interior), included in his annual report for 1868 a lengthy account of the climate perils of deforestation.[95] To the contrary, planting trees in the prairie was moderating temperature and improving rainfall. Such lessons should be "seriously heeded" and necessitated governmental action.[96] "The redemption of sterile and desert lands is one of the growing ideas of the times," and a law should be enacted requiring settlers to plant trees on their homesteads.[97] Wilson predicted that "if one-third the surface of the great plains were covered with forest there is every reason to believe the climate would be greatly improved, the value of the whole area as a grazing country wonderfully enhanced, and the greater portion of the soil would be susceptible of a high state of cultivation." The geologist Ferdinand Hayden, who led geological surveys of the west for the federal government, also advanced the notion of tree planting in western lands.[98] Hayden believed that the prairie was forested long ago in ancient times. The past forests could be restored; doing so would make the region wetter and more prosperous. Hayden's surveys included testimonies that rainfall was already increasing with westward settlement of the country. This thinking led him to claim that "the planting of ten or fifteen acres of forest trees on each quarter section [160 acres] will have a most important effect on the climate, equalizing and increasing the moisture."[99] Wilson seconded these ideas in his annual report for 1867.[100] Notices of forest effects on rainfall also appeared in the annual reports of the commissioner of agriculture.[101]

The US Congress agreed and enacted the Timber Culture Act of 1873 to promote tree planting in the western prairie. The law allowed settlers to claim 160 acres of land if they planted and kept healthy 40 acres of trees spaced "not [. . .] more than twelve feet apart" for ten years.[102] The author of the act, Senator Phineas Hitchcock from Nebraska, intended the legislation "to encourage the growth of timber, not merely for the benefit of the soil, not merely for the value of the timber of itself, but for its influence upon the climate."[103] An 1874 revision to the law shortened the time to eight years and spelled out a schedule for tree planting: ten acres in the second year; an additional ten acres in the third year; and twenty acres in the fourth year.[104] The required planting was reduced in 1878 to a total of ten acres planted in the third and fourth years with at least 2,700 trees per acre, and 675 had to be "living and thrifty" when the land claim was filed.[105] The legislation was repealed in 1891.[106]

The year 1873 saw the emergence of Franklin Hough as a leader in promoting forests. That year, Hough read a paper at the annual meeting of the American Association for the Advancement of Science "on the duty of governments in the preservation of forests."[107] He acknowledged the action of Congress to encourage tree planting with the Timber Culture Act, but still more was needed to protect forests. Chief among his reasons was the beneficial effect on climate and water supply. Hough called on states and Congress to enact further legislation and for the establishment of schools of forestry. In response, a committee was formed, chaired by Hough, to inform Congress on forest policy.[108] The committee reaffirmed Hough's call for action and concern for climate. To back up their claims, the committee quoted extensively for the work of Jules Clavé in France and Ernst Ebermayer in Germany. One recommendation was to appoint a commissioner of forestry within the federal government to study the status of forests, including "the influence of forests upon the climate."[109] Another need was to create forest meteorological observatories, similar to those then operating in Europe, to gather data.[110] For Hough, a forested America was paramount. The solution to environmental calamities was "plain and obvious: Plant trees."[111] When, in 1876, Congress authorized preparation of a report on the status of forests in America, Hough was appointed to the position. Subsequently, he became chief of the newly formed forestry division in the Department of

Agriculture. As Hough had proposed, one of the charges from Congress was to assess the influence of forests on climate.[112]

The result was Hough's monumental 1878 *Report upon Forestry*, with a more than 100-page review of forests, climate, and water supply.[113] Rainfall was one benefit of forests, and Hough advocated tree planting in the prairie (Figure 4.2).[114] A substantial portion of Hough's report retold long-ago tales of deforestation at Madeira, St. Helena, Ascension Island, and Mauritius and relied on anecdotes from agriculturalists, horticulturalists, and other individuals. For quantitative data and observations, Hough, as in the earlier 1874 report to Congress, looked to the findings of German and French researchers. In addition to Ebermayer's work, Hugh translated and included an abridgement of Becquerel's monograph on forests and climate.[115] Marsh, too, in *Man and Nature* had relied on European scholars for his review of the climate effects of deforestation. Contemporary American meteorologists, in developing their science, favored a dynamical perspective of climate based on large-scale atmospheric circulations and dismissed the influence of forests on climate as unfounded. There were no measurements of forest climates in the United States, unlike the forest meteorological observatories that had become common in Europe. In the coming decades, the contempt of prominent American meteorologists for the forest–climate question became only more palpable. Hough's *Report upon Forestry*, rather than settling the issue, only sparked more heated scientific controversy.

The fervor for forest conservation and tree planting resonated with public imagination in many forms. Ideas about deforestation and rainfall in the Caribbean, first expressed during Spanish settlement in the sixteenth century, continued to be conveyed 300 years later. A writer in 1872 remarked on the destruction of trees on the island of St. Croix. Whereas previously "woods covered the hills, trees were everywhere abundant, and rains were profuse and frequent," now, after the forests had been cut away, the rains had stopped and "the whole island seems doomed to become a desert."[116] The influential newspaperman Horace Greeley advocated reforestation to protect against floods, ensure a reliable supply of water, and for shelter from cold and strong winds. This, he claimed, "is measurably true of every rural county in the Union."[117] Ellwood Cooper, a proponent of planting eucalyptus trees in California, extolled the benefits of forests. He described the climate of California as lacking, but "moderate the winds, increase the rain, and we have perfection."[118] The remedy was obvious: "How is this to be done? How are we to obtain this

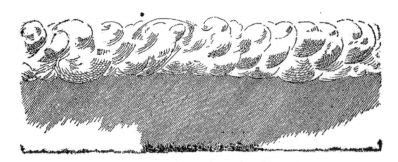

Figure 4.2 Franklin Hough's sketch of rain falling over forests in the prairie.
Hough believed that by cooling the air and increasing its relative humidity, forests favor condensation and rainfall.
A passing cloud might have scant rain, he wrote, but "I have also seen, where this cloud passed over a forest of considerable area, that these filaments of rain became a copious shower, which dried up when they came over the heated and naked fields beyond." Quotation and drawing from ref. (114).

result? By planting forest trees." Popular science magazines, too, reported on the climate benefits of forests.[119] One lasting outcome of the fervor to plant trees in the United States was the founding of Arbor Day in 1872 by the Nebraska state board of agriculture.[120]

That trees bring rain was certainly tied to efforts to entice settlement of the plains region. Money was to be made by selling land and building railroads, but first the perception of an inhospitable climate had to be overcome. Enter Richard Elliott, a railroad official and advocate of tree planting. Foresting the prairie "is a work worthy of the age, and of the nation," he claimed.[121] Elliott further asserted that the presence of the railroad itself, along with telegraph lines, was bringing moisture to the lands. To gain credibility, he brought his ideas to the attention of Henry and Hayden, who dutifully included Elliott's writings in their reports.[122] Elliott envisioned himself to be "in some measure a geologist, a botanist, a farmer, a meteorologist, a horticulturalist, and a philosopher in general," but he was also an "industrial agent" for the railroad, and his ideas of climate betterment provided "a means of advertising the road!"[123] Even if he was wrong and planting trees did not improve the land, an entirely acceptable outcome, he admitted, was that he "could at least demonstrate that it had a railway in it!"

Some years after enthusiasm for foresting the prairie had fallen into eclipse, Elliott still extolled its benefits.[124] "A time is rapidly coming when in all parts of the vast plains ... there will be farm buildings, towns, fields, orchards, and groves of forest trees, all having more or less effect on the atmospheric conditions; and, so far as climatic modification is concerned, the progress of settlements can have only ameliorating influences."[125] It was this type of rhetoric that led the editors of *Science*, in reporting on the forest–rainfall controversy in 1888, to say that as "great corporate or private interests are to be affected by its decision, much of the discussion of it, unfortunately, has not been of a purely scientific character."[126] The editors specifically called out the railroad companies for their duplicity: "But the agents of the railroad companies reply that the climate of the Far West has changed; that the planting of trees upon what was once arid lands has increased the amount of rainfall ... and they quote figures to prove it." Elliott himself acknowledged that he "made up a 'learned' report" on tree planting for Hayden in consultation with railroad officials.[127]

Rain Follows the Plow

Advocacy for tree planting to increase rainfall in the arid plains of the United States was followed with further ideas to improve the climate by cultivating the prairie. Nebraska in the 1870s was a hotbed of ideas related to climate change.[128] In a lecture given in 1873, Samuel Aughey, a professor at the University of Nebraska, asserted that the rainfall in Nebraska had been increasing as a result of planting trees, and "the climate, now good, will be still further ameliorated" with more trees.[129] Aughey, however, is more remembered for promoting the notion that plowing the prairie sod increases rainfall (Figure 4.3).[130] Although "the growth of forests exercises the happiest influences on climate," still other factors contributed to increases in rainfall in the region, Aughey argued.[131] In particular, cultivation allows the soil to retain more rainfall, which evaporates and rains back onto the land in a feedback whereby wet soils produce more rainfall. "After the soil is broken, the rain as it falls is absorbed by the soil like a huge sponge. The soil gives this absorbed moisture slowly back to the atmosphere by evaporation. Thus year by year as cultivation of the soil is extended, more of the rain that falls is absorbed and retained to be given off by evaporation, or to produce springs. This, of course, must give increasing moisture and rainfall."

Figure 4.3 Plowing on the prairies beyond the Mississippi (Theodore R. Davis, 1868).
The sketch depicts a farmer breaking soil for the first time on a Kansas prairie with a team of oxen. Reproduced from ref.
(130).

Charles Wilber popularized this concept with the phrase "rain follows the plow," evoking the image of the industrious farmer and his plow turning over the prairie sod.[132] The farmer makes "a new surface of green, growing crops instead of the dry, hardbaked earth covered with sparse buffalo grass. No one can question or doubt the inevitable effect of this cool condensing surface upon the moisture in the atmosphere." Wilber elaborated on Aughey's conjecture and further advocated a process by which "a large area of saturated soils, extending over several counties appears to have, from this cause, the power to perpetuate rainfall," but "a large tract of dry land extends and perpetuates dryness."[133] Wilber, like Humboldt, Marsh, and others, invoked the imagery of the desert: "The desert thus has a tendency to establish itself. The greater the area of sand surface the greater in quantity is the radiant heat produced. It heats a still greater amount of atmosphere, and with added force forbids the condensation of moisture, or the forming of clouds."

As with forests, the premise that plowing the prairie increases rainfall engendered caustic retorts. Nearly sixty years after Aughey proposed the idea, the US Department of Agriculture released a technical bulletin aimed at discrediting the theory.[134] The author of the report, a climatologist with the Soil Conservation Service, asserted that evapotranspiration from land does little to provide moisture for rainfall. If local evapotranspiration contributes to rainfall, "droughts will perpetuate themselves" and "periods of ample and abundant precipitation will tend to foster a continuance of such rainy spells" – ideas, the author claimed, that were lacking in proof and readily dismissed.[135] It is a "myth," he went on, that plowing the prairie increased rainfall, as subsequent droughts proved.[136] Yet feedbacks among soil moisture, evapotranspiration, and precipitation are central to the modern science of droughts and heat waves, and though droughts in the central United States have their origins in atmosphere–ocean interactions, dry soil and loss of vegetation amplify the lack of rain.

■

Advocation of forest conservation for political, cultural, or economic reasons is the narrative most commonly used by historians of science to describe the forest–rainfall controversy. In that telling, proponents of forest influences on rainfall are faulty, even duplicitous, in their thinking. The science, whatever there was, was inadequate, and the heroes are those individuals who forcefully rebutted the inflated claims. It is a thinking evidenced in the journal *Science*, which reported on the controversy as it rose to a head in the debates at the Philosophical Society of Washington in 1888. Certainly, planting trees in arid lands did not cause rainfall to increase. Nor did plowing the prairie bring rain. Recurring droughts in arid lands deflated the enthusiasm for planting trees. But flawed science and its misuse is only one part of the narrative. The science was, in fact, blossoming and new discoveries were being made.

5   Making a Science: Forest Meteorology, c. 1850–1880

Planting trees to increase rain was the grand climate controversy of the nineteenth century, having great social and economic repercussions, and with the emergence of meteorology as a science, meteorologists entered the debate. Their views, like the science itself, were complex and messy, and there was not a united position: some voiced support, others expressed contempt, and still more were tentative and qualified in their opinions. There are three elements to the story. One thread is the rise of forest meteorology. Interest in forests and rainfall was rampant in all corners of Europe throughout the nineteenth century and took on all manners of view.[1] The American meteorologist Mark Harrington remarked that "every variety of opinion can be found there."[2] Some scientists – in France, Germany, and Austria and with diverse backgrounds in physics, meteorology, forests, and soils – developed a new science of forest meteorology that blended meteorology, forest ecology, and forest hydrology. They sought answers in direct measurement of forest influences on climate and installed meteorological observatories in forested and open lands to obtain the necessary data. It was an understanding developed at the microscale and extrapolated to the macroscale.

Counter to the science of forest meteorology, prominent meteorologists in the United States held that climate was, in fact, unchanged despite more than two centuries of forest clearing. In examining macroscale patterns of temperature and precipitation, they could not find a signal of forest influences, and their theories of large-scale atmospheric dynamics, likewise, did not accommodate forests. In this, the second thread of the forest–climate question, the voices of meteorologists prevailed, and by the early twentieth century, the great debate about forests and climate, after several hundred years of thought, was over. Yet the science was only in hiatus. In part three of the story, the science arose again in the latter half of the twentieth century as atmospheric scientists considered anthropogenic climate change. The science of forest–climate influences, so resoundingly rejected at the turn of the twentieth century, has now been reinvented, first by atmospheric scientists and then by other disciplines as well, as forests are again seen as a means to improve climate. This chapter considers the first part of the narrative and the rise of forest meteorology as a science. It begins with an assessment of climate anxiety in France, from which arose the groundbreaking studies of the physicist Antoine-César Becquerel on forests and their climate influences. The German Ernst Ebermayer, too, launched his research on the physical effects of forests on air and soil, as did the Austrian Josef Roman Lorenz von Liburnau with his study of forests, climate, and water.

France and Her Forests

Deforestation, reforestation, and climate change were a focal point for scholarly, political, and public thought in late-eighteenth and early-nineteenth century France; indeed, forests were central to the rise of environmental awareness in France.[3] State claims to forests and the enactment of a national forest policy were intertwined with climate change. Pierre Poivre had previously called for the protection of forests on France's island possessions, and Jean-Antoine Fabre had warned of the torrential floods brought by deforestation. As the nineteenth century progressed, Alexander von Humboldt, Jean-Baptiste Boussingault, and Alexandre Surell studied the effects of deforestation on climate and streamflow. François-Antoine Rauch, too, told of the devastating effects of deforestation on climate.

Another entrant in the discourse was the botanist Charles-François Brisseau de Mirbel, who wrote in 1815 of forest influences in his instruction on plant physiology and botany. "Forests stop and condense clouds; they give off torrents of aqueous vapors into the atmosphere; the winds do not penetrate their enclosure; the sun never warms the land they shade," he explained.[4] Deforestation, instead, warms and dries the land. As a consequence, the famous lands of antiquity were now "only scraggy rocks and arid sands inhabited by miserable villages."[5] In vain, the traveler to ancient lands "seeks several rivers whose names history has preserved, they are erased from the earth." In this, Mirbel was associating the decline of Mediterranean civilizations with deforestation. Rauch had similarly linked the fate of ancient cultures to deforestation in *Harmonie hydro-végétale et météorologique*. This thought was an enduring refrain throughout the forest–climate controversy, picked up by George Perkins Marsh in *Man and Nature* and carried forth in the call to plant trees for rain.

Climate change preoccupied France, so much so that the French minister for the interior requested, in 1821, a survey of the changed climate.[6] Views on the matter were mixed – whether climate was changing at all, but if so, in what way, and did deforestation cause the change – but one response to the survey was telling of a position that only hardened over the ensuing decades. To one skeptic, an understanding of weather and climate required study of the atmosphere not of "the little action of our deserts or our forests."[7] Meteorology, the skeptic was saying, was for meteorologists, not botanists or foresters.

The forest–climate question seized the Royal Academy of Brussels, which sponsored a competition in 1825 to study the problem. A comprehensive report by Alexandre Moreau de Jonnès, who later became a senior French official in charge of governmental statistics, assessed the state of knowledge at the time.[8] He brought together in one document the actions of forests, as well as the consequences of deforestation, not only on temperature but also rainfall, the humidity of the air, winds, and water supply in rivers and springs. Moreau framed the factors determining climate in a manner that would be common in later studies.[9] Geographic location, atmospheric circulations, proximity to oceans, elevation of the land, and the presence of mountains are the primary determinants of climate, but surface conditions also matter. The type of soil, whether it is sandy or clayey, affects temperature and so, too, do forests decrease temperature compared with open lands.[10] The cooling influence is greater in tropical climates than in temperate climates. In this latter comparison, Moreau recognized the dominant role of evapotranspiration in the tropics, but also its lesser importance in temperate forests.[11] The modern science reveals a similar understanding. Moreau obtained his findings from an analysis of temperature observations – some 30,000 observations in various climate regions, or so he claimed.[12]

Moreau's analysis of rainfall was equally replete with data, from which he concluded that the influences of forests differ between low lying plains and mountains. Preserving mountain forests is necessary for rain and to feed springs and rivers, whereas lowland forests "are incomparably less useful."[13] It may be that "nature ties, by secret links, the destiny of mortals to that of forests," but to Moreau, some forests were more useful and worthy of protection than others.[14] Forests can have detrimental effects, and there is an optimal amount of forested land for a country: large enough for the benefits, but not too large to be harmful.[15]

To maintain a desirable climate, forests are a public investment no less necessary than navigable rivers, canals, or roads.[16] For his report, Moreau received the gold prize in the competition. Yet the report was not the definitive answer to the forest–climate question, and a French commission was subsequently proposed in 1836 to further study the issue.[17]

If Moreau waxed philosophically on the fate of humanity presented by deforestation, the author of the silver prize in the competition was no less reflective.[18] The difficulty in the forest–climate question, the author recognized, is to uncover the true causes for that which is so readily experienced. In this, the author was expressing the commonness of forest influences, observed in everyday life but still unexplainable, and he was quoting from the scientific philosophy of the prominent French botanist, naturalist, and proponent of evolution Jean-Baptiste Lamarck.[19] Lamarck also wrote on meteorology. In his understanding, one fundamental principle – one of the bases of reasoning for meteorology – was that the influence of sunlight on the surface differs between forests and bare soil.[20] Lamarck planned, but did not complete, a trilogy of works on a theory that foretold the modern concept of Earth system science by dividing "the physics of the earth into three essential parts" of meteorology, hydrogeology, and biology.[21] In Lamarck's thinking, humankind was destined to make the planet inhabitable by denuding the land of vegetation.[22]

Climates and the Influence of Wooded Soils

This is the setting in which the science of forest meteorology arose, and first on the scene was the physicist Antoine-César Becquerel. Becquerel was known for his studies on electricity, for which he received the prestigious Copley Medal (1837) awarded by the Royal Society of London. He headed what would become a dynastic family of physicists: his son Edmond was already a prominent physicist in his own right; his grandson Henri would receive the Nobel Prize for physics in 1903; and Henri's son Jean would also enter the family profession. Becquerel studied meteorology and wrote extensively on forest–climate influences. His early views relied on Samuel Williams, Thomas Jefferson, and Alexander von Humboldt.[23] In subsequent monographs, Becquerel provided his own contributions to the science. The commission of 1836 had debated the effects of deforestation in France but failed to reach consensus.[24] Becquerel was charged with ascertaining the facts, "putting aside any preconceived ideas."[25] During the 1850s and 1860s, he strove to provide the necessary observations and theories to settle the forest–rainfall controversy then coursing through France. He sought to identify the physical mechanisms of forest–atmosphere interactions, and he brought direct experimental evidence to bear on the problem. Becquerel's studies were published in US governmental reports and were highly influential among American foresters.[26]

Becquerel's 1853 publication *Des climats et de l'influence qu'exercent les sols boisés et non boisés* set the framework for the interdisciplinary study of forests and climate.[27] Becquerel intertwined principles of physical meteorology; the physical properties of soil as they affect soil temperature, soil moisture, and infiltration; observations on plant growth and the seasonal timing of flowering, fruiting, and leaf emergence and senescence; the geographic distribution of vegetation; and forest history. His multifaceted study reflected the close connection between meteorology and botany at the time before the advent of disciplinary sciences. Indeed, Becquerel wrote that "to properly study a climate, it is also necessary to consider the vegetation of a country."[28] In so doing, Becquerel formulated a comprehensive understanding of the land–atmosphere system, including its vegetation and soil, which underpins present-day theories and mathematical models of climate.

Becquerel realized that forest–atmosphere interactions are exceedingly complex and depend on many factors that are often ignored, but he drew some general conclusions.[29] His analysis relied on Humboldt and Boussingault, and Becquerel likewise concluded that forests in tropical climates reduce temperature.[30] In the midlatitudes of Europe and America, however, the data were insufficient to reliably determine the influence of forests. Becquerel also saw the possibility of remote atmospheric connections between different regions – what atmospheric scientists today refer to as teleconnections. Becquerel believed "without doubt" that covering the Sahara with trees would lessen the desert heat.[31] This, he thought, would change atmospheric circulation and thereby alter the climate of western Europe. "Thus," he concluded, "the deforestation of a large region can act on the climate of regions that are more or less distant." Present-day climate scientists are still concerned with changing the surface of the Sahara and its effects on climate.

A later work studied the temperature of trees to understand how forests affect air temperature.[32] Previously, forests were understood to cool the air because of evapotranspiration. A further mechanism uncovered by Becquerel was that radiation originating from the sun and sky heats the leaves, branches, and trunks during the day and cools them at night.[33] Becquerel articulated the fundamental dilemma of spatial scale inherent in studies of forests and climate, which still challenges scientists today. He measured the temperature of individual trees, but it is the forest that is of interest. Becquerel understood the difference between these spatial scales.[34] To extend his findings to the larger forest, Becquerel knew that the height of the forest, the surface area of all the branches, and the mass of all the leaves must be considered. He further knew that generalities were difficult because additional factors such as the type of soil, whether the soil is dry or wet, and the winds affect the specific outcome.

In *Mémoire sur les forêts et leur influence climatérique*, published in 1865, Becquerel summarized the climate influences of forests.[35] In considering the many conflicting ideas, Becquerel recognized the complexity of the problem.[36] The coupling between forests and the atmosphere varies with geographical location and in different climate regions as determined by latitude, proximity to oceans, and other large-scale features of climate. Within this overall macroscale geography, the precise outcome depends on local factors such as the spatial extent of the forests, their elevation, and the soils on which they grow; the orientation with respect to winds; and the age of the forest and the type of species, notably the difference between evergreen and deciduous trees. Forest influences also differ depending on seasons. Becquerel was particularly aware of the importance of soil texture and the physical state of the soil. Different climate effects are found for forests growing on sandy or clayey soils and wet or dry soils. Becquerel stressed the importance of roots. The roots of trees make the soil more porous, thereby promoting infiltration. They also draw water from deeper in the soil than can the shallow roots of herbaceous plants.[37] The deep roots of trees allow them to obtain more water than is accessible to grasses – recognized today as a key feature underpinning forest–atmosphere coupling.

Becquerel gave considerable attention to the storage of heat in biomass. His previous work had shown that the radiative heating and cooling of leaves, branches, and large trunks influences air temperature within and above the forest canopy.[38] To better understand this, Becquerel called for more temperature observations "at various heights near and at the periphery of trees." As Becquerel found, the nocturnal release of energy stored during the day limits nighttime cooling of the air.[39] Biomass heat storage is still poorly understood today. And Becquerel returned to the problem of spatial scale: "Whatever the effect of a forest, we can imagine that it is related to its expanse, because a tree or a clump of trees does not act as a large mass does."[40] The specific features of a forest and the conditions under which it grows preclude simple generalizations gained from the study of a small number of trees. Scientific disagreement over forest influences on climate arises because past investigators have only considered part of the problem, not the full suite of possibilities.[41] Studies in one location or one time of year do not necessarily inform studies in another place or another season. Modern-day scientists would be well served to heed Becquerel's cautions.

To resolve the conflicting ideas, Becquerel and his son Edmond established meteorological observatories to measure temperature and rainfall in forested and open lands.[42] This is the origin of modern eddy covariance flux towers that observe and monitor forests and other types of land, now in a worldwide network of several hundred stations. In these studies, Becquerel considered the physical principles by which forests could increase rainfall. He speculated that the cooler temperature and higher relative humidity of forests compared with those of fields might, with the right conditions, favor condensation. Rain occurs when air saturated with moisture cools and condenses. The cool temperatures over forested lands might allow this to happen. A further mechanism of cooling and condensation might be that surface winds lift upward when they encounter tall forests.

Becquerel's observatories led to an understanding of the movements of air created by the radiative heating and cooling of forests, which he thought might also affect precipitation. Forests can create an atmospheric circulation that operates like land and sea breezes, but in this case caused by the temperature difference with surrounding fields. On warm, calm days, the hot, dry air over cleared land rises into the atmosphere. The air above the forest is cool and moist, and being dense, it sinks to the surface. In the resulting circulation, air flows outward from the forests to the fields. The opposite circulation can occur at night when the temperature contrast is reversed. Becquerel observed "currents of ascending warm air and descending cold air in the daytime" and the reverse at night because of the heating and cooling of forests.[43]

Auguste Mathieu, a professor and assistant director at the forestry school in Nancy, also undertook meteorological measurements.[44] Mathieu was an expert on the anatomy of wood, the flora of forests, and the botanical description of trees.[45] In 1878, he authored a report on meteorological observations taken over the period 1867–77 around Nancy at two sites located in a forested region and one in an agricultural region.[46] He found higher annual rainfall over the forests than at the agricultural location, from which he concluded that forests increase rainfall. His temperature measurements showed that the air within the forests was more uniform and less extreme than in the field, with a lower daytime maximum temperature and a higher nighttime minimum temperature, and especially during summer. Other measurements revealed the interception of rainwater by deciduous canopies in summer and less soil evaporation under forest canopies than at the agricultural location. One question Mathieu could not answer was whether forests transpire more water than agricultural crops. He thought, however, that transpiration rates were not dissimilar because both have a similar amount of foliage and "the leaves of trees do not have a greater exhaling power."[47] This question, so central to the climate influences of forests, is still debated today.

Mathieu's findings were widely cited in support of forest–rainfall influences. Now, however, historians cite the research as an example of the faulty science underlying the reasoning at the time.[48] The forest sites were separated from the agricultural site by 18 kilometers.[49] Mathieu did not consider the distance to be impactful, but temperature and rainfall could vary across the region for reasons other than forest cover.[50] Matthieu, himself, acknowledged the limitations of the study. The correct way to answer the forest–rainfall question, he said, would be to compare measurements in a region that was first forested and then deforested, but "this means is impracticable."[51] Meteorological measurements taken in forested and open lands, not necessarily in proximate locations and at a far greater distance than that in Mathieu's study, are still commonly used today to discern forest influences.

Foresters were receptive to these studies. An essay published in a popular magazine on culture and current affairs, *Revue des deux mondes*, in 1875 is prescient in its description of forest–climate influences compared with now. The author, Jules Clavé, a French forester, framed forest influences in terms of distinct processes that scientists will recognize today: chemical, meaning the uptake of CO_2 and release of oxygen during photosynthesis; physical, in this case the effects of forest canopies on interception, soil evaporation, snow melt, and wind; physiological, meaning transpiration by leaves; and mechanical, or the influences of roots on soils, infiltration, and runoff.[52] Clavé recognized that the influence of forests on climate "has been much disputed; denied by some, it has been admitted by others, without however the latter agreeing on the direction in which it is exercised." The conflicting opinions arise because the processes involved "are complex and often masked one by the other." Clavé, like Becquerel, sought the totality of the problem. The modern science is still uncertain in how to comprehensively assess the many, and oftentimes conflicting, influences of forests on climate.

Other academics strongly disagreed with the science. Like their American counterparts, some French meteorologists sought an understanding of climate in the large-scale features of the atmosphere, and forests were not part of that doctrine. Émilien Renou expressed just this opinion in his 1866 theory of rain. The idea that forests affect rainfall "rests on absolutely nothing" and without "a single indisputable observation to support it."[53] It "is like all prejudices, repeated for a considerable time, without having the least proof."[54] It is wrong to ascribe meteorological phenomena to local influences. Rather, as the science advances, "we recognize the size and the generality of atmospheric phenomena."[55] In case his opinion was not clear, he put the matter more succinctly: "Forests have no influence on rain."[56] The meteorologist Hippolyte Marié-Davy, likewise, was faulted for being too preoccupied with atmospheric general circulation to see the influences of forests.[57]

In his reprinting of Alexandre Surell's 1841 study of mountain floods, the water engineer Ernest Cézanne evidenced a disdain for forest influences on climate.[58] Unlike Bequerel, who was advocating for the science of forest meteorology to inform public policy, Cézanne sought to dismiss the science. The public rushes to blame any perceived change in climate on deforestation, he thought. The readiness with which people fault forest clearing for climate evils is simply an "old grievance" that "will not be extinguished: it will be constantly reborn, maintained by the inexhaustible fund of popular tradition."[59] He disputed Mathieu's measurements of increased rainfall over forests, suggesting instead that the greater amount of rainwater collected was an artefact of the winds.[60] Deforestation, Cézanne complained, is the cure-all explanation for all meteorological problems, "truly the *Deus ex machinà* of meteorology."[61] The judgement of Cézanne was that "we can now relegate the action of forests to the infinitely small ones of meteorology."[62]

The Physical Effects of Forests on Air and Soil

Unease about deforestation and climate change arose in Germany, too, and naturalists told of the collapse of ancient civilizations as a result of deforestation. In his 1832 instruction on forest science, the forester Heinrich Cotta related how the air becomes too dry and the rains stop if there are not enough forests.[63] Proof was found in that Sicily, Sardinia, Persia, the steppes of Russia along the Volga and Don, and elsewhere once had plentiful forests but now the barren land was arid. Carl Fraas, who studied botany and agriculture, believed that the once bountiful forests of ancient Greece could not be restored because the climate had become too dry.[64] In asking why climate has changed since antiquity, Matthias Schleiden, a professor of botany at the University of Jena, was "directed to the disappearance of the forests."[65] In this, Schleiden saw the decline of civilization. To Schleiden, the track of humanity across the planet is well marked. Ahead lies "original Nature in her wild but sublime beauty"; behind, humankind "leaves the Desert, a deformed and ruined land." In describing the deterioration of the Mediterranean region, the chemist Justus Liebig, remembered today for his pioneering work on agriculture and soil chemistry, told of how wise rulers had once protected the forests because of their influence on climate.[66]

The well-known Prussian meteorologist Heinrich Dove revealed his thoughts in an 1855 essay on rainfall in temperate latitudes.[67] His writing suggests a tentativeness. The issue was of utmost importance to society, but the data could not confirm or deny forest benefits with any certainty. He believed that evidence for a reduction in precipitation was well-established in the tropics. The evidence was less clear in temperate climates, but America, where deforestation had so greatly changed the landscape, would provide the answer. Dove, too, invoked the specter of irreversible climate change and a decline in civilization, suggesting that "the continually increasing population of the earth, in its effort to maintain itself, will plant in nature the germ of a period of death, when vapors should no more condense into clouds over the treeless earth, and even the seed in the soil refreshed only by dew would lose its power of budding, or if it should shoot up, would slowly wither and die."[68] A similar concept is seen in today's notion of climate tipping points.

Some German researchers, like Becquerel, identified the mechanisms by which forests influence the larger climate. One way, as explained in an 1865 study by a "Dr. Berger in Frankfurt," was through atmospheric circulations.[69] Berger found, as had Becquerel, that the temperature of forests is cooler than that of open fields during the day and warmer at night. This, he explained, generates local daytime and nighttime currents (Figure 5.1).[70] At night, air flows into the forest from the field, but "during the day, in reverse, the cooler forest air will pour over the open air, there itself warming also – only much higher – rise,

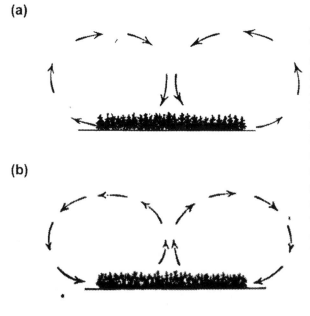

(a)

(b)

Figure 5.1 Atmospheric circulations can occur when a forest is surrounded by open fields, as illustrated by Josef Roman Lorenz von Liburnau in *Wald, Klima und Wasser* (1878). Shown are (a) the daytime circulation in which cool, moist air from the forest replaces rising warm, dry air above open land and (b) the reverse nighttime circulation. Reproduced from ref. (70).

descend over the forest to cool down and start the cycle anew."[71] The smoke from a smoldering fire, even a lit cigar, carried in the air currents provided proof. Berger further described how the heating and cooling of forests growing on mountainsides can likewise alter daytime and nighttime valley circulations.[72] These forest circulations of rising and descending air can influence precipitation, especially on the plains where forests and fields alternate across the landscape.[73] Precipitation will increase if "the uniformity of an extensive forest or that of an extensive area of land free of vegetation is replaced by the alternation between the two."[74] Local deforestation could, consequently, increase rainfall, but large-scale clearing reduces rain. To Berger's thinking, "the aim of agriculture must therefore be to produce this alternation" by both clearing some woods and foresting other exposed lands.[75] Modern science confirms the presence of atmospheric circulations in landscapes of forests interspersed among dry, open lands.

Becquerel's investigations into forests and climate were a masterpiece of interdisciplinary science. Comprehensive research, not narrow-mindedness, was required; quantitative measurements rather than qualitative conjecture would solve the problem. Ernst Ebermayer, too, embraced this concept in establishing several paired meteorological stations in forested and nearby open sites.[76] Previous studies were conflicting, and only sustained observations over several years could resolve the question, Ebermayer reasoned.[77] He installed the necessary meteorological stations in Bavaria and began taking measurements in 1868.[78] His 1873 publication on the physical effects of forests on air and soil – *Die physikalischen Einwirkungen des Waldes auf Luft und Boden* – has since been hailed as foundational to the establishment of forest meteorology as a science.[79] He measured the temperature and humidity of air; the temperature of the soil at several depths; the temperature of tree trunks; the flows of water in precipitation, interception, and evaporation; and the percolation of water through the soil.[80] His study was aimed at obtaining the influence of forests on climate and soils; the effect of snow cover and leaf litter on soil temperature; the difference in temperature between dry and wet soil; differences related to the amount of exposure to the sky; the effect of forests, and particularly leaf litter, on soil moisture; and the consequences of forest clearing – aspects of the science that are still being studied today.[81]

Ebermayer was familiar with the work of Becquerel and Mathieu, and his findings were similar.[82] Like Becquerel, he examined the heating and cooling of leaves, branches, and trunks; the vertical profiles of temperature in the forest canopy; and the storage of heat in the forest.[83] He agreed that the deep roots of trees sustain transpiration in times of drought.[84] Building on the findings of Becquerel and Berger, Ebermayer also described the local circulations generated by forests and fields.[85] A novel contribution was Ebermayer's finding that leaf litter on the forest floor further reduces the already low soil evaporation in forests compared with open sites.[86]

He concluded that deforestation increases temperature during the summer, enhances soil evaporation, and dries out the soil.[87] These changes are greater in the hot climates of southern Europe – Italy, Spain, and Greece – than in northern countries. With regard to rainfall, Ebermayer thought that forests can affect precipitation because the cool, humid air favors condensation (i.e., is closer to saturation).[88] He recognized that care must be taken to separate geographic influences from those of forests themselves, but nonetheless thought that forest influences can be consequential in mountains, though not in low-lying plains, and are most important in summer months for hot southern countries and interior continental regions.[89] Consequently, forests are more necessary for rainfall in Germany and Russia than in Ireland and Great Britain. Yet Ebermayer admitted that his observations were not sufficient to substantiate "these theoretically determined effects."[90] Regardless of changes to rainfall, deforestation produces a drying of the land, less so in some places such as Ireland and Scotland with damp climates, but more evident in Germany, Hungary, and the southern European lands of Greece, Italy, Sicily, southern France, and Spain.[91]

Other paired meteorological stations were subsequently established in France, Germany, Austria, and elsewhere in Europe.[92] An international meteorological congress held in 1873 at Vienna called for observations of forests, and a second congress at Rome in 1879 furthered the need for meteorological observatories in forests and open fields.[93] The German network of observatories grew to seventeen paired stations, and a newly planted site provided an opportunity to directly measure changes in rainfall with forest growth.[94] Franklin Hough highlighted the European forest observatories in his *Report upon Forestry*, as did others in later reports; similar observatories were lacking in the United States.[95]

Forests, Climate, and Water

The forest–climate problem ensnared foresters and meteorologists in Austria and across the lands of the Hapsburg Empire.[96] The forestry professor Gottlieb von Zötl, in 1831, explained that foresters manage not only the forests, but also the climate of a region.[97] In the latter half of the century, the issue came to the forefront with the consideration of legislation to regulate forests. The forester Adolph Hohenstein furthered the call to manage Austria's forests because of their importance for climate.[98] Yet, at least in 1860, direct meteorological observations to compare forests with open fields were lacking.[99] Even then, rain gauges needed to be carefully placed to obtain reliable measurements.[100] Nonetheless, Hohenstein told the people of Austria, forest owners, in cutting the woods, cannot wantonly deprive them of "the advantages with which nature to this day has given you."[101] One of those advantages was climate. "What are the forests for our creation, for the humidity of the atmosphere, for clouds and rain, for supplying rivers and springs?" he asked.[102] It was the duty of foresters "employed by the high administration of creation, the balance of nature, to maintain the world healthy in its adornment."[103]

The Austrian meteorologist Julius Hann brought expert knowledge, not sentimentality, to bear on the problem, but his writings, like Dove, evidence a hesitancy and unsureness. In an 1867 publication, Hann rejected the premise that forests increase rainfall, but he also outlined how this could occur. He found an analysis of long-term precipitation measurements by the American Lorin Blodget, which showed the stationarity of rainfall, despite widespread deforestation throughout America, to be especially persuasive.[104] Echoing Blodget's viewpoint, Hann suggested that forests merely grow in regions of high rainfall rather than directly affecting rain. In his consideration, the amount of rainfall is determined by factors other than forests. Yet Hann also acknowledged an influence of forests on climate, though he thought it had been overstated.[105] The enhancement of precipitation by forests was fairly certain in the tropics, but more difficult to prove in temperate regions.[106] Becquerel's studies would provide much-needed data, Hann thought, and more such observatories were vital "if one is to arrive at reliable conclusions."[107] Reporting on the earlier study by Berger, Hann advocated a dynamical framework of local circulations created by a mosaic of interspersed forests and fields interacting with larger-scale atmospheric motions.[108] Hann later acknowledged his unease with the state of the science and that there was no easy answer to this issue of pressing societal concern. A farmer could "embarrass" a meteorologist, he wrote, by asking if deforestation will decrease rainfall and harm the harvests; the science could not yet provide a precise answer to the "theoretically justified fears."[109]

A report on riverflow issued by Gustav Wex, a water engineer in Vienna, in 1873 reveals a broad interest in the forest–rainfall question.[110] Wex analyzed the flow of several rivers in central Europe including the Rhine, Elbe, and Danube, and found a decrease in riverwater, which he attributed to forest clearing and a decline in rainfall. Wex recommended quick government intervention to protect forests and safeguard the water. John Croumbie Brown warmly reviewed Wex's work in his 1877 book *Forests and Moisture*, and Wex's manuscripts were translated into English for an American audience, where they were featured in an 1897 study on forests and climate conducted by the US National Academy of Sciences.[111]

Into this heightened awareness of deforestation, climate change, and forest conservation strode Josef Roman Lorenz von Liburnau. Lorenz, like Becquerel, was an interdisciplinary visionary. He was both a meteorologist and an expert in agricultural sciences: president of the Austrian Meteorological Society and cofounder of the first university for agriculture in Vienna.[112] Lorenz thought that understanding the impact of forests on climate was of pressing scientific and social importance.[113] Doing so required a new science of "agricultural and forestry meteorology" that would provide the necessary synergy between meteorology and forestry – foresters trained in meteorology and meteorologists trained in forestry. Lorenz tried to bridge the gap between the two disciplines with a textbook on climatology with special regard to agriculture and forestry.[114] He delved into the science, relying greatly on Ebermayer's findings, but Lorenz also brought a keen meteorological perspective. He framed the forest–climate question as a problem of spatial scale. He recognized the broad climate zones related to solar heating of the planet (e.g., equatorial, subtropical, temperate, polar). Within this zonation, continents and oceans influence climate at the planetary scale; large water bodies, mountains, and forests are evident at the regional scale; and the same features influence the local climate through small lakes or a grove of trees.[115] As he explained later, forest influences on climate are a matter of scale: "It depends on the point of view of the investigator."[116] Lorenz, as he would do throughout his career, stressed the importance of air movements. The air is a fluid constantly in motion, conveying with it temperature and humidity across the land. It is through atmospheric circulations that forests influence surrounding lands.[117]

Like Becquerel's monographs, Lorenz's subsequent treatise on forests, climate, and water – the 1878 book *Wald, Klima und Wasser* – outlined the fusion of meteorology, forest ecology, plant physiology, hydrology, and soil physics that is seen in today's science. Lorenz wanted to inform the public of the "great forest question."[118] He acknowledged that the science is complex and not easily communicated to the public. While the public and governments may demand action, the science defies generalities and is "impossible to characterize in a few words."[119] The descriptor "the forest" has broad meaning that encompasses many types of trees and in different stages of growth so that very little can be said about forests in general.[120] Yet effectively outline the science and its implications for forest policy is what he did.

Lorenz explained the hydrology of forests: the physics and physiology of transpiration; the interception and evaporation of rainwater by leaves and branches in the canopy; soil evaporation; the unique characteristics of forest soils and the factors influencing soil temperature and moisture; and the movement of water across the land. Associated with the hydrologic cycle is an exchange of energy with the atmosphere. Forest biomass, with its high water content, stores energy during the day and heats up slowly. Additional energy, much larger than biomass storage, is consumed during transpiration. Lorenz explained the differences between forests and open land in light of these water and energy flows. Evergreen forests differ from deciduous forests, and forest influences vary throughout the year depending on whether the trees are actively growing, experiencing summer droughts, or in winter dormancy. Lorenz, like Becquerel and Ebermayer, emphasized the prominent vertical layering within forests. To understand the effects on climate, one needed to look at the forest "layer by layer," from the forest floor through the lower layers in the canopy that consist mostly of woody trunks and thick branches and into the upper layers of foliage and thinner branches in the tree crowns.[121]

Lorenz again demonstrated a keen awareness of spatial scale. He contrasted the local effects of deforestation, such as a small clearing in a forest, with the influences of wider deforestation over large regions – a challenge still for today's science. He understood that forests everywhere do not always have the same effects, either within the forest itself, in the immediate vicinity, or the wider surroundings.[122] He concluded the book by considering the broader implications: If forests have a beneficial influence on climate, what are the legal implications of deforestation?[123] The climate influences of forests can span large regions – even a continent – and so cutting or planting forests requires cooperation among countries. However, the influence of a small woodland on a neighboring field presents a different legal situation. As he had throughout the book, Lorenz was distinguishing between forests in general and a particular forest.[124] It is a distinction as relevant today – for both scientific study and climate policy – as it was then.

Lorenz presented a dynamical framework to study forests and climate, especially rainfall, that built upon the ideas of Berger. The influence of a forest on the surrounding land is not understood simply by extrapolating the forest microclimate to the larger region.[125] Movements of air extend forest influences beyond the immediate forest. One such circulation happens when a forest is surrounded by open fields (Figure 5.1).[126] On warm, sunny, summer days with little wind, the cool, moist air of forests flows into the warmer fields, is warmed and rises in height, and flows back to replace the sinking air over the forest and close the cycle. Furthering Berger's ideas, Lorenz emphasized the modification of air currents passing over forests, and he provided several case studies, some of which are shown in Figure 5.2.[127] Lorenz presented an example of an isolated forest surrounded by cleared fields (Figure 5.2a). The forest cools the overlying air and enriches it with water vapor. The air is carried downwind from the forest, but the effects on rainfall differ depending on the characteristics of the upwind air onto the forest. Mountains have different rainfall on their windward and leeward slopes, and forests can alter the expected patterns depending on how they are situated with respect to the prevailing winds (Figure 5.2b). Lorenz also considered the effects of heterogeneity in the landscape (Figure 5.2c). Air encountering an extensive forested land is modified differently compared with the same air moving over a mosaic of forests and fields. He urged the establishment of meteorological observatories to gather rainfall data so as to prove what he knew to be "fundamentally correct."[128]

Foresters in the Hapsburg Empire did not enthusiastically embrace Lorenz's fusion of meteorology and forestry.[129] One critic was Emanuel Purkyně, a Czech specialist in plant physiology and plant geography and a professor of forestry.[130] Purkyně was also interested in meteorology. His measurements in the lands around Prague identified the presence of local microclimates within the broader climate of the region. It was this observation – that the climate experienced where people live, grow crops, and harvest wood is locally heterogeneous rather than broadly uniform – that shaped Purkyně's perspective, and his own measurements convinced him of the difficulty in obtaining reliable data. He grew into a strident challenger to the science, and he published a sharply worded critique issued in ten installments between 1875 and 1877 and spanning almost 300 pages.[131] Unlike Becquerel, Ebermayer, and Lorenz, who sought to establish a science of forest meteorology, Purkyně dismissed the scholarship outright.

Purkyně portrayed himself as a truthsayer in a wilderness of unsubstantiated assertions, facing hostility and sacrificing his reputation to expose myths and fantasies.[132] He admitted that he was not a meteorologist,

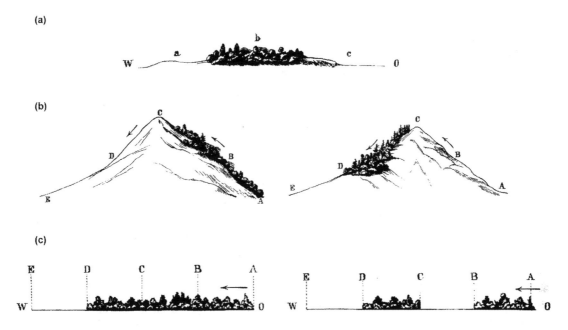

Figure 5.2 Forest influences on rainfall vary in different landscapes, as illustrated by Josef Roman Lorenz von Liburnau in *Wald, Klima und Wasser* (1878).

Shown are (a) a forest surrounded by open land, (b) mountainous terrain with forests located on the windward (left panel) and leeward (right panel) slopes with respect to the prevailing winds (shown by the arrows), and (c) a large continuous forest (left side) contrasted with a change from forest to open field (right side) with winds from east to west (right to left). Reproduced from ref. (127).

but meteorology, despite being of utmost importance, was neglected in general education, so much so that the public knew little of the science.[133] Purkyně offered as proof the "adventurous theories" about forests, which everyone considered justified without "fear of embarrassment." His task was to correct the wrong, and he appealed directly to "the reader, if he is serious about refined, independent opinion."[134] Purkyně rightly rebutted the excesses put forth by some, such as that Bohemia would transform into a dry, steppe climate because of deforestation,[135] but his single-mindedness blinded him to think beyond a narrow perspective. Lorenz, himself, described Purkyně as "well-meaning" but "impatient," naïve in his understanding.[136]

Purkyně's criticism was wide-ranging. His central argument, spanning much of his essay, was that the climate was stable, not changing.[137] Conditions vary naturally from year to year so that drought in one year cannot be ascribed to deforestation. If clearing a forest really does decrease precipitation, rainfall would have to remain low until the forest regrows, but this is not the case because a dry year is often followed by a wet year.[138] Purkyně also looked to broad geographic patterns and saw no influence of forests. The rains continue in the eastern portions of the United States despite much deforestation whereas other regions of the country with extensive forests are less rainy.[139] Heavy rains fall across poorly forested areas of Western Europe, while rain is sparse in forested eastern lands and central Siberia.[140] It is a further mistake to discern a signal of forest influences in the contrasting climates of forests and steppes.[141] They reflect the prevailing climate and soil, but vegetation "never influences meteorological and geological processes."[142]

Purkyně dismissed Ebermayer's study of forested and open lands as unoriginal, well known by "every forester"; but he also cited the work as confirmation that forests have no broad climate effects while condemning those who misunderstood this point.[143] The coolness under a forest canopy fools people into assuming a larger influence beyond the forest edge.[144] This reasoning, Purkyně declared, is as flawed as thinking that the cellar of a building alters the temperature above the rooftops in a city.[145] Other aspects of the science fared no better in Purkyně's mind. That forests create local circulations is based on "an ignorance of the simplest physical laws."[146] Nor could Purkyně conceive how forests could alter air currents high in the atmosphere.[147] Air moving across a land can dampen or dry out forests, but forests have no influence on the air itself. Some years later, after establishing meteorological observatories to gather data, Lorenz indirectly responded to Purkyně criticisms. He remembered Purkyně's analogy with cellars and replied that precise observations, not "amateurish assumptions and similes," are needed to advance the science.[148]

Purkyně's writing reveals a parochial interest – to safeguard forestry from the influence of outsiders – that likely shaped his thinking. Forests are "always only poorly protected" when in hands other than those of foresters.[149] That forests need to be preserved for their climate benefits presents "serious dangers for forestry" if the public, fearing the consequences of cutting down trees, demands action.[150] Foresters can properly manage forests and should not be forced to chase after "fairy tales" in which they must preserve primeval forests, "otherwise everything is lost."[151] It was Purkyně's duty to free the practice of forestry from "the burdensome and inhibiting influences" of the public, who wrongly believe they own the forests.[152] If the harm to forestry was not enough, Purkyně warned that regulations posed a threat to freedom, property, and the sovereignty of countries.[153] French worries over deforestation and floods were ill-founded, used by the government only to manipulate public opinion and to expropriate private land "without grumbling."[154] A similar seizure could happen if Germany, in an effort to protect the flow of the Elbe, asserted its right to forests at the river's headwaters in Bohemia.[155] All landowners should be fearful of government interference. If forests have a "real influence" on climate, the height of trees, their density, and the compactness of the canopy would need to be policed to "save the world from great disaster."[156]

CO_2, Water Vapor, and the Greenhouse Effect

Becquerel's, Ebermayer's, and Lorenz's books were the beginning of the science of forest meteorology, and much of today's science considers similar questions. One central aspect of modern climate theory – the planetary warming imposed by water vapor, CO_2, and other greenhouse gases – eluded scholars; they attached a much different significance to CO_2. A concept of the cycling of carbon between plants and the atmosphere had emerged based on knowledge of photosynthesis and respiration.[157] Its importance, however, was to purify the air. Benjamin Franklin had voiced this benefit in 1772, as did the French botanist Mirbel in 1815. To Mirbel, plants maintain an essential balance in nature, seen in that the animal kingdom consumes oxygen and respires CO_2 while plants utilize CO_2 and restore oxygen. With this balance, "all is linked in the vast system of our world."[158] The Royal Academy of Brussels silver medalist (1825), too, extolled the purification of air by forests.[159]

By the early part of the nineteenth century there was a well-developed conceptualization of the carbon cycle. Boussingault described the cycle in 1834, when he wrote that plants remove CO_2 from the atmosphere during photosynthesis while the respiration of living organisms and decomposition of organic material give back CO_2.[160] He suggested that photosynthesis and respiration are in balance "if the amount of living matter remains the same." By 1845, the French chemist Jacques-Joseph Ébelmen had developed an understanding of the biological and geochemical processes controlling long-term changes in atmospheric CO_2 and the associated oxygen cycle.[161] Ébelmen understood that living organisms, through photosynthesis and respiration, control the concentration of these gases in the atmosphere. He ended his paper with a provocative question: "Has the composition of our atmosphere reached a permanent state of equilibrium?"[162] The answer, he thought, would come from future generations.

Ten years after Ébelmen outlined the workings of the carbon cycle, Shirley Hibberd, a successful British nature writer and gardening author, wrote a book of essays that combined science and nature. In "The Poetry of Chemistry," Hibberd captured the grandeur of nature evident in the carbon cycle.[163] He described the cycling of an atom of carbon over the ages, absorbed from primeval air by the leaf of an ancient fern only to be buried in a block of coal. Liberated thousands of years later and burnt for fuel, "it shall combine with a portion of the invisible atmosphere, ascend upward as a curling wreath to revel in a mazy dance high up in the blue ether; shall reach earth again, and be entrapped in the embrace of a flower." The atom becomes an apricot that nourishes the human body and soul, "educing the thoughts which are now being uttered by the pen. It is but an atom of charcoal, it may dwell one moment in a stagnant ditch, and the next be flushing on the lip of beauty; it may now be a component of a limestone rock, and the next an ingredient in a field of potatoes; it may slumber for a thousand years without undergoing a single change, and the next hour pass through a thousand."

Later naturalists romanticized the role of forests to purify air. The German botanist Hermann Schacht considered this to be their greatest importance, more than climate benefits.[164] In Adolph Hohenstein's call for forest conservation, it is the consumption of CO_2 and production of oxygen by trees that sustains life, and does not, then, life itself depend "on the faithful professional fulfillment of all foresters?"[165] Jules Clavé, the French forest enthusiast, published a series of essays in 1862 in which he outlined the many contributions of trees to humanity. "More than any plant, the tree merits our gratitude."[166] One reason was that forests had prepared the necessary conditions for the emergence of humankind by transforming primeval air enriched with CO_2 "into breathable air."[167] In his tribute to farming, the American philosopher Ralph Waldo Emerson praised the harmony of nature in which "plants supply the oxygen which the animals consume."[168] Ebermayer, too, lauded forests for being "the greatest natural oxygen factory."[169] Perhaps not unexpectedly, Emanuel Purkyně dismissed the idea that "human breathing depends on the forest."[170]

The climate significance of CO_2 was as yet unknown. Becquerel thought the concentration of CO_2 to be an important aspect of meteorology, but did not know if it has "an influence on the general economy of nature."[171] He knew, however, that its concentration in the air varies between daytime and nighttime and seasonally according to the growth of plants, and he even estimated the amount of carbon removed from the atmosphere in forest growth.[172] John Tyndall was formulating the principles by which CO_2 and other gases in the atmosphere warm climate through the greenhouse effect, but there was as yet no cause for alarm about an increase in atmospheric CO_2 from the rapid industrialization fueled by burning coal.[173] Another gas – atmospheric water vapor – drew considerably more interest. Tyndall identified this as a climatically important gas that, like CO_2, absorbs the longwave radiation emitted by the land, preventing its loss to space and thereby warming the climate. Without water vapor, the atmosphere would be much colder. This principle is now known as the greenhouse effect, and water vapor, along with CO_2, is a key greenhouse gas. For Tyndall, "this aqueous vapour is a blanket more necessary to the vegetable life of England than clothing is to man."[174] Without it, the land would be "held fast in the iron grip of frost."

Tyndall's findings opened a new dimension to the forest–climate question, and his analogy with frost was compelling. Ebermayer grasped the significance and realized that forests, by making the air humid, protect against nighttime frosts.[175] James Meschter Anders, a medical doctor and proponent of the climate and health benefits of forests, saw this too. Known today for his medical contributions, Anders also lectured in botany at the Wagner Free Institute of Science (Philadelphia). He studied transpiration, which he considered to be one of the greatest benefits of trees by increasing the humidity of air and enhancing rainfall.[176] In Tyndall's work, Anders found further value to forest transpiration. The warmth provided by water vapor is especially evident at night, when moist or cloudy skies retain heat but temperatures plummet with clear, dry air. Echoing the analogy of Tyndall, Anders explained to fellow biologists that "aqueous vapor is a blanket."[177] While Anders admitted that much of the water vapor in the atmosphere comes from ocean evaporation, he asserted that forest transpiration can contribute sizable quantities of vapor. For this reason, "the practice of forest culture as a means of improving atmospheric conditions, cannot be too highly commended."[178] Anders subsequently published a book on the health benefits of plants.[179] He included a chapter on forests and climate along with a plea to keep large portions of the country forested. Ebermayer and Anders were concerned with nighttime frosts. Today, water vapor is more broadly recognized as a greenhouse gas that heats the planet. By altering the amount of water vapor in the atmosphere, deforesting large portions of the world, or conversely planting trees, can change planetary heating.

■

The forest–climate debate, and in particular whether forests increase rainfall, reached its zenith in the latter half of the nineteenth century as nations considered legislation to protect forests and men lent their scientific expertise to inform public policy. Antoine-César Becquerel, Ernst Ebermayer, and Josef Lorenz heeded the call for better scientific knowledge. They established forest observatories to compare forested and open lands, made the necessary measurements, and developed an understanding of the meteorological, hydrological, and ecological processes by which forests influence climate. In this way, they advanced a new science of forest meteorology. They explained forest influences in the laws of physics, fused with interdisciplinary knowledge of meteorology, forest hydrology, and forest ecology, and gathered the data to further their theories. It was an understanding based on observations of microclimates, but upon which was layered a dynamical framework applicable to macroclimates. Many of the findings have withstood the test of time, and the questions posed are still relevant to today's scientists. The existence of mesoscale circulations induced by the contrast between forests and cleared land is today verified. Heterogeneity in the landscape can, in the right conditions, generate air currents and induce rainfall. The atmospheric teleconnections that Becquerel thought might arise with planting trees in distant lands are today studied by climate scientists. The vertical structure of forest canopies and variation in temperature, seen by Becquerel, Ebermayer, and Lorenz as so essential to the science, still challenges researchers. The dismissal of forest influences by critics – French meteorologists and water engineers; the Czech forester Purkyně – proved to be too rigid. Yet an even more vocal contrarian view, one that came to dominate the science and cement opposition to the forest–climate theory, was arising from meteorologists across the Atlantic.

6   American Meteorologists Speak Out, c. 1850–1910

Meteorology was blossoming as a science in the latter half of the nineteenth century, and meteorologists, assured in their knowledge, unabashedly entered the forest–rainfall controversy. Two contrasting approaches to study forest influences on climate had been developed by the middle of the nineteenth century. Antoine-César Becquerel, Ernst Ebermayer, Josef Lorenz, and others advanced a new science of forest meteorology framed in meteorological stations located in forests and open fields. The alternative sought answers in analyses of temperature and precipitation measurements spanning several decades. Did the observations show a change in climate as forest cover decreased or increased? American meteorologists adopted the latter approach. Finding no evidence of climate change in regions of the country that had undergone deforestation – on the contrary, climate was found to be stable – they dismissed the idea that forests influence climate.

The Americans, following the reasoning of the geologist Charles Lyell, expressed a simpler explanation for the abundant rainfall over forests. In his *Principles of Geology* (1834), Lyell described a reciprocal relationship in which climate influences vegetation and vice versa, but Lyell was guarded in his thinking.[1] Forest influences are overstated, he cautioned; forests are not the chief determinant of rainfall. Rather, forests grow because of the rains. It was this idea that American meteorologists embraced in the latter half of the century, and much more forcefully and harshly as the forest–rainfall controversy exploded onto the public consciousness. Meteorology was the domain of physical scientists, not foresters, and they unreservedly crushed any notion that forests should be protected for climate benefits.

Rational Climatology

By the mid-1800s, meteorological observations had become widespread and of long duration, and the question of whether the American climate was stable or changing was put to quantitative testing.[2] Samuel Forry had first proposed that climate was stable based on his analysis of observations at military posts. Lorin Blodget, using a more extensive network of stations, followed Forry with his own *Climatology of the United States* (1857). Blodget acknowledged the common impression that deforestation had altered temperature and precipitation in the United States and elsewhere, but, he argued, climate was, in fact, unchanged over the ages.[3] He dismissed the premise that forest clearing or cultivation could change climate. He thought that climate affects vegetation, not vice versa: "The great differences of surface character which belong to the deserts, woodlands, and other more striking features, are believed to have their *origin* in climate, and not to be agents of causation themselves."[4] The presence of deserts, Blodget further asserted, is determined by geographic location and "they are not self-creating." Over the next two decades, other meteorologists, while not directly addressing the issue of deforestation, agreed on the constancy of climate.[5]

Thirty-one years after Blodget's publication, and following years of advocacy by George Perkins Marsh, Franklin Hough, and other forest conservationists and even federal legislation to promote tree planting with the Timber Culture Act of 1873, prominent meteorologists raised their voices in outcry. The tree planting mania had consumed politicians, the public, and scientists alike (Figure 6.1).[6] Even Blodget had joined the enthusiasm and favored federal legislation to forest the plains.[7] That year, in 1888, Henry Gannett of the US Geological Survey issued a forceful rejection of forest influences on rainfall.[8] His analysis of precipitation

Figure 6.1 Depiction of tree planting as a celebration of Arbor Day. Reproduced from ref. (6).

in regions of the United States that had undergone increases and decreases in tree cover showed no change in rainfall. Not only did the data not substantiate claims for an effect of forests, but, Gannett asserted, "a satisfactory explanation of this supposed phenomenon has never, as far as I am aware, been offered."[9] A second paper by Gannett published in the same year further claimed there was no evidence that tree planting had increased rainfall in the Central Plains.[10] It was this repudiation of forest–rainfall influences that sparked the heated debate with Bernhard Fernow, chief forester at the US Department of Agriculture, during the meeting of the Philosophical Society of Washington later that year.[11] In that debate, Gannett argued that "an effect has been mistaken for a cause, or rather, since it is universally recognized that rainfall produces forests, the converse has been incorrectly assumed to be also true."

In addition to the constancy of climate, a second prominent line of argumentation for American meteorologists was that measurements of precipitation are unreliable. Precipitation is highly variable from one year to the next and from one location to another, Gannett noted in his studies. Cleveland Abbe, the chief meteorologist at the US Weather Bureau, likewise drew attention to the quality of rainfall measurements. In his address to the Philosophical Society of Washington during the forest–rainfall debate, Abbe stressed that the setting of the rain gauge affects the quality of the measurements.[12] Wind speed affects the amount of water collected in a rain gauge, so different exposure to wind, not forests themselves, accounts for any reported increase in rainfall measured above forests. In asking whether climate has changed, Abbe also highlighted the intrinsic variability from one year to the next.[13] Statistical analyses must be used to determine if two sets of meteorological measurements are truly different. The combination of measurements, analysis, and theory – what Abbe called "rational climatology" – provides the answer to the forest–climate problem. "It will be seen that rational climatology gives no basis for the much-talked-of influence upon the climate of a country produced by the growth or destruction of forests," Abbe confidently declared.[14]

Abbe did not dislike forests or disavow forest science. In fact, he was a life member of the American Forestry Association and believed that "the forest is essential to health and happiness."[15] If civilized humankind has destroyed forests, then "a higher civilization teaches us to restore them."[16] He just did not believe that forests are an essential part of meteorology. The surface of the planet covered by forests is so slight, he informed members of the American Forestry Association, that they "exert a relatively small and not a preponderating influence on the general condition of the atmosphere."[17] To Abbe, the presence of forests "depends upon certain broad climatic conditions over which they have no control." Abbe did acknowledge an influence on local climate and water supply, and he advocated reforestation, but for reasons other than climate benefits.

His views may have been informed by his travels to St. Helena and Ascension Island, which had featured so prominently in earlier reports on forests and rainfall.[18] His meteorological studies on St. Helena showed that the land had little influence on the island's rainfall, and harkening back to the old observation that trees furnish rainwater on Ascension Island, Abbe found that this occurred when clouds and fog envelop the trees; the water droplets coalesce and drip off the foliage. Abbe was unwavering in his dismissal of forest–climate influences. Late in his career, he reaffirmed that the influence of forests on the atmosphere is "of minor importance in dynamic meteorology."[19] That climate determines the distribution of forests and the growth of trees, on the other hand, "affords a very important illustration of climatic influence." Meteorology, Abbe was saying, is a necessary science for foresters, but forestry has no bearing on the study of meteorology.

When forest enhancement of rainfall was not rejected outright, it was dismissed as only being minor. William Ferrel, who is remembered today for advancing dynamic meteorology and his theory to explain atmospheric circulations, offered his perspective.[20] He argued that large-scale circulations, not forests, control where rain falls. While not denying a contribution of forests to atmospheric water vapor, Ferrel believed that any effect of forests on precipitation is not felt locally in the vicinity of the forests, but rather, if at all, manifests far from the evaporative source. The water vapor is transported by winds and precipitates elsewhere so that any increased rainfall "would perhaps be entirely insensible to observation."[21]

Fernow's Forest Influences

The 1888 debate at the Philosophical Society of Washington was a confluence of the American forestry and meteorological communities with their opposing views on the forest–climate question. Yet there was no resolution, and the discord continued. Forest influences had preoccupied Fernow's immediate predecessor in administering to the forest service.[22] Fernow, likewise, featured the topic in his first three annual reports as forestry chief.[23] Unsatisfied with the outcome of the Philosophical Society debate, his report for the year 1888 included a lengthy criticism of Gannett's methodology and reasoning.[24] Gannett had analyzed three regions of the country that had, according to him, seen "radical changes in the forest-covering" – prairie states once grass-covered but now forested; the forests of Ohio long since cleared; and the New England states reforested.[25] Yet, Fernow pointed out, Gannett had no data on the extent or history of forest changes in those regions. To the contrary, Fernow provided an extended rationale for how forests affect rainfall. Even then, Fernow contemplated a fuller report on forests.[26]

The ensuing report, *Forest Influences*, appeared in 1893 and was prepared in collaboration with Mark Harrington, chief of the Weather Bureau. Fernow recognized that the public, and some scientists, had overstated the arguments for and against forest influences on climate. He sought to reframe the rhetoric "from the battle-field of opinions, scientific and unscientific, to the field of experiment and scientific research, and from the field of mere speculation to that of exact deduction."[27] He praised European investigators who had set up meteorological stations "with a view to settle the question by scientific methods and careful systematic measurements and observations."[28] Although Fernow had called for similar forest experimental stations in the United States, none had yet been established.[29] He again recommended "systematic observations bearing on the subject of forest influences," to be collected in a partnership between the Weather Bureau and the Department of Agriculture.[30]

Fernow relied on Harrington for meteorological expertise. Harrington was not only a meteorologist but had also studied biology, and he was previously a professor of botany at the University of Michigan.[31] Harrington wrote a manual to aid in the analysis, description, and identification of plants, and he published on tropical ferns.[32] He introduced his manual by explaining that botany is a "mental discipline" comparable to mathematics, and "even superior" in that it provides training in "the powers of observation and critical judgement."[33] In his review of forest–climate influences, Harrington blended his botanical and meteorological backgrounds.[34] He began by acknowledging that the characteristics of the land – its vegetation, soil, and the presence of snow – affect the climate at a location. The trees of forests, the grasses of prairies, and the barrenness of deserts contribute to the climates of those regions. He recognized the diversity of forests throughout the world. Forests differ in the density of trees, the height and openness of the canopy, composition (broadly defined as deciduous, evergreen, or mixed forests), the angles at which leaves are arranged (e.g., horizontal or vertical), the leaf litter on the ground, the soil on which the trees grow, and the elevation and slope of the land. These affect the climate at a particular location, and they defy broad generalities when considering forests and climate.

Harrington reviewed the data, mostly from European scientists including Becquerel, Mathieu, Ebermayer, and Lorenz, to contrast forested and open lands. Compared with open fields, forests moderate temperatures under the canopy (cooler during the day and warmer at night; most noticeably during summer). Similar effects are seen in soil temperature. Air temperature also varies with height in the canopy, generally cool in the understory and warmer in the overstory. Harrington recognized the significance of the large surface area of leaves in a forest canopy: "The foliage seems especially arranged for the exchange of heat. Its surface is very large for its mass."[35] Forests transpire large quantities of water but reduce soil evaporation. The foliage and branches additionally intercept some of the precipitation, which readily evaporates from the canopy. Together, the three streams of water – transpiration, soil evaporation, canopy evaporation; collectively called evapotranspiration – moisten the air. Harrington estimated the magnitude of the three fluxes to ascertain the total water flux.[36] Despite differences in evapotranspiration, the amount of moisture in the air, or the absolute humidity, does not greatly differ between forests and open fields, but the relative humidity (which depends on temperature) is higher in forests because of their lower air temperature. Evidence for more rainfall at wooded sites compared with open locations was less conclusive, though Harrington thought further study was warranted. Forests also block winds and, following Lorenz and others, generate circulations with adjacent open lands. The chemical influences of forests manifest through CO_2 and oxygen, though Harrington calculated this to be too small to affect the concentration in the atmosphere, even more so for oxygen, which is three orders of magnitude more abundant in the air.[37]

Harrington's contribution to *Forest Influences* is notable in several ways. Much of his review examined what scientists now recognize as forest microclimates. The topics Harrington raised are still studied today, albeit with much more advanced instruments and technologies and aided by modern-day theories and mathematical models. Harrington delved into forest macroclimates – a cooling of temperature in some regions or an increase in precipitation – but he felt that the data were less reliable than the local climates observed with paired meteorological stations in forested and open lands.[38] Harrington knew so because he had tried to discern a change in rainfall with westward expansion of settlements across the United States.[39] Notable, too, is that the studies Harrington cited were mostly European. The science of forest meteorology was being practiced in European countries but not in the United States. American meteorologists, in developing their understanding of climate, had pushed the emerging field of forest meteorology away from the mainstream science. Harrington called for more observations above forest canopies, especially of precipitation, to complete "the theory of the action of forests on climate."[40]

Fernow saw Gannett and others as aiming to "divide and discourage" the work of foresters in their study of climate.[41] His own contribution to the report focused on forests and water.[42] He was unguarded in advocating forest influences on rainfall, but Fernow allowed for balanced perspectives. He acknowledged inaccuracies in measurements of precipitation – a topic that had come up earlier in his annual report on forestry.[43] To that end, Fernow included Abbe's remarks on rain gauges, which had provided the substance of Abbe's lecture during the debate at the Philosophical Society. Abbe's argument was straightforward: "We must be satisfied as to the degree of reliability of our data" before the forest–rainfall question can be answered.[44] Fernow also appended a review on precipitation by George Curtis, a meteorologist with the Smithsonian Institution and known for having compiled a new edition of the famous *Smithsonian Meteorological Tables*.[45] Curtis dismissed forest influences on rainfall as being at best minor, "an amount too small to be of any considerable hydrographic or economic importance," and lacking in "rational explanation."[46] The subject had received "a disproportionate and undue amount of emphasis."[47] His analysis of tropical rainfall is telling of a meteorological bias.[48] He thought that deforestation would increase evaporation and therefore increase rainfall. That much of the water vapor is, in fact, conveyed to the atmosphere during transpiration by leaves eluded him. Deforestation could reduce rainfall, he admitted, but only by increasing runoff and thereby reducing the amount of water in the soil that can evaporate.

The Disagreement Deepens

In highlighting the influence of forests on rivers and water supply, Fernow was continuing a line of thought promoted by earlier forest conservationists, notably Marsh and Hough, but the attention on water only deepened the controversy. The dispute reached the halls of the august National Academy of Sciences in 1896. The Secretary of the Interior requested a study to consider the need for a national policy on forest conservation, including the influence on "climate, soil, and water."[49] The committee recommended that a forest policy be developed and that more forest reserves be created, but hesitated whether the climate benefits of forests should guide that policy. "This influence is potent and beneficial," but the committee acknowledged that data are lacking, and their conclusion was based on "general considerations" rather than precise observations.[50] The committee was firmer in its belief in the benefits of forests for river flow, backed by the study by the Austrian water engineer Gustav Wex.

Charles Sprague Sargent chaired the committee. Sargent was the director of the Arnold Arboretum at Harvard and an avid forest preservationist, but not for rainfall. In his thinking, "forests do not produce rain; rain produces the forest."[51] The idea that forests influence rain had been carried "far beyond its legitimate limits."[52] Forest clearing "will not diminish the amount of rain falling upon it; nor can the increase of forest area in a slightly wooded or treeless country increase its rain-fall."[53] If forests do not produce rain, they do protect the supply of water by hindering runoff to streams and rivers, by delaying snow melt, and by reducing evaporation. This, to Sargent, was the public good provided by forests. In this, Sargent was reframing the forest–climate question from rainfall to the surer footing of forest hydrology, but even that provided little relief to the arguing.

Two dueling reports to the federal government highlight the continuing dispute. Fernow's *Forest Influences* was written with a spirit of cooperation and to motivate, not hinder, further interdisciplinary study. Willis Moore, who replaced Harrington as chief of the Weather Bureau, had a different intent. In a 1910 report to the US House of Representatives, Moore parochially dismissed forest influences on climate with the comment that "while much has been written on this subject, but little of it has emanated from meteorologists."[54] And emanate, Moore did. He systematically attacked any notion that forests affect temperature, rainfall, or the flow of rivers. He left no room for misunderstanding or nuance. "Precipitation controls forestation," Moore declared, "but forestation has little or no effect upon precipitation."[55] Moore invoked the authority of Cleveland Abbe to support his position. He related how Abbe had remarked that "it is a pity that the errors of past centuries should still continue to be disseminated long after scientific research has overthrown them."[56] Furthermore, Abbe continued, "the idea that forests either increase or diminish the quantity of rain that falls from the clouds is not worthy to be entertained by rational, intelligent men."

Moore's report was written with a tone of authority, but his arguments were mostly devoid of specifics and instead relied on generalizations. The climate was wetter in the past, as evidenced by changes in forest geography, but for reasons other than deforestation, and any decrease in rainfall should "be regarded as the cause rather than as the result of the barren condition of the soil."[57] And in reasoning more histrionic than scientific, Moore claimed that "the fact that dead forests stand long after the streams have receded" is evidence that "their removal did not precede the drought."[58] Another argument employed by Moore was that biases in rain gauge measurements invalidate comparisons between forested and open locations. In this, Moore again relied on Abbe, who is reported to have said: "Those who wish to restore the good old times before the forests were cut down, when rain and snow came plentifully and regularly, have only to lower and shelter their rain gauges and snow gauges and, *presto*, the climate has changed to correspond."[59] Moore put forth the tried-and-true argument that precipitation is variable and official measurements show no change related to deforestation: "These facts are important and can not be successfully disputed."[60] Besides, rainfall over the United States is controlled by large-scale features of the atmosphere, not surface conditions.

A substantial portion of Moore's report was to also deny that forests affect river flow. He uncritically accepted studies that claimed no influence of deforestation and castigated studies that argued for an increase in floods. Moore presented his own analysis of a thirty-eight-year record of precipitation and river flow for the Ohio River at Cincinnati, which he considered to be "one of the most important contributions" of his paper.[61] The analysis, he confidently asserted, showed that changes in forest area had not altered runoff to the river. Any another study "founded with less care" would be less definitive. Moore's calculation in which he normalized runoff by the amount of precipitation is still used today. Yet one critical piece of data was missing from his analysis: Moore had no knowledge of the area deforested over the thirty-eight years. He could only "presume deforestation has been as great as in any other part of the country during recent time." It was not just Moore who rejected forest control over rivers. Engineers, too, dismissed the premise.[62] Indeed, similar to climate, river flow was part of a political debate over managing the nation's waterways that pitted foresters against water engineers.[63]

Moore's report engendered a sharp review by Fernow.[64] He thought the report to be "an important, a useful, and at the same time a mischievous publication." Fernow appreciated that governmental policy should be informed by science, but Moore was selective in his use of data, and it is "more than once evident that he is biased." While Fernow was "in hearty sympathy" with many of Moore's claims, and indeed acknowledged that some arguments for forests "were skating on thin ice," Fernow favored a more balanced approach – neither "the fictions" of forest advocates nor "the mischief" of Moore. Moore's report did nothing to advance the interdisciplinary science that Fernow thought was needed. Instead, Fernow saw the report for what it was: "His conclusions will be accepted as dogmas and as proved facts" by those wishing to discredit further study.

Raphael Zon, chief of silviculture with the Forest Service, responded with a report to the US Senate in 1912. Zon advocated protecting forests for their climate benefits. These influences, he told the Society of American Foresters, are far-reaching and extend beyond the forest itself.[65] Copious evapotranspiration makes forests "the 'oceans of the continent,'" and the moisture-laden air is transported to other regions.[66] The forests in the southeastern region of the country along the Atlantic coastal plain extending into the Appalachian Mountains, for instance, need to be preserved because they provide moisture to the central plains. If Moore's report was lacking in specifics, Zon's Senate report was a testament to thoroughness.[67] The nearly 100-page report included more than 800 references to scientific studies. Zon presented two frameworks to investigate forest influences on climate and water. Both approaches – watershed studies and the water balance – are still used today. The watershed approach compares streamflow measurements in watersheds that differ in forest cover. Zon's review of previous studies found many conflicting results, but, while acknowledging the importance of climate, geology, and topography, Zon concluded that a signal of forests is found in streamflow. Building on a newly initiated experiment in Colorado and another in Switzerland, Zon outlined the protocol of paired watersheds – one to be experimentally deforested or otherwise manipulated, the other as an unaltered control.

The second method is to examine various processes that control streamflow and the ways in which forests affect those processes. In this, Zon put forth the notion of a water budget, or water balance, that accounts for all the flows of water on land. Some of the precipitation, be it rain or snow, is intercepted by foliage and branches and evaporates. This water never reaches the ground to replenish soil water. The amount of interception depends on the intensity and duration of the rainstorm, the openness of the canopy, and whether the forest is deciduous or evergreen. Some water drips down from the canopy or flows down the trunks of trees. This water, along with the precipitation that directly reaches the ground, either infiltrates into the soil or runs off over the soil surface. Soil texture, litter on the forest floor, and other factors determine the amount of infiltration. Water in the soil returns to the atmosphere as evaporation from the soil or as transpiration from foliage. Both are controlled by meteorological conditions and the availability of soil water. Soil evaporation additionally varies with the presence of leaf litter, and transpiration depends on the physiology and surface area of leaves in the canopy and the depth of roots in the soil. The amount of water that is available to supply streams is the precipitation input minus the combined losses of water from interception, transpiration, and soil evaporation. Forests also influence the accumulation of snow on the ground and delay its melting in spring.

Disciplinary Splintering

As the nineteenth century closed and the twentieth century dawned, the forest–rainfall problem continued to vex scientists. The geographer and climatologist Eduard Brückner, then a professor at the University of Bern, was certainly flummoxed. In a lengthy review on climate change published in 1890, he noted the divergence of opinions: "As often as the link between deforestation and rainfall is attested to, as numerous are the voices opposing it, and it is difficult to decide who is right."[68] The science is inconclusive, and "only one thing is evident: We are still very much in the dark as to the link between forests and rainfall."[69] It puzzled him "that over and over again the forest is singled out as the scapegoat for a variety of frequently contradictory changes."[70] Other processes are at work that can cause climate to change. The arguing has not been productive, and there is "lack of any progress towards a solution."

Forests flustered Austrian meteorologist Julius Hann. In the second edition of his influential textbook on climatology, Hann expressed his hesitation. Hann included in his instruction on climate a short chapter on forest influences.[71] Forests reduce the temperature during warm months; they increase the humidity of the air; they protect against strong winds; they regulate the hydrologic cycle; they retain the winter snowpack; and they prevent torrential floods in the mountains. Their influence on rainfall is, outside of the tropics, less certain, and Hann showed his unsureness.[72] On the one hand, "rainfall is to be looked upon as the cause, and the condition of the cover of vegetation as the effect." Yet in the next sentence, Hann continued: "Extended forests, even in middle and higher latitudes, certainly have some influence in increasing the frequency of rainfalls, but it is naturally almost impossible to determine the extent of this influence by observation and measurement." The dilemma, he thought, is that the influence of forests on rainfall has been exaggerated, "and the natural reaction has led to the present tendency altogether to deny such an influence." Nonetheless, forests transpire considerable amounts of water and keep the air cool and moist. Extensive forests "must have a tendency to increase the amount and the frequency of precipitation" compared with dry, open country.

And in this closing chapter to the forest–climate question, the controversy took more oddity when planetary science entered the fray. The astronomer Percival Lowell, remembered today for his Lowell Observatory in Arizona, published two sensational books on Mars.[73] In those, he claimed that the planet was inhabited and with forests, but civilization was dying because of desiccation. The same fate awaited Earth. A writer in French forestry journal *Revue des eaux et forêts* voiced the mania of the time, both for alien life and forest–climate influences, in a satirical (it is hoped) commentary that related the inhospitable atmosphere of Mars to the destruction of its forests.[74]

The turn of the century marks one ending to the forest–climate question. After several centuries in which philosophical reasoning gave way to scientific methods and romantic notions of forests yielded to utilitarian rationalizations to conserve forests for climate and water, the study of forest influences largely receded from the minds of scientists, politicians, and the public. A writer to the scientific journal *Nature* in 1912 succinctly summarized the state of affairs at the time: "The literature on the subject is somewhat bewildering."[75] In the United States, calls for the creation of forest experimental stations were finally heeded, but with a focus on silviculture,[76] and a large experimental deforestation was begun in Colorado in a partnership between federal foresters and meteorologists, but primarily to study forest influences on water.[77] There was still the occasional report on forests and rainfall,[78] but historiographies of the forest–climate controversy in France, Austria, and Russia also point to this time as the resolution of the debate.[79] Press coverage on forests and rainfall in Australia and New Zealand, common at the height of public and scientific interest, declined in the early twentieth century.[80] Yet the controversy would touch future generations of scientists in how they studied nature, because the science of forest–climate influences was diminished in the eyes of both meteorologists and foresters.

The field of numerical weather prediction was gaining traction, and formal collegiate education in meteorology was needed to develop the necessary expertise. In an address to the American Association for the Advancement of Science at its annual meeting in 1890, Cleveland Abbe envisioned meteorology as a branch of "terrestrial physics, or physics of the globe" – what has since become known as geophysics.[81] Abbe evocatively invoked the wonder of plants in his rationalization for meteorology, not as an essential component of the science but rather by analogy with the growth of the science. He likened the new science of meteorology to "a young plant whose strong healthy roots have taken hold of the soil."[82] With the right care, the prospect for further advancement of knowledge was good: "Its growth is favored by the sunshine and the rains of heaven, the winds rustle its leaves; it is struggling for life and existence; give it shelter and nourishment that it may grow and drop its flowers and fruit into the laps of its benefactors." But the science must be recognized in its own as a separate academic discipline: "Don't poke it away in the corner of a building devoted to other objects, where it will die of thirst and starvation." With the right conditions, the science will thrive "and it will become as beautiful as the Cedars of Lebanon." Abbe closed his address with unabashed sentimentality. Of planet Earth, he called on the audience to "love it as you would your own personal home. Search for its beauties as you would search for the sweet nooks in your own garden."[83] Earth's garden may hold secrets still to be discovered, but it had no role in the physics of the globe.

Abbe subsequently proposed a four-year graduate-level instruction in meteorology.[84] He outlined the necessary preparation in physics and mathematics, including calculus and statistics; in fluid motions and heat transfer and thermodynamics; in statistical climatology and the practical study of weather prediction. The final year of instruction was devoted to advanced theoretical considerations. The physiology of plants and their geographic distribution were worthy of coursework by degree candidates, but only in that climate relates to other disciplines. The study of meteorology is "an application of mathematical physics" that "outranks all other branches of science in its universal importance and its difficulty."[85] Abbe rejected forest meteorology as central to the science and pushed it at best into a minor subdiscipline compared with the favored dynamic meteorology. In time, forest meteorology receded to the narrower context of the microclimates of forests and clearings, seen notably in the work of Rudolf Geiger.[86] Fernow, himself, seemingly anticipated this outcome in his *Forest Influences* report. Though the effects on climate may be local, he explained, they are nevertheless important "for upon them rest success or failure in agricultural pursuits and comfort or discomfort of life within the given cosmic climate."[87]

American foresters, too, turned away from the study of forests and climate, but not so in Europe. The French national forestry school at Nancy was born out of the forest–climate controversy. Jacques-Joseph Baudrillart, an early proponent for the school, remarked in 1823 that forests "exert the happiest influence on the atmosphere" and warned of the perils of deforestation.[88] If left unchecked, France would become "a vast desert."[89] Prospective French foresters were taught over subsequent decades that it is "of the highest importance for the climatic state of a country that the forests are distributed in a suitable manner."[90] In Germany and Britain, likewise, forest influences on climate were considered an essential aspect of instruction in forestry at the turn of the century.[91] The US Forest Service was born, in part, from federal legislation to ascertain "the influence of forests upon climate."[92] Franklin Hough, in his *Elements of Forestry* (1882), claimed the science of meteorology and "the various questions of atmospheric influence and of climate" as an application of forestry.[93] In explaining what forestry is, Fernow included the influence on climate and water as an "incalculable benefit" of forests.[94] By 1905, however, Gifford Pinchot, the reigning forestry chief, rejected climate influences. Pinchot advocated a "practical forestry" designed "to make the forest render its best service to man."[95] He did recognize the concept of a "protection forest," preserved for its environmental benefits at the headwaters of streams or planted as windbreaks in the western plains, but not so for climate. "This branch of forestry has had little definite fact or trustworthy observation behind it," he wrote.[96] "The friends and the enemies of the forest have both said more than they could prove." Pinchot, himself, claimed that "the best evidence at hand fails to show a decrease in rainfall over the United States in the last hundred years, in spite of the immense areas of forest that have been burned and cut."[97] That same year, Pinchot tasked the Forest Service with managing forests for "the greatest good of the greatest number in the long run."[98] While recognizing the microclimates within forests and their hydrologic influences in his 1905 primer on forestry, the greatest good did not involve the management of forests for the climate of the country.

By 1930, Edward Munns of the Forest Service felt a need to remind foresters of the origins of their federal service from the scientific controversy and lamented the "marked let-down in the interest felt by American foresters in forest influences as contrasted with forest products."[99] Ten years later, while lecturing on forestry and meteorology at the Massachusetts Institute of Technology, Edward Kotok, also of the Forest Service, similarly paid homage to the meteorological roots of forestry, but not necessarily for the good of forest science. Advocating the climate benefits of forests showed only that "foresters drew on meteorology to prove the necessity for their work."[100]

American foresters had opportunistically promoted a narrative of forest–climate influences to create public interest in forestry, and "out of this situation grew some classic, bitter controversies between foresters and American meteorologists." Most of his lecture concerned not the influence of forests on climate, but rather the effect of climate on forests. In that, "the forester must depend upon the guidance of the meteorologist."[101] Joseph Kittredge, a professor of forestry at the University of California – Berkeley, likewise dismissed the earlier work as "propaganda" that "stressed the evil effects of deforestation on climate because that carried a strong popular appeal" but which "were least likely to be disproved."[102] His own book, aptly titled *Forest Influences*, redirected the study of forest influences to erosion and flood control.

Yet the conflict and divergence of opinions did not entirely fade away. In the 1956 proceedings of a symposium on human modification of the planet – an update to George Perkins Marsh's celebrated *Man and Nature* – the influential climatologist Charles Thornthwaite, known for his pioneering studies of evapotranspiration and the water balance, rejected much of Marsh's writings on the climate influences of forests.[103] He acknowledged an effect of forests on microclimates, but denied that people have any significant outcome on the climate of a region. In an example of the perils of scientific hubris, Thornthwaite boldly asserted that "no responsible scientist seriously believes that man can alter the general atmospheric circulation in any significant way."[104] In the same volume, however, the ecologist John Curtis, in describing human-caused changes to forests, acknowledged the controversy over forests and rainfall but suggested that changes to the albedo, or reflectivity, of the land as a result of deforestation are a more likely cause of climate change.[105] In the next decades, the science would prove Curtis right and Thornthwaite wrong.

A Renewed Science

The seeds for a new understanding of forest–climate influences were planted unknowingly by Abbe and other meteorologists, who at the dawn of the twentieth century were developing a mathematical framework to forecast the weather. Abbe and the Norwegian meteorologist Vilhelm Bjerknes defined the fundamental equations that still underpin the weather forecast and climate models in use today.[106] The theory was sufficiently developed by 1922 that the British meteorologist Lewis Richardson, in his book *Weather Prediction by Numerical Processes*, set down the mathematical equations to forecast the weather and devised the computational methods to solve them.[107] Richardson astutely described the relevant processes as they relate to vegetation, and he formulated the necessary equations to model them: "Leaves, when present, exert a paramount influence on the interchanges of moisture and heat. They absorb the sunshine and screen the soil beneath. Being very freely exposed to the air they very rapidly communicate the absorbed energy to the air, either by raising its temperature or by evaporating water into it."[108] Furthermore, "a portion of rain, and the greater part of dew, is caught on foliage and evaporated there without ever reaching the soil." Vegetation also influences surface winds and turbulence because "leaves and stems exert a retarding friction on the air." Among the equations put forth by Richardson were ones for the evaporation of water from land. Over vegetated land, the primary water loss is through stomata. Richardson, in formulating the fundamental physical equations for weather prediction, also needed mathematical equations for stomata.

Richardson read studies on the biophysics and physiology of leaves by the chemist Horace Tabberer Brown and colleagues.[109] The basic conceptualization put forth by Richardson is instantly recognized by scientists today:

> The air in the intercellular spaces of the leaf is supposed to be saturated in equilibrium with pure water at the temperature of the leaf. The vapour diffuses out through the stomata at a rate proportional to difference of vapour density between inside and outside. The transpiration also depends on the size, shape and number of the stomata. By analogy with electric conduction, the rate of transpiration may be said to be inversely as the "resistance" of the stomata. The resistance consists of two parts in series. The larger part is due to the air in the intercellular spaces and in the constricted passages of the stomata; this part is unaffected by wind. The remainder of the resistance is due to the air immediately outside the stomata and is reduced by wind.[110]

To extend the calculations to a canopy of leaves, Richardson used a layer of "vegetation film" immediately below the atmosphere.[111] Additional equations were needed as well for soil temperature and soil moisture.[112] From this initial conceptualization began a new chapter in the study of forests and climate.

It took several decades to refine and perfect Richardson's equations, but by the 1960s global models of the atmosphere, known as atmospheric general circulation models, had come to the forefront of atmospheric science.[113] Progress in atmospheric modeling paralleled a growing concern about the modification of climate through fossil fuel combustion and other pollutants emitted to the atmosphere. By the late 1960s and into the 1970s, workshops were held and reports issued on the scope of the problem and the need for a comprehensive understanding of all the processes affecting climate.[114] Calls quickly emerged for the models to include the oceans and land, not just atmospheric general circulation. There was a recognition that anthropogenic alteration of the land through urbanization, cultivation of crops, overgrazing, deforestation, and irrigation could modify climate.[115] An interdisciplinary understanding of climate was the ardor of the day. Edward Lorenz, famed for his contributions to dynamical meteorology, weather prediction, and chaos theory, expressed this view in 1970, when he wrote of a "super-model" of climate that includes "the distribution of vegetation."[116]

The development of global climate models brought a new scientific tool to bear on the study of forests and climate. One emerging concern was the threat of tropical deforestation. Scientists defining the scope of the new climate science at an academic conference in the summer of 1970 were confronted with just that possibility when, in the midst of their debates, the July 7, 1970, front page of *The New York Times* reported that "the Brazilian nation is battering, clearing and burning its way into the heart of South America to open and settle what may be the last unknown frontier in the world – the immense Amazon Basin."[117] A graphic photograph of the logging accompanied the article. Alarmed by the devastation to the forests, the atmospheric scientists promptly considered the impact of tropical deforestation on climate through reduced evapotranspiration and increased surface heating and called for further study.[118] Prominent studies in the journal *Science* subsequently identified the importance of surface albedo, evapotranspiration, and the roughness imposed by tall vegetation in regulating climate at the macroscale.[119] The term "biogeophysics" was adopted to describe the interrelationship of biological and geophysical processes in determining climate.[120]

Initially, the land was seen simply as acting as a boundary condition imposed on the lower atmosphere, but, as the models matured and scientific understanding of climate grew, land–atmosphere coupling became seen as an integral climate process, not external to the climate system (Figure 6.2).[121] Comprehensive models of the land surface, its hydrology, and its vegetation were needed to couple with

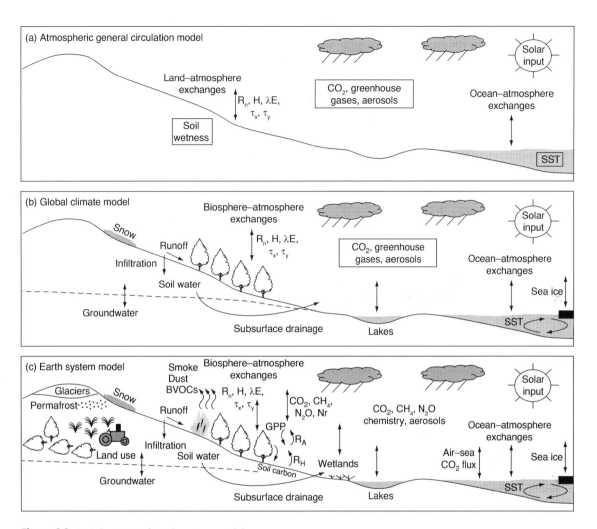

Figure 6.2 Development of Earth system models.

(a) Atmospheric general circulation models circa 1970s. These models used prescribed inputs of atmospheric CO_2, other greenhouse gases, and aerosols. They calculated land and ocean physical flux exchanges (R_n, net radiation; H, sensible heat flux; λE, latent heat flux; τ_x, τ_y, momentum fluxes) using prescribed soil wetness and sea surface temperature (SST). (b) Global climate models circa 1990s. These models added the hydrologic cycle on land and plant canopies. They included ocean general circulation to calculate sea surface temperature, sea ice, and ocean dynamics. (c) Earth system models circa 2010s. These models added the carbon cycle and other biogeochemical processes, anthropogenic land use, wetlands, glaciers and cryosphere processes, atmospheric chemistry, and aerosols. Atmospheric CO_2 concentration is calculated based on gross primary production (GPP), autotrophic respiration (R_A), and heterotrophic respiration (R_H) on land and air–sea CO_2 exchange. Other chemical fluxes on land include methane (CH_4), nitrous oxide (N_2O), reactive nitrogen (Nr), biomass burning, mineral dust, and biogenic volatile organic compounds (BVOCs). Modified from ref. (121).

models of the atmosphere.[122] The scientific advancement of climate models required concomitant advancements in land surface hydrology, the physiology of leaves, and the biophysics of plant canopies, as represented by sophisticated models of biosphere–atmosphere coupling.[123] An early incarnation of these models, developed for the National Center for Atmospheric Research (NCAR; Boulder, Colorado) global climate model, included stomata and, in a prescient anticipation of the carbon cycle science still to come, included CO_2 uptake during photosynthesis because of its close relationship with stomatal conductance.[124] Yet disciplinary biases did not die easily. Plants, with their leaves, stomata, and diversity of lifeforms, do not conform to the mathematics of fluid dynamics. In building their models, atmospheric scientists reverted to their expertise in physics and fluid dynamics and some neglected the biology. As late as 1993, a version of the NCAR model still did not include vegetation.[125]

One of the first applications of the new models was to study the climate effects of tropical deforestation.[126] In addition to biogeophysical processes, there was growing awareness of the contribution of deforestation to the observed increases in atmospheric CO_2 concentration.[127] The current generation of models now also includes the terrestrial carbon cycle and carbon cycle feedbacks with climate change, as well as other biogeochemical processes.[128] The influence of forests on climate has reemerged in light of anthropogenic climate change, as both a cause of climate change due to deforestation and as a means to mitigate planetary warming by restoring forest cover in previously forested land (reforestation), planting new forests (afforestation), and preventing further deforestation (avoided deforestation). The science, first begun centuries ago based on qualitative reasoning, has transformed into the rigor of formal mathematical modeling. The interdisciplinary vision of Alexander von Humboldt, Antoine-César Becquerel, Ernst Ebermayer, Josef Lorenz, and others was finally realized.

Yet the new science, with its call for interdisciplinarity, has been hard to define. Climate scientists initially spoke of the climate system and later gravitated to the Earth system. A report prepared by the US National Aeronautics and Space Administration (NASA) committee for Earth system sciences in 1986 formally outlined the scope of the science (Figure 6.3).[129] The diagram has been hailed for its interdisciplinary vision and as foundational in the development of Earth system science,[130] but it also shows the difficulty climate scientists have in placing ecology within the system. The committee conceived of a fluid Earth and a biological Earth represented by the physical climate system and biogeochemical cycles, respectively. In the conceptual diagram, terrestrial ecosystems are associated with biogeochemical cycles, separate from terrestrial energy and moisture fluxes and separate from the physical climate system. In doing so, the committee members defined ecology in terms of biogeochemistry, not energy and moisture fluxes, and to the exclusion of the physiology of leaves and plants, organisms and their life histories, populations and community assemblages, and myriad other central tenets of ecology. The diagram does not depict the essential interdependence among energy, moisture, and biogeochemical fluxes as mediated by ecosystems and the common ecological processes that control them. Nor does it convey the centrality of ecology and terrestrial ecosystems to climate; ecosystems are only relevant to the extent that they regulate biogeochemical cycles. Meteorologists in the nineteenth century defined climate to the exclusion of ecology, and climate scientists today are still defining the ecology of the Earth system.

■

More than five centuries of discourse and reasoning and arguing about the climate of Europe, tropical and temperate America, island territories, India, and other lands – sometimes conflicting and oftentimes erroneous – evolved from philosophical reasoning to scientific investigation. It produced an understanding of forests and climate that, though dismissed in the forming of meteorology as a science, underpins one part of our present-day knowledge of climate and how, through deforestation, human activities alter climate. Concern for anthropogenic climate change from greenhouse gas emissions and other pollutants has

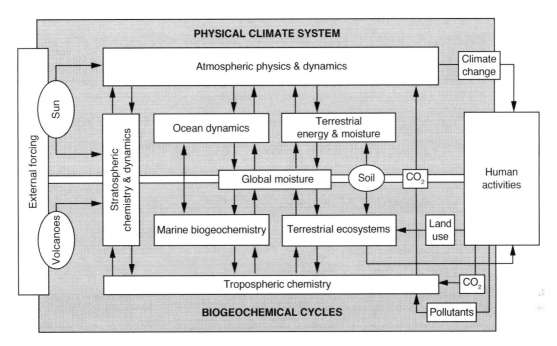

Figure 6.3 Representation of the Earth system in the so-called Bretherton diagram.
The physical climate system (top shaded box) interacts with biogeochemical cycles (bottom shaded box). Both components of the system have multiple subcomponents. Arrows denote connections among subcomponents. Francis Bretherton (National Center for Atmospheric Research, Boulder, Colorado) led the committee that prepared the report. Redrawn from ref. (129).

reignited scientific and public interests in forests and climate. Forests are again seen as one part of the solution to the climate problem. The current science, like the old, emphasizes the climate services of forests. Reforesting vast tracts of land that have been cleared of trees, afforesting other lands, and preventing deforestation are seen as necessary steps to mitigate planetary warming. The Earth system science that underlies this call to action is changing how we study the planet and is breaking down the disciplinary barriers that have for so long limited our perception of nature and the environment.

Yet one sees in the history of the forest–climate question a concept of purposely engineering climate through human actions, and today's call to plant trees likewise embraces that notion. In this, scientists, policymakers, and the public further the utilitarian view of forests. Forests are valued because they aid in protecting against climate change. As it has over the centuries, the science of forests and climate change is complex and defies generalizations. There is no single correct answer for the efficacy of forests to mitigate planetary warming. Are we to devalue some forests if their climate services are not as beneficial as others? Past scholars struggled with the same issue during the forest–climate debate. Some sought a wider perspective and extolled the aesthetics of trees in addition to their climate benefits. As the next chapter considers, forests enrich the human experience in countless ways other than their climate services.

7   Views of Forests

Climate is but one way to perceive forests. In our comparatively short time on Earth, we humans have forged an inextricable bond with forests. Trees have inspired human imagination for tens of thousands of years.[1] Forests are central to the cultural evolution of humankind and have molded our beliefs and values and societies.[2] Our world would be impoverished without forests. The forest historian Michael Williams wrote that the forest–climate question was of such common knowledge during the eighteenth and nineteenth centuries that "deforestation and consequent aridity was one of the great 'lessons of history' that every literate person knew about."[3] If so, then those same people likely also opined on the aesthetics of trees and their contributions to the human experience. Indeed, in his call to reforest England, John Evelyn delved into forest aesthetics, and he included in the second edition of *Sylva* (1670) an essay on the sacredness of forests so as to "have in some sort vindicated the honour of trees, and woods."[4] Later forest advocates followed his example. How we view forests, and how we value them, must be considered from a multifaceted standpoint that recognizes their many contributions to humanity and planet Earth.

In the Beginning, There Were Trees

In the story of planet Earth, humankind is but a recent actor. Our species, *Homo sapiens*, emerged distinct from ancestral species several hundred thousand years ago, but behaviorally modern humans arose only over the past 50,000 years, and the history of humanity is still shorter, extending back in time just several thousand years.[5] Trees have had a far longer journey on Earth – almost 400 million years.[6] Today, an estimated 3.04 trillion trees inhabit the planet,[7] covering 30 percent of land area excluding Antarctica.[8] Plants – mostly trees – comprise more than 80 percent of the total biomass on Earth.[9] Humans pale in comparison to trees. There are an estimated 422 trees for every one person on the planet and we comprise a paltry 0.01 percent of total biomass,[10] but in our comparatively short time on Earth, our story has been inextricably interwoven with that of forests.

Tropical forests were the cradle from which ancestral hominins emerged in Africa. The rise of hominins six to seven million years ago in East Africa corresponds with decreased forest cover and expansion of savanna during a drier climate, forcing our ancestors to develop bipedalism, flexible diets, social structures, and other uniquely human traits needed to survive in the open.[11] Subsequently, early *Homo sapiens* reoccupied and utilized tropical forest environments, and this adaptability may have aided proliferation across the planet.[12] Christianity, Judaism, and Islam speak to the importance of trees in the origin of humankind, and cultures throughout the ages have mythologies that forests preceded the human world. The literature professor Robert Pogue Harrison, in *Forests: The Shadow of Civilization*, used this insight to set the stage for his analysis of the symbolism of forests.[13] The imagery of forests preparing the world for humankind, making it habitable, appears in writings on forests and climate. The French forester Jules Clavé used just such imagery in 1862.[14]

Today, we gaze upward in awe and wonder at the enormity of trees – towering coast redwoods reaching over 100 meters into the California sky on trunks more than 5 meters wide; kauri trees of New Zealand, stately giants whose trunks have a diameter in excess of 5 meters. We marvel at the longevity of ancient bristlecone pines growing in the mountains of California, some dating to a distant past several thousand years ago, bystanders to long-ago events. The famed baobabs of Madagascar question our perception of reality: Their flat-topped crown at the top of a long, smooth, cylindrical stem suggests an upside-down tree with its roots in the sky. Some trees we choose to name – the acclaimed Californian giant sequoia General Sherman (84 meters tall and 11 meters in diameter at its base) and its like-sized compatriot General Grant; or the Methuselah bristlecone pine, over 4,600 years old.[15] We honor trees that record history and spin legends of the events they may have witnessed – the Ankerwycke yew in England, near where the signing of Magna Carta occurred in 1215 and where, three hundred years later, Henry VIII is said to have met Anne Boleyn;[16] or a large, hollowed-out oak in Nottinghamshire, England, known as the Major Oak, thought to have sheltered the famed Robin Hood in Sherwood Forest.[17] Some tell a more somber tale, such as the ginkgos that survived the atomic blast at Hiroshima or the horse chestnut that Anne Frank watched through a small window in her secret hiding place.[18] The mystery of ancient trees compels scholars to germinate 2,000-year-old seeds of date palm from the Judean Desert to rediscover a tree from long ago.[19]

Trees have molded humankind in countless ways since the earliest hominins sought them for shelter, fuel, food, and weapons. Religions throughout the world speak to the sacredness of trees. Forests provide countless medicinal and spiritual health benefits. Forests have shaped our laws and governments and our understanding of the fundamental rights of humankind. Trees have inspired the visual arts and literature. Forests are interwoven in our culture and economy. We extol the environmental benefits of forests for water quality, air pollution, habitat, and now, climate. Yet it is estimated that over 15 billion trees are cut down annually and that the number of trees worldwide has declined by almost 50 percent since the dawn of civilization.[20] Our views of forests are conflicted; there is an ambivalence in our attitudes. There is no one way to represent a forest, to voice its meaning.

What Is a Forest?

A longstanding thought question asks whether a tree falling in a forest makes a sound if no one is present to hear it.[21] The question poses both a scientific and a philosophical problem, grounded in whether sound is a physical phenomenon or a human experience. A similar question might be extended to trees themselves: in the absence of humans, would a tree exist? Irish philosopher George Berkeley posited just such a question in 1710.[22] One might also ask the same question of forests, because people have been defining what a forest is for over a thousand years.

Nowadays, a forest is thought of as a large area of land covered with trees, but the scientific definition of what constitutes a forest is very precise. The United Nations Food and Agriculture Organization, for example, defines a forest by area and the stature of trees.[23] The area of land covered with trees must be at least 0.5 hectares (5,000 square meters). The trees must be taller than 5 meters in height and their foliage must cover at least 10 percent of the ground or be able to reach those thresholds. A forest may be temporarily unstocked with trees due to management or natural disturbance but must be regenerated within five years. Rubberwood, cork oak, and Christmas tree plantations are forests, but tree orchards (e.g., fruits, nuts, olives) and palm plantations (e.g., oil, coconuts, dates) are not.

The origin of the word "forest," however, had nothing to do with trees and was instead a legal term. It first appeared in the early Middle Ages in reference to royal game preserves.[24] Forest referred to land claimed by the king in a royal decree, reserved for royal use, placed off limits to noblemen and commoners, and subject to special laws. The Normans brought this notion of forest from France to England. A particular royal use was hunting, and forests in twelfth century England and later were lands reserved for and under protection of the king to provide a safe abode for woodland animals.[25] In the reasoning of the time, "a forest is a place full of woods," but "it doth not therefore follow that every wood is a forest."[26]

This concept of a forest, and the right to access and use the land, was instrumental in shaping modern-day notions of individual liberties, rule of law, and representative governments. During the twelfth century, English kings expanded their land holdings by proclaiming vast tracts of countryside to be royal forest, a procedure known as "afforestation." Magna Carta, the iconic document that established limits to the power of English monarchs in 1215, was, in part, a rebellion by noblemen against royal usurpation of forested lands.[27] It required that land claimed as royal forest be "disafforested," by which the reigning king gave up possession of the land. This was followed two years later with the Charter of the Forest, which required further disafforestation and reestablished access to forest land by common people. Still, English monarchs developed an elaborate system of laws – so-called forest law separate from common law – and an administrative system to enforce the laws.[28]

In time, England's forests became less a preserve for the monarch and more an instrument of timber production.[29] A vast supply of timber was needed to provide firewood, to erect dwellings and other structures, and to construct and sustain the English naval and merchant fleets. Trees throughout the British Isles were felled, and the royal forests of earlier monarchs deteriorated. Alarm from overharvesting likewise emerged throughout continental Europe.[30] From this concern arose a new view of forests as an object of scholarship – the science of planting, growing, and managing stands of trees for timber.

John Evelyn has come to symbolize the birth of forest management and scientific conservation of forests in England.[31] Evelyn was a seventeenth-century English intellect, founder of the Royal Society, acclaimed horticulturalist, and prolific writer. He was also a voice for the strategic importance of forests – "we had better be without gold, than without timber" – and *Sylva*, published in 1664, offered a practical guide to the propagation, management, and uses of various trees.[32] Evelyn wrote the book with the intent to preserve existing forests and repair cleared lands, but the work was more than a nationalistic call for timber production and Evelyn also delved into forest aesthetics. He issued a total of four editions during his lifetime, and several more were posthumously produced. A 1776 edition edited by Alexander Hunter after his death included richly detailed engravings of foliage and trees.[33] A new *Sylva* for modern times, published on the 350th anniversary of the first printing, continues Evelyn's legacy.[34]

Evelyn's advocation of trees was a step toward a new concept of forests. Sustainable forest management arose in the Germanic states as well, part of the rich knowledge of forest science perfected during the eighteenth and nineteenth centuries. This evolving definition of forests is further evidenced in a 1757 encyclopedic entry by Charles-Georges Le Roy, "lieutenant des chasses" in the park of Versailles and in charge of forest administration.[35] In *Encyclopédie*, one of the great French cultural contributions during the eighteenth century, Le Roy defined a forest as simply a wood that covers a large expanse of land. He envisaged a forest as nothing more than an assemblage of trees of various species and ages, managed rationally based on knowledge and experience to achieve a desired purpose. Indeed, much of the lengthy *Encyclopédie* article is not a thoughtful essay on what a forest is, but rather a practical guide to forest management. Le Roy's understanding of a forest, even more so than Evelyn, who combined his sensible advice with romantic prose, can be seen as formulating a utilitarian view of forests that advanced the care and management of forests so as to provide materials and monetary profit.[36] Le Roy's countryman Henri-Louis Duhamel du Monceau, too, prepared an eight-volume work (1755–1767) on all aspects of forests, including the various species, their culture, and their uses; the anatomy and physiology of trees; forest propagation and management; and the timber industry – a work he would not have undertaken had he known how large it would become.[37]

Europeans landing on the shores of North America certainly saw opportunity in the boundless forests they encountered. The British government looked to the far-reaching forests of its American colonies to feed a voracious appetite for naval stores.[38] A policy that declared large pines found in the New England colonies to be the property of the king, marked for royal use, sparked political resentment. To colonial Americans, this was an infringement upon their rights and required acts of rebellion. They, and later generations of Americans, approached the newfound forests from a different position. As the populace spread across the continent, forests presented a hostile wilderness to be tamed and civilized, but also an abundance of riches for social and financial gain.[39] The land was to be worked and improved, the trees cleared, and the vastness of the forests inured the colonists to their wanton destruction. The future president John Adams captured this sentiment in a diary entry during the year 1756. Prior to European settlement, he wrote, "the whole continent was one continued dismal wilderness," but after more than a century of colonization "the forests are removed, the land covered with fields of corn, orchards bending with fruit, and the magnificent habitations of rational and civilized people."[40] Thomas Cole's painting *The Oxbow* (1836) aptly expresses this sentiment (Figure 7.1).[41] It was this exploitive view of forests that gave rise to the optimistic belief that clearing forests was improving the climate, only to be followed by growing concerns about the effects of deforestation on rainfall.

The intellectual debate about forests and climate played out on a broader stage of European colonialism and the accompanying destruction of forests, culminating with the birth of forest conservation in the nineteenth century.[42] Extensive clearing of trees in the westward settlement of the United States sparked awareness of the environmental devastation brought by deforestation, expressed most prominently by George Perkins Marsh in *Man and Nature* (1864).[43] In China and Japan, too, environmental damage wrought by deforestation, especially flooding and erosion on denuded mountainsides, brought forth forest conservation.[44] Nineteenth-century awareness of forest–climate influences arose from this deforestation, and from this came a new connotation of forests. William Schlich, a British professor of forestry, defined a forest as "an area which is for the most part set aside for the production of timber and other forest produce, or which is expected to exercise certain climatic effects, or to protect the locality against injurious influences."[45]

From this initial crusade for environmental consciousness, modern concerns for the environment have given rise to a new awareness of forests. Current environmental thought recognizes forests as an interconnected ecosystem of living organisms and the abiotic environment in which they live. This is the ecological view of forests and regards the forest per se, worthy of scientific curiosity in and of itself. Nonetheless, conservation philosophy still perceives forests through the lens of the human experience, in terms of services to humankind. Modern conservation thinking values forests for their environmental and socioeconomic benefits, and it establishes the economic value of the goods and services provided by forests.[46] This is an extension of the utilitarian view by which forests are valued on their usefulness. To a long list of forest

Figure 7.1 *View from Mount Holyoke, Northampton, Massachusetts, after a Thunderstorm – The Oxbow* (Thomas Cole, 1836).
The painting captures colonial American views of forests as a dangerous landscape to be conquered. Image provided courtesy of the Metropolitan Museum of Art (New York).

services – timber and wood products, wildlife habitat, recreational uses, scenic beauty, health benefits, water quality, and many more – we now also think of forests as nature's technology to fight climate change.

Visual Art

There are myriad ways to see forests, and a scientific understanding of forests is incomplete without considering their many cultural readings. Forests mean different things to different people, not only in terms of how we define forests, but also in how we experience forests. Forests imprint upon human imagination and human souls in countless ways, and nowhere is that more evident than in the diverse approaches artists use to express trees. The humanities has a language of forests independent of, but complementary to, science. In paying homage to the landscape painter Frederic Edwin Church, the evolutionary biologist Stephen Jay Gould wrote that "artists dare not hold science in contempt, and scientists will work in a moral and aesthetic desert ... without art."[47]

Large, ancient trees have long fascinated humankind. The Cowthorpe oak, a much revered tree in England, stood 26 meters tall and its trunk measured in excess of 4.5 meters in diameter at hip-height above the ground when described by Alexander Hunter in 1776 (Figure 7.2a).[48] Of the imposing tree, Hunter exclaimed: "When compared to this, all other trees are but children of the forest." A tribute to the oak was published in 1857, again in 1904, and the tree was mentioned in the scientific journal *Nature* in 1905.[49]

Figure 7.2 Drawings of (a) the Cowthorpe oak and (b) the Greendale oak in Evelyn's *Silva* (1776). These illustrations show a reverence for large, ancient oak trees in England, but also their disfigurement for human pleasure.

A photograph taken three years later showed a tree alive but decaying, its thick, sagging limbs propped up with poles.[50] Not all large trees were treated with veneration. Evelyn described the Greendale oak, another ancient British oak, in his 1664 edition of *Sylva*.[51] The tree had already been disfigured – three vertical arms had broken off, though eight still remained. An illustration in Hunter's 1776 edition revealed a grotesque remnant of the once grand tree (Figure 7.2b).[52] Only one arm remained, and a large arch had been cut into the trunk so that a carriage could pass through. Yet the tree survived and was spotlighted in 1890 and again in 1904.[53]

Jacob Strutt's *Sylva Britannica* (1822) celebrated the magnificence of trees with exquisitely engraved prints. Among them was a large oak at Fredville, which survived to have its photograph taken in 1994 (Figure 7.3).[54] Trees are "silent witnesses of the successive generations of man," and Strutt spoke of "the gratification arising from the sight of a favorite and long-remembered tree."[55] The latter half of the eighteenth century extending into the nineteenth century was a period of marked interest in drawing trees, and the artwork conveys changing British mores toward trees and forests.[56] Today, photographs taken throughout Britain and the world show the beauty, diversity, and individuality of trees – in their large stature, spreading crowns, and oddly shaped branches; the nuanced textures of their bark, some smooth, others roughly furrowed; the artistry in the many shapes and sizes of their foliage – and the accompanying essays tell their stories.[57]

Depictions of forests in the arts stir expectations of awe and wonder or feelings of apprehension and nervousness, and they bring thoughts of mystery and enchantment to light. *Scholar Viewing a Waterfall* (Figure 7.4), by Ma Yuan from late-twelfth to early-thirteenth century China, contrasts the knotted, twisted contortions of a pine tree in the foreground with the serene image of a pensive, white robed, scholarly man gazing downward into the rushing stream of a waterfall, listening to the roar of the cascading water in the

Figure 7.3 The Fredville oak in Strutt's *Sylva Britannica* (folio ed. 1822).
Jacob Strutt's engravings capture British veneration of large, ancient trees.

background and contemplating his place in the world. Several centuries later, the impressionist painter Claude Monet skillfully captured the imagery of trees in the French countryside, along the seacoast, and in city parks and gardens.[58] Monet used trees to explore the atmosphere of the moment as expressed on the foliage and branches through the nuances of sunlight, gusts of wind, and enshrouding mists and fog. A viewer is left with different impressions upon seeing the trees of Vincent van Gogh.[59] His trees capture emotions, too, but of psychological drama, anxiety, unsettledness, and even terror. Poplars were a common subject of both artists, and their contrasting impressions of trees are evident (Figure 7.5).

In the United States, the Hudson River School was known for portraying the American wilderness and its forests. *The Oxbow* (Figure 7.1), completed by Thomas Cole in 1836, has been interpreted as conveying the views Americans at that time felt toward forests.[60] The untamed forest, with storming clouds spewing a deluge of rain, is a threatening wilderness, while the pastoral farmland, awash in sunshine, is serene. It is an image of American progress realized in the westward expansion across the continent. Yet the painting also, whether intentionally or not, speaks to the commonly held opinion in colonial America that deforestation was improving climate. Another work from the Hudson River School intentionally blended art with science. One of the leading scholars of the day was the naturalist and geographer Alexander von Humboldt. In *Cosmos*, arguably his most important work and which Stephen Jay Gould described as "the greatest of all testaments to the essential humanism of science,"[61] Humboldt described the interconnectedness of life with the physical world and sought to "comprehend nature as a whole" rather than through scientific specialization.[62] He also examined nature through the

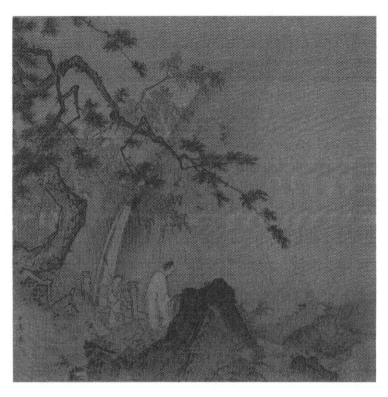

Figure 7.4 *Scholar Viewing a Waterfall* (Ma Yuan) from the Southern Song dynasty in late-twelfth to early-thirteenth century China.
Chinese landscape painting is known for its serene and humble depiction of nature. Image provided courtesy of the Metropolitan Museum of Art (New York).

humanities. Humboldt perceived a close affinity between natural science and landscape painting, the latter of which he considered to be the greatest way to convey nature.[63] Frederic Edwin Church's *Heart of the Andes* realizes that blending of art, science, and nature (Figure 2.3).[64] Exhibited to acclaim in New York City in 1859, the piece is a vast rendering of the South American Andes extending from lush tropical forest in the foreground through a grassy plain and up into the mountains to a snow-clad peak in the background. True to Humboldt's understanding of the geography of plants gleaned during his expeditions through tropical America, the work depicts the zonation of vegetation and climates that Church saw in his own travels to the region.

Other artistry also combines art with science. The botanical engravings of leaves, flowers, and fruits by John Miller (also known as Johann Sebastian Müller)[65] in *Silva* (1776) are meticulous in their detail (Figure 9.3). So, too, are the engravings of leaves in François André Michaux's *North American Sylva* (1819) with their richness of color from original drawings by Pierre-Joseph Redouté and Pancrace Bessa, two French masters of botanical art.[66] John Loudon's *Arboretum et Fruticetum Britannicum* (1838) is both a scientific guide to trees and their culture and a rendering of their artistry.[67] A modern melding of art and science is the Fondation Cartier pour l'art contemporain exhibition *Trees* presented in Paris (2019).[68] The exhibition used drawings, paintings, photography, and cinematography to intertwine artistic, philosophical, and scientific perspectives of trees. The work expresses our relationship with forests, the threats of global

(a)   (b)

Figure 7.5 (a) Claude Monet's *Poplars on the Epte* (1891) and (b) Vincent van Gogh's *Two Poplars in the Alpilles near Saint-Rémy* (1889).
The artists invoke different feelings in their paintings of poplars in the French countryside. *Poplars on the Epte* reproduced with permission of the National Galleries of Scotland (Edinburgh). *Two Poplars in the Alpilles near Saint-Rémy* provided courtesy of the Cleveland Museum of Art (Cleveland, Ohio).

deforestation, and yet again changes the meaning of forest. Those who read the exhibition's catalog cannot look at trees with indifference.

Public art installations, too, capture the ecological aesthetic of trees. *Tree Mountain – A Living Time Capsule* by Agnes Denes is one such piece.[69] Built from 1992 to 1996 in Finland, the work covers a hill with 11,000 trees arranged in a spiral pattern planted by 11,000 volunteers. Each person has ownership of their tree for 400 years, transferable to subsequent generations. It is a hopeful expression for the future, built in the permanence of a healthy, thriving forest. Bonsai, a long-standing Chinese and Japanese form of art, expresses a similar sentiment, though on a much smaller scale, to create living displays that elegantly capture the grandeur of trees in the wild (Figure 18.1). Maya Lin's *Ghost Forest*, constructed in New York City, stirred a different relationship with nature.[70] The temporary exhibition installed forty-nine full-grown but dead cedar trees in Madison Square Park to explore the contrast between living and dying trees and to visualize the consequences for forests of climate change.

The complex geometric shapes formed by tree branches spark artists and scientists alike. In *Remarks on Forest Scenery* (1791), William Gilpin, an Anglican priest, found beauty in the ramification of branches on a large, aged oak tree, its thick limbs extending outward from a central trunk, each subdividing into a complex geometrical shape of interwoven smaller branches that become still smaller twigs and terminating in budding shoots (Figure 7.6a).[71] Two hundred years later, the Italian architects Cesare Leonardi and Franca Stagi

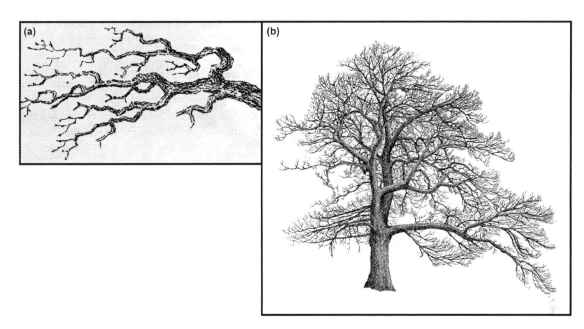

Figure 7.6 Artistic visualizations of the branching of trees.
Shown are (a) the ramification of an oak branch in *Remarks on Forest Scenery* (William Gilpin, 1791) and (b) *Castanea sativa* Miller (chestnut) in *L'Architettura degli Alberi* (1982; © Cesare Leonardi, Franca Stagi) and reproduced with permission of Archivio Architetto Cesare Leonardi (Modena, Italy).

expressed a similar beauty in their illustrations of the architecture of trees (Figure 7.6b).[72] Charles Darwin, in *On the Origin of Species*, used the metaphor of a tree to describe his theory of evolution: "The limbs divided into great branches, and these into lesser and lesser branches, were themselves once, when the tree was small, budding twigs; and this connexion of the former and present buds by ramifying branches may well represent the classification of all extinct and living species in groups subordinate to groups."[73]

In some movie productions, trees become humanlike beings with human emotions. Trees are the central characters in the animated short *Flowers and Trees*, Walt Disney's Academy Award winning story of love and vengeance.[74] In this production, released in 1932 and the first cartoon short produced in color, two male trees – one young and healthy; the other a hideous, old stump, gnarled and leafless – compete for the affection of a young, female tree. His affection spurned, the elder tree starts a fire that threatens to destroy the forest, but which in the end engulfs him in flames. The 1939 movie production of *The Wonderful Wizard of Oz* by Metro-Goldwyn-Mayer envisioned festive, dancing trees, but the fanciful frolicking was cut from the final production. The deleted sequence, featuring the song "The Jitterbug," was set in the Haunted Forest as Dorothy and her companions encounter an insect who causes the foursome to franticly dance.[75] The surrounding trees sway to the song's rhythm and clap their branches in accompaniment.

Literature

Trees and forests appear in many genres of literature, and they present a profound literary device perfected by writers throughout the ages. Many works of literature purpose trees to construct moral and environmental thoughts. Forests are the stage upon which to search for human meaning.[76] The ecological workings of forests cause us to reconsider our anthropocentric perception of intelligence and social networks.[77] Nor can one indifferently walk through the woods after reading of the mother tree.[78] The natural history and cultural uses of various trees provides a means to relate what we will miss if they disappear because of disease, insects, or climate change.[79] Stories of forests worldwide – living and petrified; primeval and urban; towering giants in the wild or tamed bonsais in a museum – introduce a select group of trees and the insights they provide to understand our world.[80] A worldwide tour of notable trees uses science, natural history, and storytelling to convey in short vignettes the biography of eighty trees and our relationship with them.[81] It tells of countless trees exploited for human uses, their value soaring and ebbing in relation to our interests.

A particular type of literature uses forests as a muse, relating inspiration and meaning found during woodland wanderings in what Wallace Stegner described as "rumination hung upon the framework of an outdoor excursion."[82] William Gilpin's *Remarks on Forest Scenery* (1791) is an early form of this type of writing, in which outings in the forests of his English countryside enthused Gilpin's quest to discover the essence of picturesqueness.[83] He wrote of the beauty and aesthetics seen in the form of an individual tree, its branching, and the arrangement of trees in parks and woodlands. He described various species and their growth habits, and he critiqued each on their picturesqueness. Too, he lamented the loss of forests in Britain and elsewhere at the hands of mankind. Two hundred and twenty five years later, the paleontologist Richard Fortey used his walks over the course of a year in English woodlands to contemplate the ecology, history, and peoples of the region.[84] David Haskell's daily tramps in the woods of southeastern Tennessee become a lesson on biology and the ecology of forests, the mysteries and complexities of nature.[85] Sara Maitland's tour of British forests inspired not lessons in natural history and ecology, but rather helped her to consider the interwoven relationship between forests, people, and fairy tales and to reimagine familiar stories from the past.[86]

Henry David Thoreau was a particularly accomplished practitioner of nature writing inspired by outdoor wanderings. He studied trees scientifically, contemplated them philosophically, marveled at their picturesque stature, and critiqued their character. He was familiar with the great works on trees – Evelyn's *Sylva*, Michaux's *North American Sylva*, Loudon's *Arboretum et Fruticetum Britannicum*.[87] He read Humboldt's *Views of Nature* for its description of tropical forests,[88] and he advanced the science of forest succession, by which forests change in community composition over time and which is still widely studied by ecologists today.[89] But in his philosophical writings and his journal entries, he related to trees in a much more reflective way.[90] The outline of a leafless elm on the winter sky, its crooked, angular branches extending outward in all directions, prompted Thoreau to imagine "vast thunderbolts stereotyped upon the sky."[91] Looking upward beneath an oak tree, Thoreau described how the leaves, silhouetted in the sky, "dance, arm in arm with the light, – tripping it on fantastic points, fit partners in those aerial halls."[92] Thoreau developed a special fondness for *Pinus strobus*, more commonly known as eastern white pine. He was especially struck by the interplay of light among its long needles. One day, while out in strong winds, Thoreau saw distant pines swaying and a "fine silvery light reflected from its needles."[93] Thoreau likened the incessant reflection to "the play and flashing of electricity." It was as if the trees were speaking: "Surely you can never see a pine wood so expressive."

It was in *The Maine Woods* (1864) that Thoreau conveyed his passion for pines. Thoreau reserved harsh judgment for loggers. The lumberman "admires the log, the carcass or corpse, more than the tree," Thoreau decried.[94] "He cannot converse with the spirit of the tree he fells, he cannot read the poetry and mythology which retire as he advances." Thoreau saw pines differently: "Strange that so few ever come to the woods to see how the pine lives and grows and spires, lifting its evergreen arms to the light, – to see its perfect success."[95] Of the lumberman, he declared, "all the pines shudder and heave a sigh when *that* man steps on the forest floor." Thoreau recognized that the true essence of a tree – its spirit – is restorative: "It is the living spirit of the tree, not its spirit of turpentine, with which I sympathize, and which heals my cuts."

Some works give trees human emotions. They are sympathetic, thoughtful, contemplative, or ill-behaved as needed for the story. In the Roman poet Ovid's *Metamorphoses*, Orpheus plays his lyre on a barren hill, and all sorts of trees – poplar, oak, linden, laurel, beech, ash, hazel, fir, maple, willow, elm, pine, cypress, and more – are drawn to Orpheus, roused by the melodic tune to march across the land, so many that the ensuing forest engulfs Orpheus with needed shade.[96] Trees as humanlike beings is a recurring motif in Roman culture.[97] *Metamorphoses* has not only walking trees, but also people who become trees, such as Cyparissus the cypress.[98] Modern works, too, animate trees. L. Frank Baum, in his telling of *The Wonderful Wizard of Oz*, created not foppish, dancing trees but rather menacing, fighting trees. His trees guard the forest to keep Dorothy and her companions from passing through. "The trees seem to have made up their minds to fight us," cries the Cowardly Lion.[99]

Forests are integral to the medieval chivalric romance.[100] The twelfth century story *Tristan* tells of the fated love of Tristan and Iseult, in which the lovers escape from expectations of the royal court to shelter in a forest; only there can they be safe in their illicit love and can their passion flourish. Yet the forest is also savage, a landscape of suffering from which Tristan and Iseult must ultimately leave. The contemporaneous Arthurian romances authored by Chrétien de Troyes – *Erec et Enide* and *Lancelot* among others – relate the love of the protagonists for each other in a chivalric quest for honor and exploits among courtly knights. These Arthurian romances present forests as a landscape of adventure, a place of destiny, and a setting for idyllic love and spiritual awakening. The myriad romance tales over the ensuing centuries, each set among forests but seeing the forest in a multiplicity of ways, speak to the centrality of forests to medieval peoples and as source of a literary tradition.

Not all is serene in the woods, though. Many tales present forests as dark, impenetrable, and mysterious; strange, bewitching things await those who enter the woods. Innocents venturing off well-worn, safe paths into unfriendly forest provide the vehicle to impart moral lessons in the stories of Little Red Riding Hood, Hansel and Gretel, and others who encounter wolves, witches, and magical creatures in the imaginations of the Brothers Grimm. In their tales, the forest is foreboding or enchanting as needed. The wolf awaits Red Riding Hood in her grandmother's cottage deep in the woods, Hansel and Gretel cannot find their way out of the witch's woods, but Snow White finds safety in the forest. The symbolism of forests was important for the Grimms.[101] In Jack Zipes's analysis of their stories, "the forest is always large, immense, great, and mysterious. No one ever gains power over the forest, but the forest possesses the power to change lives and alter destinies."[102] The social compacts that guide everyday mores are disrupted in forests, and storytellers have used this insight to craft their tales. To the Brothers Grimm, the forest "is the place where society's conventions no longer hold true."[103]

Gothic novelists play upon the mysteries of a forest. For these writers, forests construct the requisite atmosphere of fear, danger, and anxiety. In gothic novels, trees become ghoul-like beasts of the forest, and no good can come to those who pass through dark woods on a moonless night. A classic gothic encounter with a tree is found in "The Legend of Sleepy Hollow," published by Washington Irving in 1820, which relates the travails of Ichabod Crane and his confrontation with a headless horseman in the countryside of Tarrytown along the Hudson River in New York.[104] On that fateful evening, late when "it was the very witching time of night," Ichabod rode home alone from a party and "all the stories of ghosts and goblins that he had heard in the afternoon, now came crowding upon his recollection."[105] In the road before him stood "an enormous tulip-tree, which towered like a giant above all the other trees of the neighborhood, and formed a kind of landmark. Its limbs were gnarled, and fantastic, large enough to form trunks for ordinary trees, twisting down almost to the earth, and rising again into the air." Many superstitions surrounded the tree, and with his imagination sparked and his nerves jittery, "as Ichabod approached this fearful tree, he began to whistle: he thought his whistle was answered – it was but a blast sweeping sharply through the dry branches. As he approached a little nearer, he thought he saw something white, hanging in the midst of the tree – he paused and ceased whistling; but on looking more narrowly, perceived that it was a place where the tree had been scathed by lightning, and the white wood laid bare. Suddenly he heard a groan – his teeth chattered and his knees smote against the saddle: it was but the rubbing of one huge bough upon another, as they were swayed about by the breeze." We recognize this primordial fear in our language: to be "out of the woods" means to be free from danger or difficulty.

The forest as a place of refuge for outlaws ready to pounce upon unsuspecting travelers is a common symbolism since the days of Robin Hood. Shakespeare and his contemporaries excelled at using woodland settings to create discomfort, anxiety, and confusion in their theatrical productions.[106] The characters drawn into Shakespearean forests encountered wild men, outlaws, exiles, and hermits. Anything and everything could and did happen in the woods. Yet all is not sinister in the forest. The author A. A. Milne presented his favored woodland as a place of innocence and adventure as Christopher Robin roamed the Hundred Acre Wood with Winnie-the-Pooh, Tigger, Piglet, Eeyore, and others.[107] This is the enigma and paradox presented by forests: innocent but perilous, enchanted but disenchanted, sacred but ungodly, protective but lawless. There is magical virtuousness to the woods but also wickedness that breeds doubtfulness upon entering the forest. As Robert Pogue Harrison perceived, western literary culture has a rich history of depicting the forest "as a place where the logic of distinction goes astray."[108]

The sounds trees create incite our fears or calm our souls. We cannot see wind, but we can see its effects in the fluttering of leaves during a gentle breeze or the swaying of treetops in a storm. We can also hear these effects in the noises created by wind in a forest. One speaks of rustling leaves or creaking branches, the whoosh of wind rushing through trees, or a whistling or howling or moaning wind. These are the sounds produced when wind passes through and around leaves and branches. Physically, the sounds are longitudinal pressure waves in the air that travel to our ears, the pitch set by how fast the wind causes the leaves or branches to vibrate. Spiritually, however, they are much more. They are the sounds that instilled dread in the hapless Ichabod Crane along the roads of Sleepy Hollow.

Others hear calming melodies in the songs of trees. The Chinese scholar Liu Chi, in 1355, described the way in which the sights and sounds of pines unburdened him: "Gazing at the pines soothed my eyes; listening to the pines soothed my ears."[109] Trees with "large leaves have a muffled sound; those with dry leaves have a sorrowful sound; those with frail leaves have a weak and unmelodic sound," but not so pine. "Listening to it can relieve anxiety and humiliation, wash away confusion and impurity, expand the spirit and lighten the heart, make one feel peaceful and contemplative." Several hundred years later, writers and poets are still trying to convey the sounds of trees. The British writer Rebecca Hey blended artwork, prose, and poetry in *The Spirit of the Woods* (1837) so that readers could "partake the enthusiasm of the writer towards the whole leafy race."[110] One of the spirits of the woods was that trees "discourse most eloquent music."[111] The "wild minstrelsy" of Scots pine on a winter day was one musical section "in the grand chorus of nature."[112] The American poet Henry Wadsworth Longfellow conveyed a similar sentiment in "A Day of Sunshine," composed in 1863:

> I hear the wind among the trees
> Playing celestial symphonies;
> I see the branches downward bent,
> Like keys of some great instrument.[113]

The sounds of forests have aroused other writers. Thoreau likened the branches of pines to "great harps on which the wind makes music."[114] The rush of wind through oak trees one blustery winter day, as the wind stirred dried leaves still clinging to branches, called to Thoreau like "the voice of the wood."[115] He recognized it as "the roar of the sea, and [it] is enlivening and inspiriting like that." It arose from "billows of air breaking on the forest like water on itself or on sand and rocks. It rises and falls, wells and dies away, with agreeable alternation as the sea surf does." Yet Thoreau also heard more gentle sounds, such as when a light breeze stirred aspen leaves "to rustle with a pattering sound, striking on one another. It is much like a gentle surge breaking on a shore, or the rippling of waves."[116] The environmentalist John Muir, too, was moved by the sound of wind among the trees. During a storm in the Sierra Nevada Mountains of California, Muir "could distinctly hear the varying tones of individual trees, – Spruce, and Fir, and Pine, and leafless Oak" and "each was expressing itself in its own way, – singing its own song."[117]

Trees appear in the mythology and spirituality of many cultures. *Cedrus libani*, commonly known as cedar of Lebanon, has had tremendous cultural significance in the Middle East spanning several thousand years.[118] Native to the mountainous regions of the eastern Mediterranean, the tree was prized by the great civilizations of antiquity to build ships, construct temples, and for other uses. Cedar of Lebanon has become a symbol of our unsure relationship with trees, revered in prose yet overexploited and threatened. The Babylonian epic of *Gilgamesh*, one of humanity's oldest written stories composed some three thousand years ago, describes the destruction of the forests.[119] In the tale, the Sumerian hero Gilgamesh travels to the famed cedar forest to kill its protector and harvest its trees in a quest for greatness and immortality.[120] The Bible, too, makes many references to cedar of Lebanon, and its passages use the tree as a metaphor for ennobling qualities.[121] The tree has been sought by generations of travelers. One was the US Supreme Court justice William O. Douglas, who sought refuge under the shade of cedars from the hot Mediterranean sun and "thanked God for trees."[122] More than immediate relief from heat, "their majesty suggested repose, tranquility, and nobility." Today, the tree is the national symbol of Lebanon, but the habitat of the once great cedar forests has diminished greatly and the remaining forests are threatened by a changing climate.[123]

The Forest for Trees

Humankind envisions forests in various ways that intertwine art, literature, and concern for the environment. In this mosaic of thoughts, the manner in which forests are championed and the breadth with which they are advocated depends on who is voicing their cause. Pieter Bruegel the Elder's *Hunters in the Snow* exemplifies how a shared circumstance – a painting of a wintry landscape – can produce divergent meanings (Figure 7.7). Completed in 1565, the work portrays rural peasant life, one of several paintings by Bruegel that examined the Dutch countryside and its inhabitants at different seasons of the year.[124] This particular scene is of winter. The sky is grayish and overcast; snow blankets the ground; the waterways are frozen; the trees are bare; it is a cold, dreary day. Three men and their dogs return to the village, weary and downtrodden from an unsuccessful hunt. Nearby, townsfolk hunch over a roaring fire to prepare a meal. Skaters play on a frozen pond in the distance. It is

Figure 7.7 *Hunters in the Snow* (Pieter Bruegel the Elder, 1565).
Bruegel's painting of a wintry Dutch landscape has been interpreted as a depiction of climate change and also a lesson on landscape ecology. Reproduced with permission of KHM-Museumsverband (Vienna).

a view of the Dutch world that Bruegel sought to convey with sincerity and sympathy, and *Hunters in the Snow* shows Bruegel as moralist, humanist, and social critic. To one art historian, his landscapes are acclaimed for their "poetic power" and for their authentic depiction of the human condition with "humour and psychological depth."[125] To another, the imagery of *Hunters in the Snow* is carefully crafted with great fidelity to nature and the natural lighting of subjects.[126] The four trees in the foreground, too, are illustrative of the way in which Bruegel provides depth and rhythm and guides the viewer into the scene.[127] Yet the snow-clad mountains rising in the background are incongruous with the Dutch countryside; they are more reminiscent of the Swiss Alps. Why did Bruegel invoke such feelings of wintry harshness? Bruegel drew the scene during a time of severe winters that were colder than usual.[128] And so *Hunters in the Snow* has been used to illustrate climate change.[129] From a climate perspective, the painting depicts Bruegel's impression of those frigid winters and records a changing climate. Ecologists have similarly used the artwork to illustrate the ecological concept of a landscape.[130] Instead of a visual record of an unusually cold climate, the ecological perspective perceives the painting as an expression of the core tenets of landscape ecology: heterogeneity of landscape elements, spatial scale, and movement across the landscape. Forests, like *Hunters in the Snow*, are an experience shared by many but perceived in different ways.

The multiple perspectives of forests are aptly captured by the phrase *seeing the forest for the trees*. The earliest record of the saying is by Thomas More during the struggle in Tudor England with the Roman Catholic Church. In *The Confutation of Tyndale's Answer* (1532–33), More, a devout Catholic, answered the heresy of church reformers in a half-million-word essay of logic, heckling, wit, and folksy tales.[131] In the particular passage, More countered the illogical distinction between the church – invisible as an institution – and its congregants – visible beings – put forth by the reformer Robert Barnes with the equally absurd assertion that the individual trees in a forest are visible but the forest as a whole is not. Barnes had attacked the authority of the church as an institution.[132] He saw the church as merely a congregation of people joined by faithful belief. That belief is personal, not set forth in doctrine and ceremony by any manner of religious authorities. In book eight of *Confutation*, More ridiculed Barnes in a lengthy attack. More first mocked Barnes by questioning how the church can be invisible if it is a congregation of visible men and women. He next equated the reasoning with the laughable claim that a woman can be invisible even though her hands, feet, or head are not. More then wrote of Barnes, "and as he might tell us, that of Paul's church we may well see the stones, but we cannot see the church. And then we may well tell him again, that he cannot see the wood for the trees. To say that the whole thing is invisible, whereof he saith we may see every part, is a thing above my poor wit / and I suppose above his too, to make his saying true."[133]

Thirteen years later, John Heywood included the saying in a book of common proverbs. Known now for its proverbs, Heywood's *A dialogue . . . of all the prouerbes in the englishe tongue* (1546) is in fact a study on marriage given by an older man – the narrator of the story – to a youth who must choose between marrying for love or money.[134] A common literary style at the time was to proffer lessons on morals and manners through proverbs. Indeed, More, the uncle of Heywood's wife, was especially skilled at this and had used the device effectively in *Confutation*. Heywood's *Dialogue* interweaves narrative, dialogue, and more than one thousand proverbs to tell how the narrator counsels the youth with tales of two unhappy marriages.[135] The first story is of a young man who marries a pretty, but poor, girl for love. The marriage fails for lack of money. The second story is of a young man who marries for the wealth of an unattractive, older widow and quickly grows unhappy – unaffectionate, wasteful with money, attentive to other women. This marriage, too, is doomed. The proverb appears in the fourth chapter of the second story, in which the narrator relates how at dinner one night the husband and wife begin quarreling in front of their guests, who are the narrator and the couple from the first tale. The rich husband is attracted to the younger wife of the poorer man and exchanges sharply worded barbs with his own observant wife. Heywood advises:

> An olde saied sawe, itche and ease, can no man please.
> Plentie is no deyntie. ye see not your owne ease.
> I see, ye can not see the wood for trees.
> Your lyps hang in your light. but this poore mā sees
> Bothe howe blyndly ye stande in your owne lyght,
> And that you rose on your right syde here ryght.[136]

With proverbs, Heywood conveys lessons on how the rich man does not value what he has and instead wants what the poorer man has. He unnecessarily torments himself (itch and ease, can no man please) by undervaluing what he has in excess (plenty is no dainty) and overlooking things that are in front of him (you cannot see the wood for trees).[137] The husband makes himself unhappy (your lips hang in your light; blindly you stand in your own light) by his actions.[138] Instead, he should consider himself lucky to have come into wealth (you rose on your right side).

. Used mockingly by More and as a moral by Heywood, how we see the forests for the trees changes with context. In its literal interpretation, the difference between the forest and the trees is expressively found in a description of the Wabash River valley in the Midwest region of the United States.[139] An observer of the pristine landscape wrote that "if the forest is viewed from a high bluff, it presents the appearance of a compact, level sea of green, apparently almost endless." In scanning the horizon, he sees "the tree-tops swaying with the passing breeze, and the general level broken by occasional giant trees which rear their massive heads." A different view – of immense trees and a rich diversity of species – is presented upon descending from the overlook. "Going into these primitive woods, we find symmetrical, solid trunks of six feet and upwards in diameter." The enormity of old sycamores, "with a trunk thirty or even forty, possibly fifty or sixty, feet in circumference, while perhaps a hundred feet overhead stretch out its great white arms," overwhelms the human scale. Our keen witness of nature finds "the tall, shaft-like trunks of pecans, sweet gums or ashes, occasionally break on the sight through the dense undergrowth, or stand clear and upright in unobstructed view in the rich wet woods, and rise straight as an arrow for eighty or ninety, perhaps over a hundred, feet before the first branches are thrown out."

The Wabash River forest and its trees illustrate the modern figurative meaning of the phrase. Today, it means to discern a broad or general understanding amid an overabundance of detail. Someone who cannot see the forest for the trees is overwhelmed in specifics and minutiae and does not grasp the larger meaning of a situation; they are lost in the details. The view of the Wabash forest from the overlook is of a uniform carpet of leaves and offers an unlimited horizon, but within the forest, massive trees restrict the vista. Studying forest–climate interactions requires striking the appropriate balance between specific details and generality. The interaction of forests with climate is the sum of many physical, chemical, and biological processes, some reinforcing and some counteracting one another. The net outcome of these interactions is not simple, nor is it evident from that of the individual components. Like the thick Wabash forest interior, the complexity and uncertainty and inability to provide exact answers to the forest–climate question can block the broader understanding of forests. *Seeing the forest for trees* symbolizes a need to understand the various component processes and the resulting emergent behavior, but also to find the broad view necessary to communicate a complex science. Unsubstantiated claims, embellishments, and hyperbole are found in writings of forest–climate influences throughout the centuries. Equally common, too, are pejorative dismissals of the science. Yet there is a larger picture to the story: our actions influence climate, and one of those actions has been to deforest the world.

More's distinction between visible trees and the invisible forest, though mocking, is equally relevant to the forest–climate question in that the distinction between the forest and the trees denotes spatial scale. The Wabash River forest canopy is endless when seen from above, but within the forest, scattered enormous trees with spreading crowns dominate the view. Scale of study is critical to understanding forests and climate because trees influence their local microclimate in a manner different from how forests influence the macroclimate at regional to continental scales. While a forest may be the visible sum of its individual trees, its interaction with climate differs from that of a single tree and is not simply the sum of the many trees. Much of the past advocacy for and rejection of forest influences on climate was a misunderstanding of scale. That confusion persists in today's science.

It is Heywood's usage, however, that is perhaps most apropos to the forest–climate question. Heywood's counsel that "you cannot see the wood for trees" is an admonishment to not overlook the things that are readily before us. It is an apt warning to humankind about forests, how we perceive them, and our relationship with them. People throughout the world and in all times have defined what a forest is: a preserve for royal pleasure; a tool for naval supremacy; a provider of wealth; a fearful wilderness to be conquered; a place to be protected for public good; a muse for art, poetry, literature, and moral philosophy. Like British forester William Schlich before, we now define forests in relation to climate – specifically whether they mitigate or exacerbate planetary warming. Perhaps we need to simply see the forest for the trees.

Part II

The Scientific Basis

This confirms the long-held idea that the surface vegetation, which produces the evapotranspiration, is an important factor in the earth's climate

– Shukla and Mintz (1982)

8   Global Physical Climatology

The issue of spatial scale runs rampant throughout the forest–climate controversy. At a broad scale, there is a distinct poleward zonation of climate defined by gradations of progressively colder annual mean temperature in tropical, subtropical, temperate, boreal, arctic, and polar latitudes. Additional climate zones are defined based on annual precipitation and the seasonality of temperature and rain. The climate at large spatial scales extending over thousands of kilometers is known as the macroclimate. It is determined by geographic variation in solar heating of the planet, which sets in motion large-scale atmospheric circulations that transport heat poleward from the tropics, and also by proximity to oceans, which similarly transport heat in ocean currents. Mountains and large lakes create a regional climate that can deviate markedly from that expected according to the macroclimate. Climate at this scale, generally up to several hundred kilometers or so, is referred to as mesoclimate. Variation in topography, soils, and vegetation can generate local climates at a spatial scale ranging from a few to tens of kilometers, known as microclimates. A south-facing slope has a different microclimate than a north-facing slope. Forests have a different microclimate compared with open land. Although this is the general conceptualization of climate, these distinctions are not rigid and there are interactions across scales. The large-scale climate, for example, sets the background for micro- and mesoscale processes that in turn impact the large-scale atmospheric dynamics. Misunderstanding of forest influences across spatial scales bedeviled past scholars and is still apparent in today's science.

Earth's Energy Balance

Earth's climate can, to first order, be understood in terms of energy inputs and outputs. One form of energy is electromagnetic radiation. We perceive this energy principally as visible light and in the warmth of the Sun's rays. Terrestrial objects also emit radiation, but at wavelengths that are longer than that of sunlight and that are not visible to the eye. This radiation is called longwave, or infrared, radiation. At low temperatures, we do not sense this energy because of its low amount. At higher temperature, such as an electric heater or burning fire, the radiant energy increases, and we feel its warmth.

   Radiative fluxes determine the balance of energy gained or lost by the planet as a whole (Figure 8.1).[1] Solar radiation heats the planet. Clouds, gases, and particles in the atmosphere absorb some of the incoming radiation, and the continents and oceans, too, absorb radiation that reaches the surface. The portion that is not absorbed is reflected to space. The absorbed solar radiation warms the surface, which emits longwave radiation, but clouds, gases, and particles absorb most of this radiation so that only a small amount passes through the atmosphere and escapes to space. The gases and particles suspended in the atmosphere also

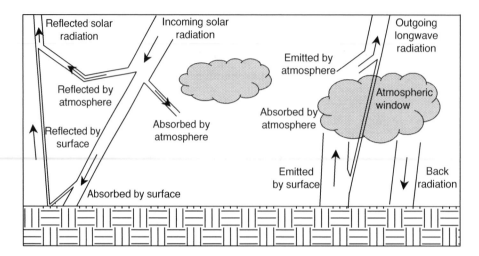

Figure 8.1 Earth's global energy budget showing solar radiation (left) and longwave radiation (right). Solar radiation heats the surface, causing longwave radiation to be emitted. Most of the emitted longwave radiation is absorbed in the atmosphere and reemitted onto the surface. Adapted from ref. (1).

emit longwave radiation, with some lost to space and some reaching the surface. The reemission of longwave radiation back to the surface is the greenhouse effect that warms the surface. The principal greenhouse gases are water vapor, CO_2, methane, and nitrous oxide. Greenhouse gases are poor absorbers of solar radiation but are strong absorbers of longwave radiation.

The planet as a whole balances annual solar and longwave radiation at the top of the atmosphere, but the geographic distribution is unequal. Latitudes near the tropics, between 30° S and 30° N, gain more energy from solar radiation than is lost by longwave radiation; they have a net surplus of energy. Poleward of these latitudes, solar radiation diminishes and there is a net deficit in radiation. The latitudinal gradient in net radiation results in an equator-to-pole temperature gradient. Low latitudes, which gain radiation, are warmer than high latitudes, which lose radiation.

Atmospheric General Circulation

The uneven geographic heating of the planet sets in motion the broad general circulation of the atmosphere and the resultant transport of heat from warm tropical latitudes to cold polar regions (Figure 8.2).[2] Warm air is less dense than cold air. Hot tropical air rises along the equator and moves northward aloft at high altitudes. The poleward-moving air cools, becomes denser, and sinks at about latitude 30°. The descending air returns to the equator along the surface. Earth's rotation deflects the surface winds to the west, forming what is known as the trade winds. This Northern Hemisphere circulation of ascending air in the tropics, poleward flow aloft, descending air at latitude 30°, and equatorward flow along the surface is known as the Hadley circulation. Two other circulations work similarly. The polar cell is a circulation between latitude 60° and the pole. Warm air flows toward the pole aloft and back as the polar easterlies along the surface

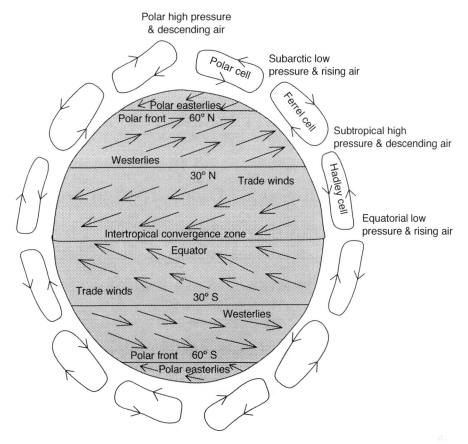

Figure 8.2 Atmospheric circulation, surface winds, and surface pressure.
Air circulates between the tropics and poles in the Hadley, Ferrel, and polar cells, with resulting trade winds, midlatitude westerlies, and polar easterlies along the surface. Adapted from ref. (2).

(southward moving air deflected to the west). A third cell in the middle latitudes between latitudes 30° and 60°, known as the Ferrel cell, connects the Hadley and polar cells. One branch of the descending air at latitude 30° returns to the tropics as the trade winds. The other branch flows poleward, being deflected to the east in middle latitudes. These are the mid-latitude westerlies. Similar circulations develop in the Southern Hemisphere.

Figure 8.2 shows several prominent features that characterize the circulation of the atmosphere. Semipermanent low surface pressure develops at the equator and latitude 60°, where air rises. Semipermanent high surface pressure develops at the poles and the subtropics at latitude 30°, where air subsides. Between latitudes 30° N and 30° S, surface trade winds converge along the equator in a broad region known as the intertropical convergence zone. Between latitudes 30° and 60° in both hemispheres, surface winds are westerly. At about latitude 60°, these winds clash with polar easterlies along the polar front.

The general circulation of the atmosphere accounts for the major climate zones in the world. Convergence and lifting of warm, moist air in the tropics lead to high annual rainfall. Many tropical regions have wet and dry seasons because of seasonal variation in the location of the intertropical convergence zone. Regions of subsidence, as occur with high surface pressure, generally have low rainfall. Many of the world's major deserts – in southwestern United States, North Africa, southern South America, South Africa, and western Australia – are located on the eastern flanks of the subtropical high pressures near latitudes 30° N and 30° S. Rainfall is also high, though not as much as in the tropics, in midlatitudes where warm, moist air clashes with cold air along the polar front.

Temporal change in solar radiation over the course of a year drives seasonal changes in temperature and precipitation. The tropics and subtropics, between latitudes 30° S and 30° N, have the least seasonal variation in temperature. Temperatures in this region are relatively constant throughout the year. In contrast, middle and high latitudes have pronounced seasons, with cool to cold winters and warm to hot summers. In the tropics, the intertropical convergence zone, where the convergence of the trade winds produces rising motion and heavy rainfall, migrates from south of the equator in boreal winter (austral summer) to north of the equator in boreal summer (austral winter).

The general circulation of the atmosphere occurs because warm air rises, creating low surface pressure, and is replaced by air flowing from regions of high surface pressure. The same physical mechanism creates regional monsoon circulations. Monsoons arise from thermal differences between continents and oceans. Land heats and cools faster than water. In winter, the continents are colder than oceans and have high surface pressure; oceans are warmer and have low surface pressure. The high pressure over land drives air offshore to the oceans. The opposite atmospheric circulation – onshore winds – occurs in summer when the continents warm, the surface low pressure intensifies, and moist air flows from oceans (high surface pressure) to warmer land (low surface pressure). Such monsoons can be seen in Asia, Africa, Australia, and southwest North America and bring heavy seasonal rain. Onshore sea breezes that develop along coastlines on hot, sunny days are a local manifestation of this type of circulation.

Ocean Circulation

Oceans exert a strong influence on climate. Water has a large heat capacity, meaning much energy must be absorbed or lost to change the temperature of one cubic meter of water by one degree. The large heat capacity of water bodies is seen in the moderating effect of oceans on climate. Seaside locations have less extreme seasonal fluctuations in temperature than in interior regions of the continents. In addition, water is a fluid and carries heat from one region to another. Like the atmosphere, the general circulation of the ocean transports heat from the tropics to the poles.

Wind-driven surface currents operate in all the ocean basins. Along the western boundary of the Atlantic and Pacific Oceans, for example, poleward-flowing surface currents transport warm water from the tropics across the basins in the middle latitudes. The Gulf Stream current in the North Atlantic off the US coast is a prominent example of a western boundary current. As a result, water off western and northern Europe is warmer than water off eastern North America at similar latitudes. A similar current operates in the Pacific off Japan so that water off southeast Alaska and western Canada is warmer than corresponding water north of Japan. Other currents operate on the eastern boundaries of the basins but transport cold polar water toward the tropics.

In addition to surface currents, deep ocean currents occur because of density differences in water. Cold, salty water is denser than warm, fresh water, which creates what is known as the thermohaline circulation.

In the North Atlantic, warm surface water flowing northwards becomes more dense as it cools and evaporation increases surface salinity. The dense water sinks and returns southward along the ocean bottom. Near the southern tip of Africa, this deep water joins similar cold bottom water from Antarctica. The river of bottom water spreads into the Indian and Pacific Oceans. The deep water slowly rises in these oceans and begins a long journey along the surface back to the North Atlantic. Changes in the thermohaline circulation, by altering the transport of heat by oceans, can cause large and rapid climate changes.

The Hydrologic Cycle

The hydrologic cycle describes the cycling of water among land, ocean, and air. Evaporation is the physical process by which liquid water in the oceans or on land changes to vapor in the air. It occurs when unsaturated air comes into contact with a moist surface. Evaporation provides the atmospheric moisture that returns to the surface as rain or snow. Evaporation also consumes an enormous amount of heat, which helps to cool the evaporating surface. Once in the atmosphere, water condenses, forming clouds; and if conditions are right, the water falls back to the surface as precipitation. Heat is released as water vapor condenses and changes from vapor to liquid. This heat is a source of energy that drives atmospheric circulation and fuels storms.

Additionally, the discharge of freshwater from rivers into oceans prevents the oceans from becoming saltier, which in turn influences ocean heat transport. Annually, the same amount of water that falls from the atmosphere as precipitation returns as evaporation. Although the planet as a whole balances water, oceans and land differ in precipitation and evaporation. More water evaporates annually from the oceans than is added from precipitation, producing an annual deficit of water. Runoff from land replenishes this imbalance. Over land, precipitation exceeds evapotranspiration. The surplus water runs off to streams and rivers where it flows to the oceans. About 65 percent of the water reaching the land surface as precipitation returns to the atmosphere as evapotranspiration, and 35 percent runs off to the oceans.

Climate Zones

Earth has five major climate zones defined by temperature and precipitation, each with several subzones.[3] This scale is referred to as macroclimate. The geographic distribution of forests, grasslands, shrublands, and deserts corresponds with these climate zones.

Humid tropical climates occur where temperatures are warm throughout the year. Rainfall determines two subzones. A tropical rain forest climate occurs where rainfall is abundant throughout the year but may vary seasonally with the position of the intertropical convergence zone. Tropical rain forest climates occur in hot, wet equatorial regions of South America, Africa, Southeast Asia, and Indonesia. A tropical savanna climate occurs in tropical regions that are warm year-round but with a pronounced dry season. Tropical savanna climates occur in Central America, to the north and south of the Amazon Basin in South America, to the north and south of the Congo Basin in Africa, eastern Africa, parts of India and Southeast Asia, and northern Australia.

Dry climates, which occur where rainfall is sparse throughout the year, are divided into semiarid and arid climates. The semiarid climate develops in temperate regions, most prominently in the Great Plains of the

United States, the steppes of Central Asia, and parts of southern South America, southern Africa, and Australia. Temperatures are hot during summer months when the clear sky and intense solar radiation heat the surface. The arid climate of deserts is not only dry but also intensely hot as solar radiation readily penetrates the clear, dry skies. Desert climates occur on the eastern flanks of the subtropical high pressures near latitudes 30° N and 30° S, but also in continental areas of the middle latitudes that are far removed from sources of atmospheric moisture such as central Asia, central Australia, and the Great Basin of the western United States.

Moist subtropical midlatitude climates occur in regions with distinct summer and winter seasons, where summers are warm to hot and winters are mild. They are divided into Mediterranean, humid subtropical, and marine zones. Mediterranean climates develop where a summer dry season is pronounced, such as in southern California in the United States and along coastal areas of the Mediterranean Sea. The climates have mild, moist winters and hot, dry summers. Humid subtropical climates form in the southeastern United States, eastern China, Japan, and along the southeastern coasts of South America, Africa, and Australia. The climates occur on the western edge of subtropical high-pressure areas, which drives warm, moist tropical air toward the middle latitudes, and consequently have hot, humid summers. Winters are mild and precipitation is abundant throughout the year. Moderate to pronounced seasonality is a dominant feature of climate. Marine climates occur in the Pacific Northwest region of the United States, western Europe, and western South America in middle latitudes where oceans moderate climate. Marine climates have mild winters, with temperatures rarely below freezing, cool summers, and abundant precipitation year-round.

Moist continental climates occur in the northern regions of North America, Europe, and Russia. Large seasonal variation in temperature, with moderate to cool summers and cold winters, characterizes the climate. The humid continental subzone is divided into warm summer and cool summer regions. Farther north in Alaska, northern Canada, northern Europe, and northern Russia, where the winters are bitterly cold and the summers are cool and short, the climate is subarctic.

Polar climates develop in high latitudes or mountaintops where winters are extremely cold and summers are chilly. Polar climates are categorized as tundra or ice cap based on temperature. Tundra climates arise where temperatures are sufficient such that plants can still survive in the short summers and long, cold winters. In the extreme cold of Greenland and Antarctica, characteristic of an ice cap climate, little vegetation grows and permanent glaciers cover the land.

The natural vegetation of the world has a distinct geographic pattern that corresponds to climate zones. Forests grow in tropical rain forest, humid subtropical, marine, humid continental, and subarctic climates (Figure 8.3).[4] In these regions, precipitation is abundant year-round. Trees cannot survive in the bitter cold of tundra climates. Instead, small shrubs, herbaceous plants, and mosses grow. Extensive grasslands occur in semiarid and savanna climates. Short, dense woody bushes form chaparral (also known as Mediterranean) vegetation in Mediterranean climates. Deserts, with sparse or widely spaced scrubby plants, establish in arid climates. The close correspondence between climate and vegetation is readily apparent, and climate zones such as tropical rain forest, tropical savanna, Mediterranean, and tundra are named after vegetation.

Climate Variability

The actual temperature and precipitation in any year can deviate markedly from the long-term climatology. Some years are warmer or colder than normal; some are wetter or drier than normal. This is a realization of

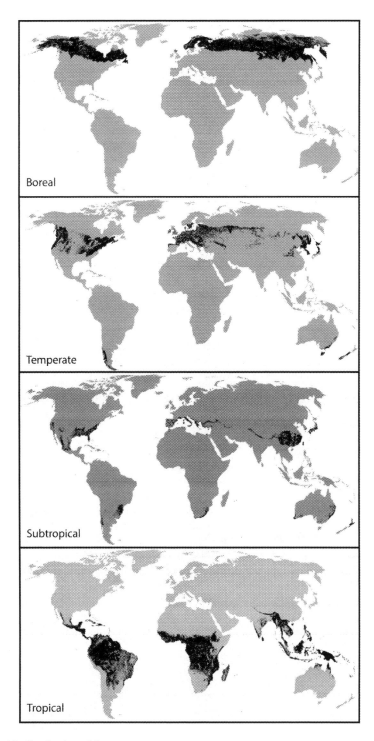

Figure 8.3 Geographic distribution of forests.
Boreal, temperate, subtropical, and tropical forests correspond to climate zones. Adapted from ref. (4).

climate variability at seasonal-to-interannual timescales. There are many modes of climate variability arising from atmosphere–ocean interactions. The most prominent is the El Niño/Southern Oscillation (ENSO), in which a large-scale warming of sea surface temperature in the eastern tropical Pacific in conjunction with a change in atmospheric winds across the Pacific alters temperature and precipitation worldwide. The period of higher-than-normal sea surface temperature is often followed by a cold phase (La Niña) in which temperature in the eastern topical Pacific is below normal temperature. El Niño (warm episodes) and La Niña (cold episodes) are opposite extremes of the ENSO cycle. El Niño refers to the warming of surface waters, while Southern Oscillation refers to changes in atmospheric circulation. The consequences of ENSO are felt throughout the world, so much so that the occurrence of El Niño or La Niña is a good prediction of temperature and precipitation in many regions. Other modes of variability arise from atmosphere–ocean interactions in the North Pacific, North Atlantic, and other basins.

Mechanisms of Climate Change

Climate has changed over the course of Earth's history. Just 18,000 years ago, the planet was in the grips of a prolonged cold period in which ice covered vast tracts of the Northern Hemisphere. Over the past few million years there have been numerous such ice ages separated by shorter, warm interglacial periods. Ice cores extracted deep below the surface in Greenland and Antarctica reveal the climate history over the past several hundred thousand years. One ice core from Antarctica covers the past 800,000 years and shows eight glacial cycles (Figure 8.4).[5] During glacial periods, temperature was several degrees colder than present. There is a close relationship between greenhouse gases and temperature over the glacial cycles,

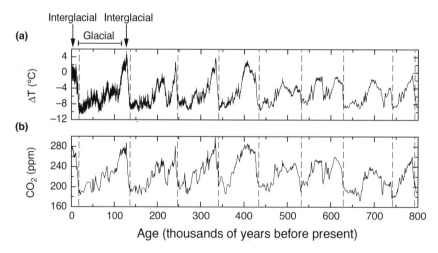

Figure 8.4 Climate history over the past 800,000 years reconstructed from the EPICA Dome C ice core in Antarctica. Shown are eight glacial cycles in terms of (a) temperature deviation and (b) CO_2 concentration in parts per million (ppm). CO_2 concentration was low during cold glacial periods and increased during interglacials. Dashed vertical lines indicate approximate glacial terminations. Adapted from ref. (5).

with low atmospheric concentrations of CO_2 during glacial periods. The last ice age reached its maximum 18,000 to 21,000 years ago, when extensive glaciers one kilometer or more thick covered northern regions of North America, Europe, and Russia. Thereafter, climate warmed, the glaciers melted, and vegetation that had been restricted to southern locations migrated northwards.

Ice cores and other data also reveal rapid climate changes since the last glacial maximum. One such climate change was the Younger Dryas cold period, which occurred from about 12,800 to 11,500 years ago. At that time, following prolonged warming that brought an end to the ice age, temperature abruptly decreased in North America and Europe. Newly emerged forests reverted to glacial tundra. Glaciers advanced southwards and down mountains. This cold period lasted for about 1,300 years before warming began again. Climate also changes over shorter timescales. For example, climate was relatively mild from about 950 to 1250 CE in an era known as the Medieval Warm Period. Then climate began to fluctuate, with cold winters interspersed among warm winters. Beginning about 1550, climate entered a prolonged period of cold temperature known as the Little Ice Age that extended to 1700 and later. Winters were long and cold; summers were short. Alpine glaciers advanced to lower elevations.

One of the longest timescales at which climate changes relates to the slow movement of continents over periods of several hundred million years. About 500 million years ago, the continents were widely dispersed along the equator. They drifted together and collided so that by 300 million years ago a large supercontinent called Pangaea had formed. Pangaea began to break apart 200 million years ago, and since then the continents have slowly moved to their current locations. The movement of continents alters the geographic distribution of land masses across the planet and changes the circulation of the oceans.

One reason for the recurring waxing and waning of glaciers is that the amount of solar radiation varies in a process known as the Milankovitch cycles. Cyclical changes in the eccentricity of Earth's orbit around the Sun, the angle of tilt (currently 23.5°), and the time of year when the planet is closest to the Sun affect the amount of solar radiation received on Earth over periods of tens of thousands to one hundred thousand years.

Oceans transport vast quantities of heat poleward from the tropics in the thermohaline circulation. A weak thermohaline circulation reduces heat transport. Increases in freshwater input to the North Atlantic due to changes in continental runoff or glacial melt affect the poleward transport of heat by making the surface water less salty, so that the water is less dense and does not sink to the ocean bottom. Climate records throughout the North Atlantic show frequent large and rapid climate changes over the past several thousand years related to the reorganization of the North Atlantic thermohaline circulation.

The Sun is often described as emitting a constant amount of radiation. In fact, however, the amount of radiation varies by a small amount in relation to the appearance of dark spots on its surface. During times of maximum sunspot activity, the Sun emits more radiation. Observations show an irregular cycle in sunspot abundance with a period of about 11 years. Sunspots were virtually absent during the period 1645–1715, during which time Earth received less solar radiation. This decrease in solar radiation, known as the Maunder Minimum, corresponds to the Little Ice Age and may explain why temperatures were abnormally cold during that time.

Aerosols are microscopic particles suspended in the atmosphere, including dust from land, salts from ocean spray, black carbon (soot) from fires, and volcanic ash. Other aerosols form when chemical reactions in the atmosphere convert gases to particles. Fossil fuel combustion and other human activities inject aerosols into the atmosphere. The presence of these aerosols affects climate by absorbing or scattering atmospheric radiation and by altering clouds. Volcanoes are a particularly important source of short-term climate change. Volcanoes inject dust, debris, and gases high into the atmosphere where they can remain for many months and are transported around the world. The airborne material reflects solar radiation to space and cools surface temperatures worldwide. Recurring volcanic activity may have contributed to cooler temperatures during the Little Ice Age.

The chemical composition of the atmosphere, with increasing concentrations of CO_2, methane, nitrous oxide, and other gases, warms climate through the greenhouse effect. These chemicals have strong geochemical and biologically driven cycles (collectively termed biogeochemical cycles), and biogeochemical cycles are an important process that affects the magnitude and trajectory of climate change. Carbon dioxide, for example, cycles among atmosphere, oceans, and continents in response to the chemical weathering of rocks. Annual carbon fluxes in the geologic cycle are small but can result in large changes to atmospheric CO_2 over tens to hundreds of millions of years. The biological carbon cycle, with carbon uptake during photosynthesis and carbon loss during respiration, superimposes large, rapid changes on the small, slow changes of the geological cycle. Ice cores show that CO_2 varies naturally from a concentration of 180 parts per million (ppm) at glacial maximum to 280 ppm during interglacials (Figure 8.4b).

Anthropogenic Climate Change

Following recovery from the last ice age, atmospheric CO_2 concentrations remained at about 260–280 ppm for several thousand years. Since the mid-1800s, the CO_2 concentration has increased. Measurements at Mauna Loa Observatory, at an altitude high in the atmosphere, show an increase from about 315 ppm prior to 1960 to above 415 ppm in 2021 (Figure 8.5).[6] The signature of the biosphere is seen in the measurements, with lower concentrations during the growing season when plants remove CO_2 from the atmosphere during photosynthesis and higher concentrations when plants are not photosynthetically active but still respire CO_2. The current atmospheric concentration is higher than observed over the past several hundred thousand years.[7] Although CO_2 naturally cycles among the atmosphere, land, and oceans, fossil fuel

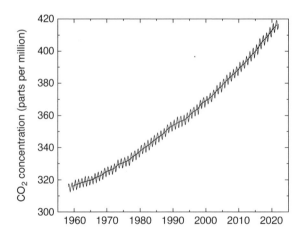

Figure 8.5 Carbon dioxide concentration measured at Mauna Loa Observatory, Hawaii.
Shown is the monthly concentration from March 1958 through December 2021 (thin line) and the annual mean concentration (thick line). For the data source, see ref. (6).

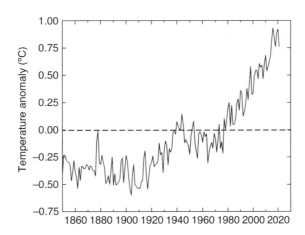

Figure 8.6 Global mean surface temperature, 1850–2021.
Shown is the annual global mean temperature over ocean and land, expressed as the difference from the 1961–90 mean. For the data source, see ref. (8).

combustion, deforestation, and agriculture emit CO_2 to the atmosphere. The concentrations of methane and nitrous oxide have also increased from human activities. The period since the mid-1800s has seen a prominent warming of Earth's surface (Figure 8.6).[8] Temperature has increased by about 1°C. The primary cause of the warming is increases in CO_2 and other greenhouse gases.[9] The greenhouse gas warming is partially offset by anthropogenic aerosols in the atmosphere. Natural changes in climate from volcanoes and solar variability are minor compared with anthropogenic changes.

To understand why climate has changed in the past or how it will change in the future, scientists use models that simulate Earth's climate. Global climate models represent a set of mathematical equations that describe the large-scale circulation of the atmosphere and ocean and their physical state, including interactions among oceans, atmosphere, land, and sea ice that affect climate. The equations are solved by discretizing the planet into a three-dimensional grid (longitude × latitude × height) so that all calculations are performed in every grid cell. A typical spatial resolution for a grid cell is 1° in longitude and latitude (equivalent to about 100 kilometers in mid-latitudes). The vertical thickness of a grid cell is variable, ranging from tens of meters to many kilometers in the atmosphere and ocean; and from several millimeters to several meters in the soil. In this way, a model has several million locations distributed spatially and vertically across the planet. The equations are numerically solved over a short time interval (typically 30 minutes or less) so that the atmosphere, ocean, land, and sea ice states are updated many times over the course of a simulated day, tens of thousands of times for a simulated year, and millions of times to recreate past climates and anticipate future climates. One of the longest continuous simulations spans the period 850 CE to 2005.[10] The most complex models, known as Earth system models, simulate physical, chemical, and biological processes that underlie climate, including the carbon cycle, terrestrial and marine ecosystems and biogeochemistry, wildfires, atmospheric chemistry, and human uses of land.

In simulations of future climate, input data on greenhouse gas concentrations, aerosols, chemically reactive gases, and land use are obtained based on descriptions of socioeconomic possibilities. These possibilities, called shared socioeconomic pathways (SSPs), depict scenarios of the future with respect to demographics, economies, lifestyles, technology, energy consumption, and land use; the resulting

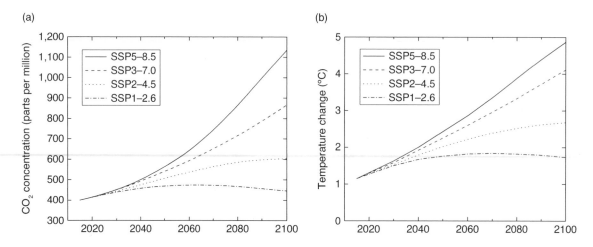

Figure 8.7 Shared socioeconomic pathways (SSP) for 2015 to 2100.
(a) Atmospheric CO_2 concentration varies among SSPs in relation to radiative forcing. (b) Global mean temperature change expressed as the deviation from preindustrial. The planet warms in relation to the radiative forcing. SSPx-y identifies the name (x) and radiative forcing (y). SSP5-8.5, for example, denotes SSP5 with a radiative forcing of 8.5 W m^{-2}. For data sources, see ref. (12).

emissions of greenhouse gases, reactive gases, and aerosols; and the chemical composition of the atmosphere.[11] The SSPs describe technological and socioeconomic changes to limit planetary warming to certain targets. The target is specified in terms of a radiative forcing. Radiative forcing quantifies the change in the planetary energy balance in units of energy flux (watts; W) per unit area (square meter; m^2) from the preindustrial state because of greenhouse gases and other changes in atmosphere composition, and it directly corresponds to the amount of planetary warming. The radiative forcing specifies the targeted perturbation to the atmosphere, and the SSP describes the societal, cultural, economic, and technological means to achieve the radiative forcing.

Four specific scenarios span a range from high CO_2 concentration and high radiative forcing that leads to large planetary warming to low CO_2 concentration and low radiative forcing with less warming (Figure 8.7).[12] SSP5-8.5, with a radiative forcing of 8.5 W m^{-2}, has high emissions of CO_2 and other greenhouse gases. The concentration of CO_2 exceeds 1,100 ppm by 2100, and other greenhouse gases also increase. As a result, global mean temperature warms by almost 5°C. SSP1-2.6, in contrast, has a radiative forcing of 2.6 W m^{-2}. It is a mitigation scenario with low greenhouse gas emissions and atmospheric concentrations. In this scenario, CO_2 concentration remains less than 450 ppm at 2100, and planetary warming is less than 2°C. To attain the low CO_2 concentration, anthropogenic emissions must decrease from the current level and, in fact, become negative by 2080, meaning that there is a net removal of CO_2 from the atmosphere. SSP2-4.5 is another mitigation scenario with decreasing CO_2 emissions, a concentration of 600 ppm at 2100, and intermediate planetary warming. SSP3-7.0 is more similar to SSP5-8.5 with increasing CO_2 concentration and planetary warming of several degrees. Taken together, the SSPs span a range of possible societal futures to show climate as it is likely to be with differing degrees of intervention, and they provide a guide for governmental policy making.

∎

Transport of heat by the atmosphere and ocean is the classic way to understand Earth's climate. The geographic heating of the planet by the Sun, the geographic distribution of oceans and continents, and other large-scale features of the land such as mountain ranges and lakes are primary determinants of climate across broad spatial scales. This is the view of climate that emerged in the nineteenth century and that set the framework within which foresters and meteorologists argued the pros and cons of forest influences on climate. It is now understood that superimposed on the broad geographic control of climate is, indeed, a signature of vegetation. Furthermore, while nineteenth century meteorologists argued for a constancy to climate, it is now recognized that climate changes from various natural and anthropogenic forcings. One of those forcings is natural and human-mediated change in forest cover. The scientific concepts that substantiate this view began to emerge in the latter half of the nineteenth century with the establishment of forest meteorological observatories, their measurements of forest–atmosphere coupling, and the study of surface energy, water, and chemical exchanges between forests and the atmosphere.

9   Forest Biometeorology

Energy transfers between the biosphere and atmosphere affect climate at spatial scales from microclimates to the globe. These include radiative exchange, seen, for example, in the warmth of sunlight; convection, experienced in the cooling influence of a breeze; evaporation; and conduction of heat in the soil. These fluxes are regulated, in part, by the biophysical and physiological characteristics of plants. In addition, plants remove CO_2 from the atmosphere during their growth and emit many chemical species to the atmosphere. Biometeorology is a field of science that studies the interactions between the biosphere and the atmosphere through energy, water, and chemical exchanges. These are regulated by the amount of water in the soil, and so the hydrologic cycle, in which some water infiltrates into the soil and some is lost as runoff, is closely coupled to the energy and chemical cycles. Seasonal changes in leaves, both foliar chemistry and the amount of leaves in the canopy, drive short-term variation in energy, water, and chemical fluxes. The growth of vegetation, accumulation of carbon in plant biomass and soil, and changes in floristic composition cause longer-term changes in biosphere–atmosphere coupling over periods of decades, centuries, and millennia.

Principles of Environmental Physics

Energy flows between the land surface and the atmosphere, along with the cycling of water and numerous chemical species (Figure 9.1). These exchanges of energy and mass are regulated, in part, by vegetation and influence climate at spatial scales from the microclimates of a forest canopy to the large-scale climate of the Amazon basin and even the entire globe. Electromagnetic radiation from the Sun combines with the longwave radiation emitted by gases and particles in the atmosphere to heat the land. This energy is dissipated as electromagnetic radiation from land or through turbulent movement of heat away from the land surface. A portion of the energy is also stored in the soil and plant biomass. Water falls to the ground as precipitation in liquid (rain) and solid (snow) forms and returns to the atmosphere as a gas (water vapor) through evapotranspiration. Plants absorb CO_2 from the atmosphere during photosynthesis, and respiration by plants and soil microorganisms returns CO_2 to the atmosphere. Associated with the cycling of CO_2 is a cycling of oxygen that is produced during photosynthesis and consumed during respiration. Plants emit numerous other chemicals to the atmosphere. One study measured more than 500 compounds actively exchanged over an orange grove.[1] Many gases, collectively known as biogenic volatile organic compounds (BVOCs), undergo chemical reactions that affect atmospheric chemistry and climate.

    Physical principles govern the flows of energy between the biosphere and the atmosphere. The Sun is the principal source of radiant energy. Direct beam radiation is solar radiation not scattered by the sky. It

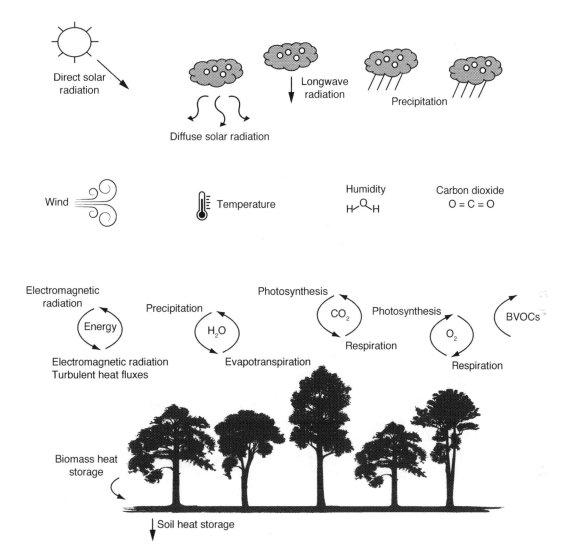

Figure 9.1 Cycles of energy, water, and chemicals with the atmosphere.
The biosphere exchanges electromagnetic radiation, turbulent heat fluxes, water vapor (H_2O), carbon dioxide (CO_2), and oxygen (O_2) with the atmosphere. Plants also emit biogenic volatile organic compounds (BVOCs), which form aerosols. The fluxes depend on characteristics of the atmospheric boundary layer (e.g., temperature, humidity, wind speed, cloud cover, CO_2 concentration) and in turn influence boundary layer dynamics.

emanates based on the Sun's angle above the horizon and its location with respect to compass direction. An object that blocks the path of direct beam radiation casts a shadow that can be seen. In contrast, diffuse radiation originates from all directions because of scattering of sunlight by the sky. With cloudy or overcast conditions, the sky becomes a uniform source of radiation and no shadows are cast. The gases and particles suspended in the atmosphere emit longwave radiation, as do objects on land. Direct beam solar radiation is the dominant flux during the day under clear, dry skies. Diffuse solar radiation and longwave fluxes are both important with overcast or cloudy skies. Longwave radiation is the sole source of radiant energy to the surface at night.

The radiant energy impinging on a body can be dissipated in several ways. All materials except for the blackest reflect sunlight. The proportion of the incoming radiation that is reflected is known as reflectance; the remainder is the solar radiation that is absorbed or transmitted through the material. Material with a high reflectance absorbs less solar radiation than material with low reflectance and sustains less solar heating. Emission of longwave radiation is a second means of radiative cooling. Similar to gases and particles in the sky, terrestrial objects emit electromagnetic radiation in the infrared waveband. An object emits longwave radiation in relation to its temperature. Objects with a higher temperature emit radiation at a greater rate than those with a lower temperature. Emission of longwave radiation is a key means to regulate body temperature.

Movement of air carries heat away from objects in a process known as convection. A common example of convection is the warmth felt as warm air rises from a radiator. This heat exchange is called sensible heat because it is heat that can be felt or sensed. An object loses energy if its temperature is warmer than the air; it gains energy when it is colder than the air. The larger the temperature difference, the larger the heat flux (if all other factors are equal). The magnitude of the sensible heat flux is also directly proportional to the movement of air molecules. This movement occurs through molecular diffusion in still air, or through the movement of air itself in windy conditions. Transport is ineffective and the sensible heat flux is small with calm conditions, but strong breezes efficiently carry heat away from an object. This is why a breeze is refreshing on a hot summer day.

An object can also lose heat by evaporation. Evaporation occurs when a moist surface comes in contact with drier air. Liquid water evaporates from the surface, increasing the amount of water vapor in the surrounding air. The moisture is transported away from the object by molecular diffusion and air movement. Considerable energy is required to change water from liquid to vapor, during which energy is absorbed from the evaporating surface without a rise in temperature. The amount of energy required to evaporate one gram of water (i.e., to change water from liquid to gas) is almost 600 times the energy required to increase the temperature of the same amount of water by 1°C. This energy is stored (i.e., is latent) and released when water vapor condenses to liquid in the atmosphere. The energy associated with evaporation is known as the latent heat flux. Evaporation, therefore, consumes an enormous amount of energy, and evaporative cooling is an important means to regulate body temperature. When water evaporates from a surface, the surface becomes cooler. Evaporation involves a transfer of both mass and energy to the atmosphere. Transfer of mass is seen as wet clothes dry on a clothesline. Heat loss is why a person may feel cold on a hot summer day when wet but hot after being dried with a towel. The rate of evaporation increases with more radiation, with stronger winds, and with drier air. We know this from experience: Clothes hanging on a line outside dry quickly on a sunny, windy day in an arid climate, but more slowly in a humid climate.

Direct contact with another object transfers heat in a process known as conduction. Conduction is the transfer of heat along a temperature gradient from high temperature to low temperature but, in contrast to convection, due to direct contact rather than movement of air. The heat felt when touching a steaming cup of coffee is an example of conduction. The rate of heat transfer by conduction depends on the temperature gradient and thermal conductivity. Thermal conductivity is a measure of the ability of a material to conduct heat. Differences in thermal conductivity create perceptions of hot or cold when touching an object. A metal spoon placed in a pot of boiling water feels warmer than a wooden spoon because it conducts heat to the hand much more rapidly than the wooden spoon. Materials with low thermal conductivity reduce heat loss by conduction and are effective insulators. Styrofoam is a poor conductor of heat. Air is also a very poor conductor of heat, so that double-paned glass windows with an inner layer of air are a very effective insulator. Water, in contrast, has a thermal conductivity that is nearly thirty times larger than that of air. This is why a body immersed in cold water loses heat rapidly.

An object's temperature responds to changes in its radiative environment and the ease at which heat is dissipated. We know this from experience. The shade of a tree is refreshing on a hot summer day because our body is subject to less radiative heating than in full sunlight. A breeze from a fan carries heat and moisture away from our body on a sweltering day and helps us to not overheat. The cooling effect of evaporation is why we perspire, and it is why a person may feel comfortable in dry climates, where low humidity results in rapid evaporation of sweat, but hot and uncomfortable in humid climates, where evaporation is not as efficient.

Leaf Temperature

The energy balance of a leaf illustrates fluxes in the biosphere–atmosphere system and their control of body temperature (Figure 9.2). A leaf absorbs solar and longwave radiation from the sky and its immediate

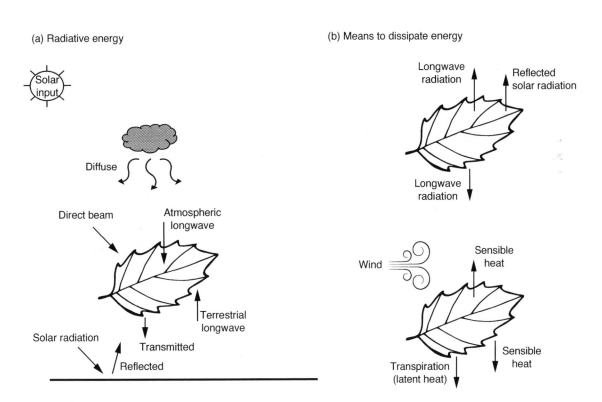

Figure 9.2 Energy balance of a leaf.
Shown are (a) sources of radiative energy and (b) means to dissipate energy. The radiative energy absorbed by a leaf from solar and longwave radiation is reflected by the leaf, carried away as sensible or latent heat, or emitted as longwave radiation.

environment, emits longwave radiation, and exchanges sensible and latent heat with the surrounding air. Energy storage in the leaf mass may also occur, but storage is generally small compared with other fluxes. In studying the leaf energy budget, it is convenient to conceptualize a leaf as a flat plate with an upper and lower surface. This is readily evident for the leaves of many deciduous trees, but even needles, despite their cylindrical shape, can be represented as a two-sided plate because their length is much larger than their thickness.

Two general categories of leaves are broadleaves, which are flat, platelike leaves commonly found on deciduous trees, and needleleaves, which are cylindrical or flat needles or scalelike leaves commonly found on conifers (Figure 9.3).[2] Broadleaves range in size from a few square centimeters to tens of square centimeters or more and vary in shape from linear to ovals to irregular shapes. The edges of a broadleaf can be smooth or deeply indented with lobes; some are serrated similar to the blade of a saw. Broadleaves can be a single blade attached to the stem (e.g., beech) or compound leaves divided into several blades known as leaflets (e.g., ash). Conifers such as pines, spruces, and firs, on the other hand, have needles that vary in length from several millimeters to several centimeters. Some conifers, such as the cypress family, have scales instead of needles.

Solar radiation is absorbed, transmitted, or reflected by a leaf. Absorption occurs on the upper and lower leaf surfaces, but the amount varies with wavelength. The Sun's radiation is broadly divided into the visible

(a)   (b)

Figure 9.3 Illustrations of leaves in Evelyn's *Silva* (1776).
Shown are the lobed broadleaves of oak (panel a) and the needles of fir (panel b). Leaves differ in size and shape, which affects energy and chemical fluxes. Reproduced from ref. (2).

and near-infrared wavebands. Visible light is the light we see with our eyes. Green leaves typically absorb more than 85 percent of the solar radiation in the visible waveband; this light is used during photosynthesis. Light in the near-infrared waveband, at longer wavelengths than visible light, is not utilized during photosynthesis, and leaves typically absorb less than 50 percent of the radiation so that they do not overheat. A leaf similarly absorbs longwave radiation impinging on its upper surface from above and on its lower surface from below. A leaf emits longwave radiation from both surfaces. The net balance of solar and longwave fluxes is the net radiation absorbed by the leaf.

Sensible heat flux occurs from both sides of a leaf. The ease at which heat is carried away from the leaf depends on wind speed, leaf size, and leaf shape. High wind speed efficiently mixes the air, transporting heat from the leaf and decreasing leaf temperature. In contrast, still air reduces heat transport, and, on warm, sunny days, leads to high leaf temperatures. Turbulent mixing also depends on the size of a leaf. A small leaf is closely coupled to the surrounding air; heat is easily dissipated by convection and leaf temperature is similar to air temperature. Mixing and heat transport is less efficient for large leaves, which are more decoupled from the air and can be several degrees warmer than air. Deep lobes, such as found in some broadleaf deciduous trees, also promote mixing and heat transport away from leaves. The ease at which heat is transported by convection is encompassed by a biophysical quantity termed the leaf boundary layer conductance. Boundary layer conductance increases with stronger wind speed. Small leaves or deeply lobed leaves have a higher conductance than large leaves or leaves that are not lobed, all other conditions being equal.

Latent heat flux occurs when water inside a leaf evaporates to the atmosphere. The process by which this occurs in plants is called transpiration. Leaves have numerous microscopic pores embedded on their surface known as stomata. These pores open so that CO_2 can diffuse into the leaf during photosynthesis. At the same time, however, water diffuses out of the saturated cavities in the stomatal pore to the drier air surrounding the leaf. Stomata open and close in response to a variety of conditions: they close in the dark and fully open with high amounts of light; they close with temperatures colder or hotter than some optimum; they close as the soil dries so as to protect the leaf from desiccation; they close if the surrounding air is too dry; and they become less open with high atmospheric CO_2 concentrations. Stomatal conductance is a physiological quantity that measures how open the pores are. Stomata are typically, but not always, located on the lower leaf surface. For sensible heat, the leaf boundary layer conductance is the total leaf conductance for heat. The leaf conductance for transpiration consists of two conductances connected in series: stomatal conductance regulates water loss from inside the leaf to the leaf surface; and the boundary layer conductance regulates water loss between the leaf surface and the air surrounding the leaf.

Measured leaf temperatures reveal rapid temperature fluctuations related to changes in solar radiation and wind speed. In one study, the leaves were 5–7°C warmer than the air when exposed to full sunlight, but intermittent clouds caused rapid cooling to 2–3°C below air temperature (Figure 9.4).[3] Wind had a similar effect on leaf temperature, and a change from calm air to blowing air produced a near-instantaneous decrease in leaf temperature of several degrees. Transpiration, too, cools the leaf. A rapidly transpiring leaf has a cooler temperature than a dry leaf with little transpiration. Many physiological and morphological traits control leaf temperature. Large leaves, for example, are at a disadvantage in hot, dry environments with high amounts of solar radiation. In addition to leaf size, the thickness of a leaf and its water content determine how much heat can be stored in a leaf. A high leaf reflectance decreases absorption of solar radiation, and many desert plants have white coatings on their leaves to reduce temperature and prevent

Figure 9.4 Measured temperature of a broad-leaved tree. Shown is the leaf temperature of a cottonwood tree (square symbols) on a summer day in Boulder, Colorado, for sunny and cloudy conditions over an 11-minute period. Also shown is the temperature of the air. Adapted from ref. (3).

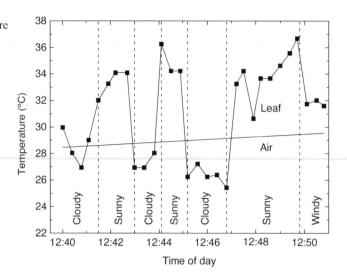

heat damage. An additional factor is leaf inclination angle. A larger inclination angle (relative to horizontal) decreases the amount of solar radiation incident on the leaf.

Forest Canopies

The physical principles that govern the energy fluxes of a leaf also regulate the fluxes between the biosphere and the atmosphere. Consider, for example, the energy balance of a forest. Solar radiation and atmospheric longwave radiation are energy inputs at the top of the canopy. Heat may be transported horizontally by winds (termed advection) into and out of the volume represented by the forest, but the net transport is generally considered to be negligible. Some of the energy added to the forest is stored in the soil and tree biomass. The remainder of the energy is dissipated through radiative and turbulent fluxes. Leaves, branches, trunks, and other materials reflect solar radiation in all directions. Some of the reflected radiation escapes above the canopy so that a portion of the solar radiation onto the canopy is reflected upward. Similar to reflectance, the proportion of the incident solar radiation that is reflected is known as the surface albedo; the remainder is the solar radiation absorbed by surface materials. The forest biomass and soil also absorb and emit longwave radiation and collectively contribute to an upward flux of longwave radiation above the canopy. Likewise, there is an exchange of sensible and latent heat between the forest and the atmosphere that depends on the biophysical and physiological characteristics of the forest and turbulent mixing with the atmosphere. Transpiration from leaves, evaporation from the soil, and evaporation of the water held externally on leaves and branches, such as the intercepted water following a rainstorm, are combined into a single term called evapotranspiration because the component fluxes are difficult to distinguish in field measurements.

Sensible and latent heat fluxes from the land impact the atmospheric boundary layer. The atmospheric boundary layer is the layer of the atmosphere directly above Earth's surface. Sensible heat from the land warms the boundary layer, and evaporated water moistens the boundary layer. Evaporated water is carried into the boundary layer, where it releases the stored latent heat when the vapor condenses back to liquid droplets and forms clouds and, if conditions are right, rain. Where soil moisture does not limit evapotranspiration, the boundary layer is generally cooler, moister, and shallower than in the absence of evapotranspiration. Dry sites have less evapotranspiration and a warmer, drier, and deeper boundary layer.

The exchanges of sensible and latent heat between the land and the atmosphere occur because of turbulent mixing of air and the resultant transport of heat and moisture. Turbulence occurs when the ground, grasses, trees, and other objects interfere with the flow of wind. The frictional drag imparted on air as it encounters rough surfaces slows the fluid motion of air near the ground. We see this force in the fluttering of leaves or the swaying of branches and treetops in strong breezes. The reduction in wind speed creates turbulence that mixes the air and transports heat and water from the land surface into the lower atmosphere. Surface heating also generates turbulence due to buoyancy. Warm air is less dense than cold air and rises in the atmosphere whereas cold air sinks. Rising air during the day with strong solar heating of the land enhances mixing and transport of heat and moisture away from the surface. At night, longwave emission generally cools the surface more rapidly than the atmosphere cools. Cold, dense air becomes trapped near the surface, with warmer air above in the lowest levels of the atmosphere. Under these conditions, vertical motions are suppressed, and transport is reduced.

The fluxes of sensible and latent heat above plant canopies are analogous to those of an individual leaf. The sensible heat flux from a canopy of leaves depends on the temperature of the effective canopy surface and that of the air above the canopy. It also depends on the ability to transport heat away from the canopy. Whereas the boundary layer conductance of a leaf relates to diffusion near the leaf surface, the comparable conductance for sensible heat flux from plant canopies, referred to as an aerodynamic conductance, relates to turbulent mixing with the overlying atmosphere. Tall vegetation is more aerodynamically rough than short vegetation, generating more turbulence and mixing. The aerodynamic conductance increases with taller vegetation so that forests have more turbulent mixing and a larger sensible heat flux compared with short grasses, all other factors being equal. At the scale of an individual leaf, stomatal control of transpiration is quantified by stomatal conductance. At the scale of a canopy of leaves, an aggregated canopy conductance represents the integration over all leaves in the canopy.

The magnitude of surface fluxes varies geographically and seasonally in relation to climate, principally due to changes in incoming solar radiation, temperature, and precipitation (Figure 9.5).[4] Annual evapotranspiration over land is high in the tropics and generally decreases toward the poles. In the tropical rain forests of Amazonia, large amounts of net radiation provide energy to evaporate water, and ample precipitation keeps the soil wet to sustain evapotranspiration.[5] In other tropical regions, there may be sufficient energy to evaporate water, but low annual precipitation reduces the availability of water that can be evaporated. Outside the tropics, temperature can drive seasonality in surface fluxes. In a deciduous forest located in the temperate continental climate of Massachusetts, for example, latent heat flux has pronounced seasonality with low rates in winter and high rates in summer.[6] At this site, sensible heat flux exceeds latent heat flux in spring before budbreak and leaf emergence. Thereafter, latent heat flux exceeds sensible heat flux during the summer as the foliage sustains high rates of transpiration. A pine forest in Oregon has a different seasonality to fluxes driven by precipitation.[7] The forest experiences wet, cool winters and dry, warm summers. In this locale, sensible heat flux is similar in magnitude to, or exceeds, latent heat flux throughout much of the year. At a boreal conifer forest in Canada, the cold winter with snow and frozen soil restricts latent heat flux until warmer months.[8]

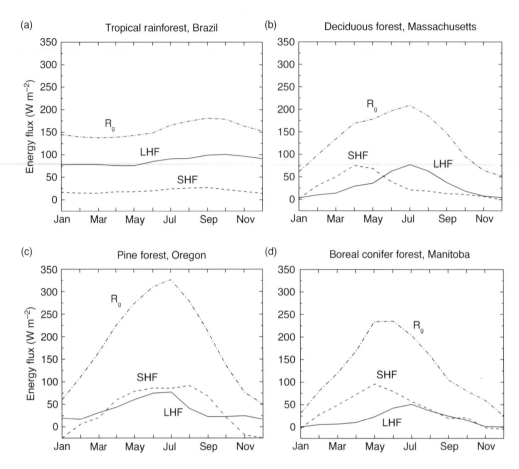

Figure 9.5 Monthly energy fluxes measured at four forest sites.
Shown are monthly solar radiation (R_g), latent heat flux (LHF), and sensible heat flux (SHF) for (a) tropical rain forest in Brazil (ref. 5), (b) temperate deciduous forest in Massachusetts (ref. 6), (c) pine forest in Oregon (ref. 7), and (d) boreal conifer forest in Manitoba (ref. 8). The scientific unit for energy flux is the amount (joule; J) per unit time (second; s) per unit area (square meter; m^2), denoted by $J\ s^{-1}\ m^{-2}$. One watt (W) is equal to one joule per second ($1\ W = 1\ J\ s^{-1}$) so that fluxes are expressed as $W\ m^{-2}$. Adapted from ref. (4).

Surface fluxes vary over the course of a day in response to the diurnal cycle of solar radiation, as shown in Figure 9.6 for an aspen forest on a summer day.[9] In early morning, the forest has a negative radiative balance, meaning that no solar radiation is absorbed but longwave radiation is lost. Sensible and latent heat fluxes are small. Stomata are closed, and the trees do not absorb CO_2 during photosynthesis. CO_2 is, however, lost during respiration, and there is a net flux from the forest to the atmosphere. As the Sun rises and the forest absorbs solar radiation, there is a net gain of radiation because solar absorption exceeds longwave loss. The forest begins to warm and some of the energy returns to the atmosphere as sensible heat. Stomata open, allowing for net CO_2 uptake (photosynthesis exceeds respiration) and water loss during transpiration. Fluxes typically are largest in early to middle afternoon and decrease in late afternoon when solar radiation decreases. This particular forest has moist soil. Latent heat exchange is an important means

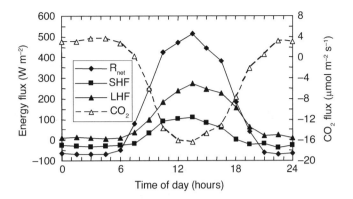

Figure 9.6 Diurnal cycle of surface fluxes.
Shown are net radiation (R_{net}), sensible heat flux (SHF), latent heat flux (LHF), and net CO_2 flux for quaking aspen forest in Prince Albert National Park, Saskatchewan, Canada. Data are averaged for the period July 19–August 10, 1994. Energy fluxes are shown on the left axis. The CO_2 flux is shown on the right axis. Net radiation is positive when there is a gain of energy and negative when there is a loss of energy. Positive values of sensible heat flux, latent heat flux, and CO_2 denote a flux to the atmosphere; negative values denote a flux to the surface. Carbon flux is expressed in micromoles (μmol) per square meter (m^2) per second (s). One micromole is equal to 10^{-6} moles. Adapted from ref. (9).

of dissipating the energy absorbed by the aspen forest and exceeds sensible heat flux during the day. On drier sites, however, sensible heat flux can exceed latent heat flux.

Several characteristics of vegetation influence surface fluxes. Vegetation with a small albedo absorbs more radiation than does vegetation with a large albedo. Albedo varies with wavelength. Plant canopies reflect more radiation in the near-infrared waveband than in the visible waveband. Albedo varies over the course of a day depending on the angle of the Sun above the horizon, and it varies seasonally depending on the amount of leaves, the presence of snow, and soil moisture. Some generalizations are possible. Albedo is largest for fresh snow; typical values range from 0.80 to 0.95. Vegetation has a much smaller albedo, typically ranging from 0.05 to 0.25 when averaged over the light spectrum, with forests having a lower albedo than grasslands or croplands (Figure 9.7).[10] Coniferous forests generally have a smaller albedo than deciduous forests. Trees mask the high albedo of snow. In seasonally snow-covered regions, the surface albedo of herbaceous grassland and cropland is much greater during winter than that of forests, as seen in Figure 9.7.

Another key characteristic is surface roughness. Surface roughness is a complex function of canopy structure.[11] It increases with canopy height so that tall canopies are rougher than short canopies. It also depends on the amount of leaves, branches, and stems and their vertical distribution in the canopy. A dense canopy with much foliage is less rough than a sparse canopy of the same height. Tall trees are aerodynamically rougher than short grasses, exert more drag on air flow, and generate more turbulence (all other factors being equal). This manifests in a larger aerodynamic conductance to heat and moisture transfer in forest than grassland; a forest is said to be closely coupled to the atmosphere. Heat is readily exchanged with the atmosphere, and the forest canopy is generally cool during the day as a result. Grassland, conversely, has a small aerodynamic conductance and is decoupled from the atmosphere. Short vegetation dissipates heat less effectively than tall vegetation so that the leaves of short vegetation experience warmer temperatures.

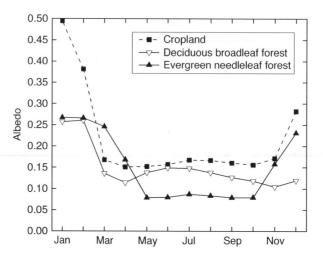

Evapotranspiration is influenced by stomata. If stomata are not open, the plant cannot photosynthesize, but when stomata are open, water inside the leaves diffuses out to the surrounding air during transpiration. If too much water is lost, the plant becomes desiccated and will die if its internal water is not replenished by water in the soil. Plants have evolved compromises between the need to open stomata to take up CO_2 and the need to close them to prevent water loss. Plants differ in their strategies to optimize CO_2 uptake while minimizing water loss. Some plants have high stomatal conductance and high rates of transpiration; other plants have a more conservative water use strategy.

An additional characteristic of vegetation is the large surface area covered by leaves. The amount of foliage in a plant canopy is described by leaf area index. Leaf area index is the projected, or one-sided, area of leaves per unit area of ground. Projected leaf area is different from the total surface area of a leaf. Thin, flat leaves have a total surface area that is twice the projected leaf area (both sides of the leaf are included). Needles have a total surface area that is more than twice the projected leaf area. One square meter (m^2) of ground covered by leaves with a combined surface area of 1 m^2 has a leaf area index of one. Because plant canopies have a vertical depth, many overlying layers of leaves can be stacked above one another. A typical leaf area index in a closed forest canopy is 4–6 m^2 m^{-2}; there is 4–6 times the leaf area as there is ground area. With small leaf area index, the vegetation absorbs little solar radiation. Instead, much of the radiation penetrates to the ground and heats the soil. The absorption of radiation increases with greater leaf area index, and surface albedo responds more to the optical properties of foliage rather than soil. A large leaf area index increases the area from which heat and moisture are exchanged with the atmosphere, and it can also influence the roughness of the canopy. When plants cover a small portion of the ground, evaporation is the dominant flux; transpiration becomes more important as plant cover increases. Photosynthetic CO_2 uptake by a canopy increases with greater leaf area index because there are more leaves to photosynthesize. However, the amount of solar radiation absorbed by the canopy saturates with leaf area index greater than about 5–6 m^2 m^{-2} so that canopy photosynthesis also saturates.

Soil moisture strongly influences the partitioning of net radiation into sensible and latent heat because of its control on stomata. Stomata close as the soil becomes drier so as to prevent excessive water loss and leaf desiccation. Over the course of a typical summer day, the majority of net radiation at a well-watered site is used to evaporate water rather than warm the surface. As a result, the air is cooler and moister than in the

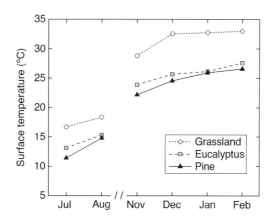

Figure 9.8 Surface temperature for grasslands and forests across Argentina and Uruguay.
Shown are monthly radiometric surface temperature for grassland, eucalyptus forest, and pine forest in the Concordia and Corrientes regions of Argentina and the Rivera region of Uruguay. Redrawn from ref. (12).

absence of evapotranspiration. In contrast, a dry site dissipates much less energy as latent heat and more as sensible heat. As soil dries, less water is available for evapotranspiration, stomata close, and more energy is dissipated as sensible heat or stored in the ground. Because most of the net radiation is dissipated as sensible heat, the lower atmosphere is likely to be warm and dry.

The characteristics of soils, too, affect surface climate. Soils with a high thermal conductivity gain and lose energy faster than soils with a low thermal conductivity. Sandy soils generally have a higher thermal conductivity than clay soils. Organic material has an extremely low thermal conductivity, and soils with high organic matter content have a thermal conductivity that is one-quarter to one-third that of mineral soils. The decomposing leaf litter on the forest floor, for example, effectively insulates the soil. The thermal conductivity of water is much larger than that of air and mineral or organic particles in the soil so that moist soil has a substantially larger thermal conductivity than dry soil. Similarly, heat capacity increases with more soil moisture. Heat capacity is a measure of the temperature change arising from a change in heat storage. Soils with a small heat capacity warm and cool faster, for a given heat flux, than soils with a large heat capacity. A moist soil requires more energy to increase its temperature than does a dry soil.

These surface characteristics combine to determine the effective temperature of vegetation. Figure 9.8 compares surface temperature for grassland and forest sites.[12] Surface temperature is defined as the radiometric temperature at which all phytomaterial in the canopy collectively emits longwave radiation and is several degrees cooler at the forests than at the adjacent grassland. Although the lower albedo of the forests leads to more solar heating, the tall trees produce better mixing with the atmosphere and efficiently carry heat away from the surface. Plentiful evapotranspiration from the forests also contributes to the cooler surface temperature. It is important to remember, however, that surface temperature is not a biological temperature. It is the effective temperature at which the canopy emits longwave radiation. In fact, a forest canopy is not an actual surface. It is a complex, three-dimensional arrangement of leaves, branches, and trunks spreading across a horizontal area and with vertical layering. The temperature of leaves can vary by several degrees depending on position in the canopy.[13] Likewise, the temperature of air varies vertically in the canopy (Chapter 11). Nonetheless, the radiometric surface temperature is useful in that it integrates the effects of albedo, surface roughness, leaf area, canopy physiology, soil moisture, and other conditions as they affect surface energy fluxes and temperature.

■

The exchanges of energy, water, and chemicals between forests and the atmosphere affect climate at spatial scales from microclimates to the globe. The reflection of solar radiation and emission of longwave radiation at the surface are part of the atmospheric radiative budget. Sensible heat flux directly influences the temperature of the air in the atmospheric boundary layer. Evapotranspiration cools the surface and moistens the boundary layer. Photosynthesis and respiration influence the amount of CO_2 in the atmosphere. Emissions of BVOCs and other chemicals determine the composition of the atmosphere and air quality. The means to study forest biometeorology arose during the forest–climate debate with the development of forest meteorological observatories. Today, there are many advanced means to monitor forests, measure their fluxes with the atmosphere, and assess their impacts on climate from the microscale to the macroscale.

10   Scientific Tools

Many scientific methods are available to study the manner in which forests affect climate. These are broadly distinguished as environmental monitoring, experimental manipulation, or use of numerical models of climate. For each technique, there are a variety of tools that vary in complexity. For example, meteorological measurements of air temperature and wind speed obtained using weather stations situated in forests and adjacent clearings are a simple means to characterize microclimates. More complex measurements of energy, water, and CO_2 fluxes obtained using principles of eddy covariance require sophisticated instruments located on tall towers extending above the forest canopy. In situ measurements of leaves and individual trees give an indication of physiological functioning and can be extrapolated to characterize an entire forest. Whole-ecosystem experimental manipulations that, for example, warm the soil or enrich the air with CO_2 provide insight to ecosystem responses to environmental change. Ecosystem studies monitor carbon and elemental stocks and fluxes, and watershed studies similarly monitor the hydrologic cycle. Such studies extend over several decades. Airborne and spaceborne instruments that acquire radiative signatures of the land surface provide an indicator of vegetation type, health, and productivity and help to bridge the gap between local in situ measurements and larger spatial scales. Atmospheric CO_2 observations at numerous locations throughout the world, too, provide information about the seasonal dynamics of biosphere–atmosphere CO_2 exchange and continental-scale fluxes. The longest such record, at Mauna Loa, Hawaii, dates to 1958 (Figure 8.5). Numerical models of terrestrial ecosystems and climate provide a means to test theories and develop understanding of the biosphere–atmosphere system.

Chamber Measurements

Conservation of mass is a fundamental physical principle. It means that in a closed system (i.e., in the absence of inflows and outflows) mass cannot be created or destroyed; the total mass of the system is conserved. In an open system with inflows and outflows, the change in mass is equal to the difference between inputs and outputs to the system. Conservation of mass is formulated mathematically as a continuity, or balance, equation and describes many physical and chemical processes in atmospheric science, hydrology, and ecology. A simple, everyday example is a kitchen sink filling with water. Provided the drain is plugged, the sink fills with water when the faucet is turned on because there is no loss of water. Even if the drain is opened, water will accumulate so long as the inflow from the faucet exceeds the outflow through the drain.

Conservation of mass provides the basis to measure gas fluxes in the biosphere using an enclosed chamber. The chamber is sealed, but air flows in at one end and out at the other end. The concentration of

Figure 10.1 Leaf chamber used to measure gas fluxes.
A scientist measures chemical emissions from the needles of a pine tree by placing them in a glass enclosure. Copyright University Corporation for Atmospheric Research. Photo credit: Carlye Calvin.

a gas of interest is measured at the inflow and outflow. The difference in concentration between the air flowing into the chamber and air flowing out of the chamber relates to sources or sinks within the chamber. For example, when a chamber is placed around a leaf, a decrease in the concentration of CO_2 in the air flow is a measure of the rate of CO_2 uptake during photosynthesis; an increase in the concentration of water vapor is a measure of transpiration. This is the principle by which leaf gas exchange is measured (Figure 10.1). The environment within the chamber can be purposely manipulated by changing the amount of light onto the chamber, the temperature of the chamber, or the relative humidity or CO_2 concentration of the inflowing air so that physiological responses to environmental factors can be studied. Chambers can even be constructed around an entire tree to measure whole-tree photosynthesis and transpiration (Figure 10.2).[1] Whole-tree chambers can be combined with experiments that add CO_2 into the chamber, warm the air, add nutrients to the soil, or alter other environmental conditions so as to understand the response to environmental change.[2]

Eddy Covariance Flux Towers

Chamber studies provide a precise measurement of photosynthesis, transpiration, and other chemical emissions by leaves and individual trees, but they do not provide a complete picture of a forest. Leaf measurements are not necessarily representative of a canopy of leaves; nor are individual tree measurements descriptive of an entire forest. Whole-tree chambers can be criticized, too, because the enclosure that surrounds a tree alters the conditions in which the tree grows, even in the absence of any purposeful

Figure 10.2 A whole-tree chamber.
Trees growing in chambers are used to measure gas fluxes at the Flakaliden research forest in Sweden. See ref. (1) for an overview of the experiment. Photo credit: Sune Linder.

manipulation. An alternative approach to measure and monitor fluxes of energy, water, CO_2, and other chemical gases in the biosphere–atmosphere system over large spatial scales is known as eddy covariance. Eddy covariance flux towers provide continuous measurements of biosphere–atmosphere exchanges of energy, moisture, and trace gases at fast timescales (sub-hourly).

Eddy covariance utilizes fundamental principles of geophysical fluid dynamics. Air can be considered to be comprised of discrete parcels with properties such as temperature, water vapor, or CO_2 concentration. As the parcels of air move, they carry with them their heat, water vapor, and CO_2. Turbulence creates eddies that mix air from above downward and from below upward and, in doing so, turbulence transports heat, water vapor, and CO_2 as the parcels are mixed. The eddy covariance method estimates fluxes by measuring rapid fluctuations of vertical wind velocity and the quantity of interest (temperature for sensible heat flux; water vapor for evapotranspiration; CO_2 for net ecosystem exchange of CO_2); the flux relates to the covariance of vertical velocity with the variable of interest.[3] Eddy covariance is technologically sophisticated and requires much hardware and software infrastructure including high-quality sensors to rapidly acquire the data; tall towers to support the sensors above forest canopies; gas analyzers to measure CO_2 and water vapor concentrations; software to process the data; devices for data storage; and computers to manage the system (Figure 10.3). The advantage is that direct measurements can be made in the field

Figure 10.3 An eddy covariance flux tower.
A scientist installs sensors on a tower at a walnut orchard in California to study turbulence within and above the canopy. Copyright University Corporation for Atmospheric Research. Photo credit: Carlye Calvin.

without altering growing conditions, and the measurements are representative of an area extending hundreds of meters. Eddy covariance has become a standard technique to measure ecosystem fluxes of energy, water, CO_2, and other chemical gases. Several hundred towers provide continuous flux measurements in forest, grassland, cropland, and other biomes throughout the world.[4] Figure 9.5 provides an example. The longest sites have been continually operating for over two decades.

Inventory Measurements

Inventory measurements examine individual trees and upscale the tree measurements to the forest based on the number of trees in the study area. Inventory measurements are commonly used to assess the carbon balance of a forest by measuring the biomass of trees. A substantial portion (up to two-thirds or more) of the mass of a tree is freshwater; the remainder is the dry mass, which is more commonly referred to as biomass. Carbon comprises about one-half of the dry mass, and this is carbon that has been removed from the atmosphere during the growth of the tree. The biomass of a forest is estimated by measuring the mass of individual trees that differ in trunk diameter. Smaller trees contain less mass than larger trees, but overall, there is a strong relationship between the diameter of the trunk and biomass. The exact relationship varies among species. By sampling trees of different sizes and different species, the relationships can be determined. The trees are felled, their foliage, branches, and trunks are dried and weighed, and a statistical relationship is obtained between tree diameter and biomass (Figure 10.4).[5] Then, knowing the number of trees by species in different size classes, the biomass of the forest can be estimated. Determining the biomass of trees is a labor-intensive exercise. Root mass is especially hard to measure because the roots need to be excavated from the soil. Long-term repeat measurements in permanent sample locations can be used to examine changes in biomass, for example, in the Amazon rain forest during drought.[6]

Forest transpiration can be estimated by measuring the flow of sap in trees. Water loss during transpiration drives soil water uptake by roots and transport through the trunk to leaves. Transpiration provides the force that pulls water from the soil, and cohesion binds water molecules together so that the pull exerted by transpiration extends down the trunk to the roots. Sap is the water in the tree trunk, and its ascent during transpiration can be

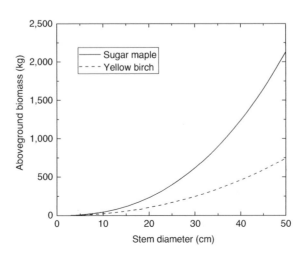

Figure 10.4 Relationship between stem diameter and aboveground biomass. Shown are relationships for sugar maple and yellow birch trees at the Hubbard Brook Experimental Forest, New Hampshire. Reproduced from ref. (5).

measured. In 1727, Stephen Hales reported on studies in which he measured the transported water in pear and apple trees by severing a root or a branch and attaching a long glass tube filled with water (Figure 3.2). Today, sap flow is measured using sensors placed into the trunk of a tree.[7] Sap flux density is the volume (cubic meter; m^3) or mass (kilogram; kg) of water moving per unit area (square meter; m^2) per unit time (second; s). Sap flow is expressed per unit area, which, depending on the particular study, is the cross-sectional area of the trunk, the area of the living sapwood that transports water, or the leaf area of the tree. From a sample of trees, sap flow is then scaled to the forest (total water flow per unit ground area) using an inventory of tree diameter.

Ecosystem Experiments

Experimentation is the essence of scientific discovery. Carefully designed experiments demonstrate cause and effect by measuring the response to a particular factor that is varied. If one wants to study how warmer temperatures affect tree growth, small trees can be grown in several chambers in which only temperature is varied; all other environmental conditions are identical for each chamber. Experiments can also be conducted in the field over large areas. A common experiment is to add fertilizer to see how nitrogen availability affects tree growth. Similar experiments can warm soil, mimic drought, or alter the forest community in some way. Soil temperature can be raised, for example, by placing heating cables underground, or air can be warmed by infrared heaters placed aboveground. Trees can be removed to study forest response to and recovery from disturbance. One of the more challenging experiments has been to study how forests respond to elevated CO_2 concentrations. Chambers provide one means to do so but are not representative of large spatial scales.

Free-air CO_2 enrichment (FACE) experiments allow trees grown in the field under prevailing environmental conditions to be exposed to elevated CO_2 concentrations.[8] In this approach, pipes are arranged horizontally or vertically on towers in a circle around the experimental area in which the trees are growing (typically an area up to a few tens of meter in diameter), and CO_2 is pumped into the air to expose the trees to a specified concentration of CO_2 (Figure 10.5). Sensors connected to computer systems regulate the flow of CO_2 to maintain the desired concentration. FACE studies have been conducted in many ecosystems including forest, grassland, cropland, and desert. They are frequently combined with other manipulations that warm temperatures, add nutrients, or mimic drought. FACE experiments are technologically intensive but can be combined with eddy covariance systems to provide a comprehensive assessment of forest responses to a changing environment.

Forest response to drought is another pressing environmental concern. Drought can be experimentally induced by preventing rain that drips through forest canopies from reaching the soil. In these so-called throughfall exclusion experiments, a barrier or some type of drainage structure installed 1–2 meters above the ground deflects the water away from the study area. Without water inputs, the soil dries out over a period of time. One such experiment in Brazilian tropical rain forest extended over an area of 1 hectare (10,000 m^2) for 10 years (Figure 10.6).[9]

Watershed Studies

Eddy covariance and sap flow measurements provide two means to estimate evapotranspiration. Another approach is to monitor water inflows and outflows from a watershed, also known as a catchment. A watershed is the geographic area that contributes to water flow in a stream or river. It can be hundreds

Figure 10.5 The Oak Ridge National Laboratory FACE site in Tennessee.
Shown are vertical pipes for CO_2 release and the supporting tower in a young sweetgum forest. Photo credit: Richard Norby.

of thousands of square kilometers for a large river or a few square kilometers for a small creek. High elevation points that form topographic divides define the area of the watershed. Water on the streamward side of the divide flows downslope to the stream; water on the other side of the divide flows into another stream. The complexities of the hydrologic cycle can be simplified to a simple mass balance equation in which the change in storage within the watershed is the balance between water input from precipitation, water loss from evapotranspiration, and water loss from runoff to streams. Over a full year, change in storage is small so that the difference between precipitation and runoff is an estimate of evapotranspiration. Precipitation is measured using rain gauges distributed throughout the watershed. Runoff is measured by observing stream flow at the mouth of the watershed.

Figure 10.6 A throughfall exclusion experiment.
Shown are the physical barriers that channel rainwater away from the soil in a throughfall exclusion experiment in Caxiuanã National Forest, Pará, Brazil. Photo credit: Patrick Meir.

One watershed where the water budget has been studied in detail is the Hubbard Brook Experimental Forest in New Hampshire (Figure 10.7).[10] Between 1956 and 1974, annual precipitation averaged 1,300 millimeters (mm) with 500 mm (38 percent) returned to the atmosphere in evapotranspiration and 800 mm (62 percent) exported as runoff in streams. Over the period studied, annual precipitation ranged from 950 to 1,860 mm, but evapotranspiration remained relatively constant from year to year, ranging from 418 to 542 mm. Evapotranspiration did not increase in wetter years or decrease in drier years. Annual streamflow, however, did increase in wetter years. Overall, there was a linear increase in annual streamflow in response to precipitation. This suggests that precipitation is first used to replenish water lost during evapotranspiration. Any excess water then contributes to streamflow. The chemistry of the stream water can also be analyzed to monitor cycling of nitrogen, calcium, potassium, and other elements in the watershed. Some studies of hydrology and biogeochemistry, such as Hubbard Brook, extend over several decades.

Satellite Remote Sensing

A crucial aspect of forest–climate interactions in the context of global change research is to upscale an understanding of forests developed for a small stand of trees to larger regional and continental scales. Satellite sensors orbiting in space provide a means to monitor Earth's atmosphere, oceans, and land across regional and continental scales and provide critical information about forest cover, productivity, and health.[11] Remote sensing is a field of science in which satellite- or aircraft-based instruments measure

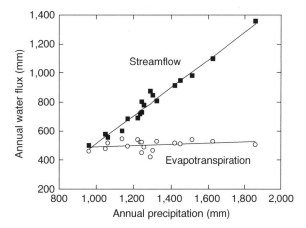

Figure 10.7 The water balance of a forested watershed. Shown are relationships of annual streamflow (closed squares) and evapotranspiration (open circles) with precipitation for the Hubbard Brook Experimental Forest during 1956–1974. Reproduced from ref. (10).

reflected or emitted electromagnetic radiation. The radiation received by the sensor is processed using mathematical algorithms that convert the received radiation signal into information on the environment.

All materials reflect or emit electromagnetic radiation. The amount varies among materials depending on wavelength, thereby providing a unique spectral signature to the materials. The same is true of water vapor and CO_2 molecules in the atmosphere, snow and ice on land, phytoplankton in the oceans, leaves in forest canopies, and many other objects. Two broad classes of satellites sense the spectral signal of terrestrial bodies. Passive sensors detect natural electromagnetic radiation that is reflected or emitted. Such instruments include radiometers, which measure the intensity of electromagnetic radiation in certain wavebands, or spectrometers, which measure the spectral characteristics of the reflected electromagnetic radiation. Passive sensors typically utilize the visible, infrared, and thermal infrared portions of the electromagnetic spectrum, but they differ in the wavelengths they are capable of measuring and the resolution with which they sample the electromagnetic spectrum. The Airborne Visible/Infrared Imaging Spectrometer (AVIRIS), for example, measures reflectance in 224 spectral channels with wavelengths between 0.4 and 2.5 micrometers, which is used to distinguish details of the chemistry of leaves in forest canopies. Most passive sensors cannot penetrate dense clouds, and thus have limited observing capabilities. Active sensors emit radiation in the direction of a target and measure the radiation that is scattered back to the sensor. Radio detection and ranging (radar) sensors emit microwave radiation. Light detection and ranging (lidar) sensors use a laser to emit a light pulse. Active sensors typically operate in the microwave band of the electromagnetic spectrum, which gives them the ability to penetrate clouds.

Most satellite sensors do not measure environmental data directly. Rather, mathematical algorithms are required to process the observed radiance acquired by the satellite to derive relevant geophysical and biological variables. For the atmosphere, variables include clouds, aerosols, water vapor, temperature, precipitation, and CO_2 concentration, among other quantities. Ocean variables include sea surface height, sea ice, sea surface temperature, and ocean color (an indicator, for example, of phytoplankton). On land, satellites provide data on glaciers, snow cover, elevation, surface albedo, surface temperature, near-surface soil moisture, total water storage, and vegetation. Vegetation is characterized by land cover type, productivity, leaf area index, canopy height and vertical structure, the chemical composition of foliage, the water content of forest canopies, and occurrence of wildfires. Some products are further processed using models and other meteorological data streams. Carbon uptake during photosynthesis, for example, relates to the absorbed photosynthetically active radiation at the surface. Evapotranspiration relates to land surface temperature, which is used in conjunction with meteorological information and models to provide an estimate of evapotranspiration. Many satellites provide global coverage of the land surface at a very high spatial resolution. One kilometer is a common scale, but some data products have a spatial resolution of 250 meters and others have a spatial resolution of tens of meters.

Global Climate Models

Scientific understanding of how forests affect climate relies on numerical models of Earth's weather and climate and the cycling of energy, water, and chemicals between the atmosphere and biosphere (Figure 10.8).[12] The models use mathematical equations to simulate the physical, chemical, and biological processes that drive Earth's interacting atmosphere, hydrosphere, biosphere, and geosphere system. The first implementations of the models in the 1960s focused on atmospheric physics and dynamics and are

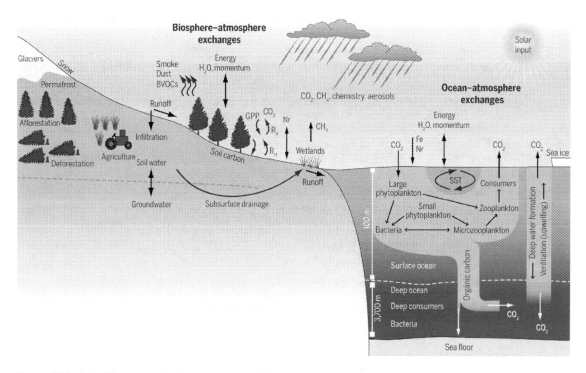

Figure 10.8 Scientific scope of an Earth system model.
The physical coupling of land and atmosphere occurs through biogeophysical fluxes of energy, water, and momentum and also the hydrologic cycle. Terrestrial and marine biogeochemical cycles are new processes in Earth system models. The terrestrial carbon cycle includes carbon uptake through gross primary production (GPP) and carbon loss from autotrophic respiration (R_A), heterotrophic respiration (R_H), and wildfire. Many models also include the reactive nitrogen cycle (Nr). Anthropogenic land use (deforestation, afforestation, agriculture) are additional processes. The fluxes of CO_2, methane (CH_4), Nr, aerosols, biogenic volatile organic compounds (BVOCs), wildfire chemical emissions, and mineral dust are passed to the atmosphere to simulate atmospheric chemistry and composition. Reproduced from ref. (12).

known as atmospheric general circulation models. Later versions of the models, beginning in the 1980s and 1990s, recognized the importance of ocean circulation, sea ice, and hydrological coupling with the land and are known as global climate models. The models require a mathematical formulation of the exchanges of energy, water, and momentum between land and atmosphere to solve the equations of atmospheric physics and dynamics. These fluxes are mediated by plants and so the models require depictions of the terrestrial biosphere, such as the type of vegetation, canopy height, the amount of leaves, the stomata on leaves, and rooting depth. The current generation of models, known as Earth system models, are the most complex yet, and additionally include atmospheric chemistry, terrestrial and marine ecology, and biogeochemistry. In addition to surface energy and water fluxes, the terrestrial biosphere component of Earth system models simulates the carbon cycle, its control by nitrogen and phosphorus, and other chemical exchanges with the atmosphere. Plants grow and die based on the prevailing climate, human uses of land (e.g., deforestation), and wildfires. In doing so, plant growth and community composition respond to and influence climate, and biome boundaries shift as climate changes. The biosphere is a dynamic component of the model. By altering fluxes of energy, water, and chemicals with the atmosphere, changes in the biosphere affect the magnitude and trajectory of climate change.

Paired model simulations, similar to laboratory or field experiments, are performed to examine the impact of vegetation on climate (Figure 10.9).[13] For example, to study the effects on climate of conversion of tropical rain forests to pastureland, a simulation is performed with intact tropical rain forests and a second simulation is performed in which the forests are replaced with grasses. The difference between

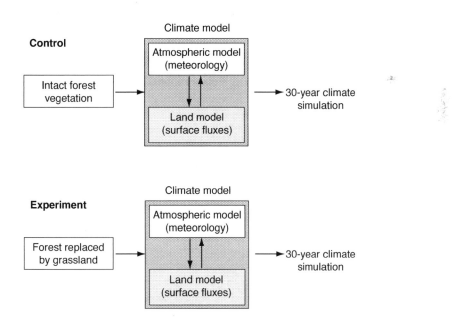

Figure 10.9 Paired climate simulations show the effect of deforestation on climate.
The atmosphere component model provides temperature, wind speed, humidity, precipitation, solar radiation, and longwave radiation to the land. The land component model returns to the atmosphere the surface fluxes of latent heat, sensible heat, momentum, reflected solar radiation, and emitted longwave radiation. The coupled land–atmosphere model is integrated for many model years. The climate effect of deforestation is the difference between the experiment climate and the control climate. Adapted from ref. (13).

the two simulated climates is the effect of tropical deforestation on climate. More complex approaches simulate transient climate change over many decades (e.g., 1850–2100) with historical land cover and projections of future land cover so as to attribute past and future climate changes to deforestation. The global climate models used today typically have a spatial resolution of about 1° in latitude and longitude (approximately 100 kilometers in midlatitudes). At that scale, they simulate large-scale deforestation rather than fine-scale heterogeneity in forest cover. Local variation in climate produced by small-scale mosaics of forests, grasslands, and croplands is not resolved amid the large-scale atmospheric processes. Regional models have a finer spatial resolution (e.g., 10 kilometers or so), but they are limited to small spatial domains and do not have feedback with large-scale atmospheric circulations.

■

Today's scientists have many advanced tools to observe, monitor, and model forests and their influences on climate and to predict the outcome of a changing environment. These include leaf and whole-tree chamber measurements, eddy covariance flux towers, inventory measurements of individual trees, ecosystem experiments, watershed studies, satellite remote sensing, and climate models. Together, these are used in scientific studies of environmental monitoring, experimental manipulation, and numerical modeling that unequivocally demonstrate the influences of forests on climate from the microscale to the macroscale.

11   Forest Microclimates

The climate influence of trees can be seen at spatial scales ranging from a solitary tree in a city park, to the contrast between closed forest canopies and open clearings, and the continental scales of tropical rain forests in Amazonia or the vast coniferous woods of the boreal forest. Some of the climate influences of forests are evident and can be experienced in our daily life. The shade of a tree offers a cool refuge from the heat of a hot, sunny day. The leafy canopy of a tree intercepts rainfall and provides shelter during a downpour. The fluttering of leaves and the swaying of branches and treetops during a strong breeze mark the absorption of momentum by tall trees protruding into the sky. Other processes affect climate less directly. The scents and odors of pines and eucalyptus are some of the countless chemical species emitted by trees, some of which alter atmospheric chemistry and air quality. Over the course of many years, the thick trunk and large crown of a mature tree are an embodiment of nature's own CO_2 removal technology; the carbon stored in biomass is removed from the atmosphere during photosynthesis.

This chapter examines the microclimates of forests, meaning the climate within a forest in contrast to a nearby open area created by a clearing or pasture (Figure 11.1a). These are the climates observed by the meteorological stations of Antoine-César Becquerel, Ernst Ebermayer, Auguste Mathieu, and others in the nineteenth century and in today's eddy covariance flux towers, and they are the focus of the science formalized by Rudolf Geiger in his landmark book *The Climate near the Ground*.[1] Differences in the microclimate between forests and clearings are greatly influenced by the size of the clearing. A small clearing on the order of a few tens of meters is more similar to under-canopy measurements than is a large clearing extending over hundreds of meters or more. Likewise, the distance from the forest edge affects measurements, both in the forest and in the clearing. As is common in scientific study of the natural world, no two forests are exactly the same, but some generalities can be found. A robust observation is that less sunlight reaches the ground in a forest than in open areas, but this depends on the size of the clearing, the height of trees surrounding the clearing, and the location in the clearing (relative to the edge) where measurements are obtained. Daytime temperature is generally lower in forest than in open areas, with greatest differences during the growing season, but this, too, is subject to edge effects and whether the canopy is sparse or dense. Wind near the ground is less than in exposed areas or above the canopy. Another key point is that there are various microclimates within a forest canopy. The environment of the forest overstory differs from that of the understory (Figure 11.1b), and this has important consequences for species conservation and protection from climate change.

Foliage

Forests alter meteorological conditions within the canopy in relation to their leaves. The morphology of leaves – whether broadleaf or needleleaf, small or large, smooth or deeply lobed – affects the exchanges of

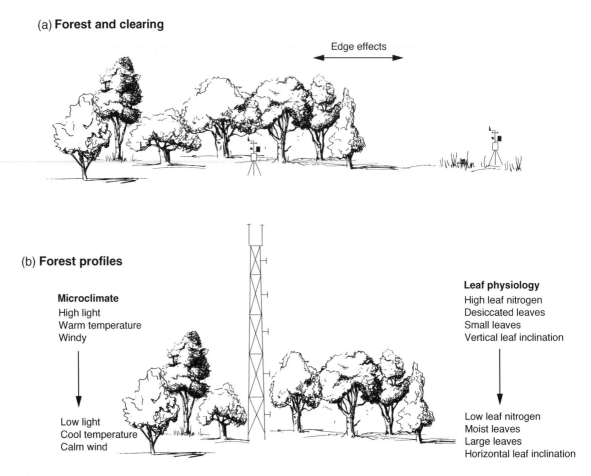

Figure 11.1 Generalized depiction of forest microclimates.
Shown are (a) a forest adjacent to open land in which differences in microclimate are measured with meteorological stations located below the canopy and in the open and (b) changes in microclimate and leaf physiology between the overstory and understory measured on a meteorological tower.

energy and chemical gases with the surrounding air (Figure 9.3). Leaves also vary in the angle at which they are oriented. Some leaves are oriented horizontally while others are vertically inclined and still others are oriented randomly so that there is an equal probability of orientation in any direction. Needles tend to have a random leaf distribution; broadleaves tend to be semihorizontal; and grasses have semivertical foliage. The orientation of leaves affects the absorption of sunlight.

Another important distinction is between deciduous and evergreen foliage. Deciduous leaves tend to be thin and have a large surface area per unit mass. They are commonly broadleaved. Evergreen foliage tends to be thick and have a smaller surface area per unit mass. Most, but not all, conifers have evergreen needles. Some larches, for example, have deciduous needles. Many tropical trees are broadleaf evergreens, with large, thick leaves. These leaf characteristics (broadleaf or needleleaf; deciduous or evergreen) are important determinants of leaf physiology and gas exchange.[2] Broadleaves tend to have a higher rate of photosynthesis than do needles. Short-lifespan deciduous leaves, which are retained for less than one year, are more photosynthetically vigorous than are needles, which can have a longevity of several years. Pronounced vertical variation in leaf morphology and physiology can be found within a canopy, too. Leaves in the overstory tend to have high amounts of nitrogen (per unit leaf area) and therefore have high photosynthetic capacity, are small, may be more vertically inclined, and are more desiccated compared with leaves in the understory.[3]

Chapter 9 introduced leaf area index as a measure of the amount of foliage in the canopy. Leaf area index describes the total area of leaves, but the amount of leaves varies with height in a forest. Cumulative leaf area index increases progressively with greater depth from the top of the canopy. Figure 11.2 shows vertical profiles of leaves in two contrasting forests. Both forests have comparable total leaf area index (5–6 $m^2\ m^{-2}$), but they differ in canopy height and the vertical distribution of leaves. Leaves in the oak–hickory forest are predominantly in the overstory 15–23 meters above ground; 75 percent of total leaf area is in this canopy

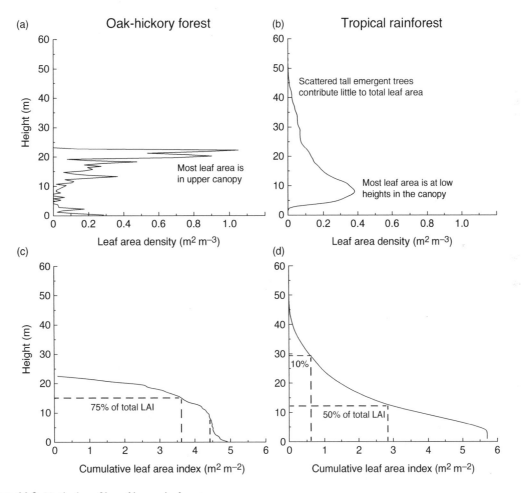

Figure 11.2 Vertical profiles of leaves in forests.
Shown are leaf area density (top panels) and cumulative leaf area index (bottom panels) in relation to height in an oak–hickory forest and tropical rain forest. Leaf area density is the area of leaves (m^2) per unit volume of canopy space (m^3) at a given height. Cumulative leaf area index is the area of leaves (m^2) per unit ground area (m^2) above a given height. (a, c) Oak–hickory forest in Tennessee. Redrawn from ref. (4). (b, d) Amazonian tropical moist forest during the middle of the wet season. Data from ref. (5).

layer.[4] The tropical forest is considerably taller, and the upper canopy has scattered emergent trees that extend to a height of 50 meters or more, but most of the leaf area is in the lower canopy; only 10 percent of the total leaf area is situated above 30 meters.[5]

Solar Radiation

Leaves, branches, and other woody material absorb much of the incoming solar radiation in forests. In a dense forest with large leaf area index, very little light, often less than 5 percent of that above the canopy, reaches the ground. The amount of light at the ground depends on many factors, including the angle of the Sun above the horizon, whether the canopy is sparse with few leaves or closed with many leaves, the angle at which the leaves are inclined, and whether the leaves are clumped together or more uniformly distributed along a branch. Direct beam radiation, which originates from the Sun's position in the sky, penetrates a canopy differently than does diffuse radiation, which emanates from the entire sky hemisphere. In general, light absorption is concentrated in the overstory in a closed canopy where leaves are predominantly in the upper canopy, but light penetrates deeper into the canopy when the leaves are sparse and the canopy is more open. Diffuse radiation penetrates deeper into forest canopies than does direct beam radiation. Consequently, photosynthesis can increase as a greater proportion of solar radiation is diffuse rather than direct beam because more leaves in the canopy are illuminated.[6]

Figure 11.3 illustrates the profile of light in a mixed species deciduous forest.[7] The total leaf area index is 6.1 m^2 m^{-2}, which is representative of a closed canopy. There is a strong gradient in light through the canopy from full sun at the canopy top to less the 2 percent of incoming radiation at the forest floor. Twenty-five

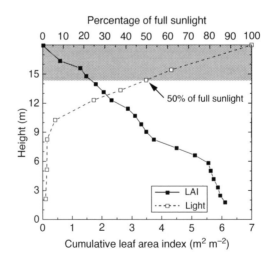

Figure 11.3 Leaf area index and light profiles in a forest.
Measurements were made in an 18-meter-tall deciduous forest in Wisconsin characteristic of North American northern hardwood forests. The bottom axis is the cumulative leaf area index (LAI) from the top of the canopy (shown as a solid line). The top axis is the percentage of solar radiation relative to above the canopy (shown as a dashed line), ranging from 100 at the top of the canopy to near zero at the ground. The shaded area is the portion of the canopy in which light levels are 50 percent or more of full sunlight. Redrawn from ref. (7).

percent of total leaf area is situated above 14 meters height, and the amount of sunlight has been reduced to one-half that above the canopy. Overall, the amount of light decreases exponentially with greater cumulative leaf area index, and this relationship provides a simple mathematical model of radiative transfer in forests.

Wind

Wind speed profiles show the drag exerted by trees in which leaves and woody material attenuate wind through the canopy. Measurements made in a 10.5-meter-tall pine forest and a nearby field exemplify the attenuation of wind by forests (Figure 11.4).[8] Wind speed above the forest at a height of 21 meters is similar to that at the same height above the field and both follow the expected logarithmic relationship with height. In the canopy, however, wind speed decreases compared with the same height in the field so that the forest understory has still air while the field experiences stronger winds. Similar reduction in wind speed is commonly found in forests (see, for example, Figure 11.6b).[9] Reduction of wind speed by a forest is the basis for planting trees as a windbreak, or shelterbelt, to reduce winds blowing across open farmlands and reduce soil erosion.

Air Temperature

Forests create a within-canopy microclimate that differs from the prevailing macroclimate. Because the understory is shaded, one would expect cooler daytime air temperature than in open locations exposed to full sunlight. Likewise, the forest canopy reduces the loss of longwave radiation at night from the soil, and nighttime temperatures in the understory can be warmer than in open areas. As a result, the diurnal temperature range, defined as the difference between daily maximum and minimum temperatures, decreases in a forest. These effects are most prominent in summer and can be found in numerous forests, though the exact outcome varies with forest type, climate regime, and factors such as soil moisture and canopy cover.

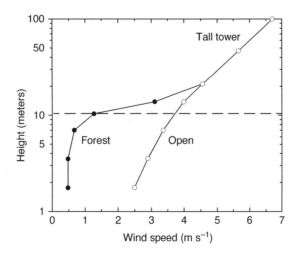

Figure 11.4 Wind speed above and within a forest.
Shown are average wind profiles measured at the Brookhaven National Laboratory, Long Island, New York, in the open, on a tall tower, and 60 meters into a pine forest when winds are directed into the forest edge. The dashed line shows the canopy height. Height is given in a logarithmic scale. Redrawn from ref. (8).

One way to assess the different microclimates of forests in contrast with open land is to compare measurements of air temperature obtained in various locations. Geiger, in *The Climate near the Ground*, published just such measurements in the forests around Leipzig, Germany, over a summer day (Figure 11.5).[10] The transect along which the observations were obtained was sufficiently short, extending over several kilometers, and elevation was relatively uniform so that topography or other factors were unlikely to have produced the observed temperature differences. The data show that in evening (1600–2000) and early morning (0400–0800), temperature was relatively uniform along the transect; the

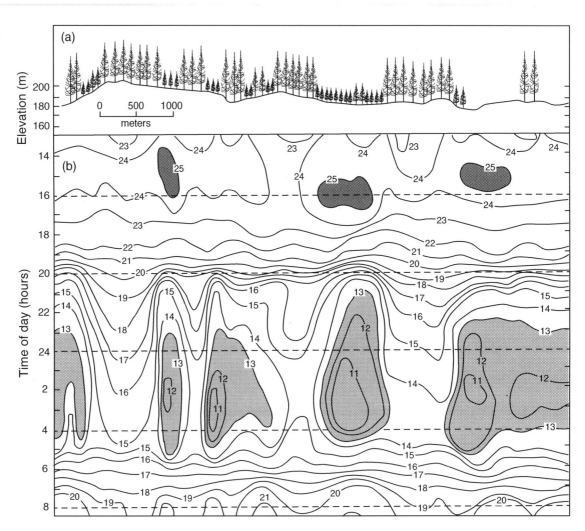

Figure 11.5 Microclimates in a landscape of forests and clearings.

Air temperature was measured along a transect in a forested region near Leipzig, Germany, on July 8–9, 1933. (a) Shown is a cross-section of elevation and forest cover along the transect spanning several kilometers. (b) Shown are temperatures (°C) at a given time of day along the transect. The horizontal axis is location along the transect. The vertical axis is time of day from afternoon (1400) to the following morning (0800). The solid lines are contours of equal temperature. The dark shading shows temperatures greater than 25°C, and the light shading shows temperatures less than 13°C. Redrawn from ref. (10).

temperature isopleths are mostly horizontal, meaning all locations had similar temperature. During the afternoon and at night, however, warm and cold pockets developed in open areas. The warmest temperatures (in excess of 25°C) occurred during the afternoon (1400–1600) in locations that were open or that had young trees. At night (2400–0400), the same locations were several degrees colder than dense forest. Figure 11.5 illustrates the general differences between forests and clearings. Many factors such as the orientation of the forest edge, the height of the trees, the amount of leaf area, and the direction of the prevailing winds influence the microclimates of forests. The size of a clearing relative to the height of surrounding trees is also critical, especially at night. The colder nighttime temperature of clearings compared with forests arises because of exposure to the cold sky in large clearings. In smaller clearings surrounded by trees, the temperature difference is less pronounced.

Other studies show similar differences between forests and nearby open locations. The assumption is that the paired sites share the same background climate so that temperature differences can be attributed to forest cover. For example, a study of 14 Swiss broadleaf deciduous (oak, beech) and needleleaf evergreen (spruce, fir, pine) forests paired with nearby open locations found that the forests generally reduce daytime warming under the canopy by several degrees, particularly during summer, compared with the open sites.[11] During the measurement period, daily maximum air temperature in forests was 2.3°C lower, on average, during summer months compared to open areas. At night, the forests can be warmer than the open sites. The cooler daytime maximum temperature, relative to open areas, correlates with leaf area index so that dense canopies with high leaf area index have more below-canopy cooling than do sparse forests.[12] In these Swiss forests, the below-canopy cooling is also most pronounced with wet soils and decreases with drier soils. In general, as reported in many studies, the moderating effects of forests on below-canopy air temperature are most evident on warm, sunny days, in dense forests, and vary depending on specific site conditions.[13] A meta-analysis of 74 studies at 98 sites worldwide from tropical to boreal and in deciduous, evergreen, and mixed forests shows that the daytime cooling and nighttime warming is robustly found at the majority of sites.[14] Forest understories across Europe are, on average, about 2°C cooler than the prevailing macroclimate during summer and warmer by a similar amount during winter.[15] The forest microclimate has important implications for habitat and biodiversity conservation and may act to buffer forest understory organisms from the severity of climate change.[16]

Forest microclimates differ not only in contrast with open land, but also vertically with position in the canopy. The vertical profile of leaf area and the associated profile of radiative absorption drive gradients of air temperature. During the day, the upper canopy in dense forests absorbs much of the incoming solar radiation, and the air in the upper canopy warms. The forest understory experiences less solar heating and is typically a few degrees cooler than the overstory. At night, the upper canopy cools from exposure to the overlying cold air and loss of longwave radiation to the sky. At the same time, the leaves absorb and reemit longwave radiation onto the soil surface, keeping the lower canopy air warm. Ebermayer and others identified these temperature gradients during the forest–climate debate. Geiger illustrated daytime profiles in the 1927 edition of his book, and many other studies have since found similar profiles.[17] Figure 11.6a, for example, shows temperature measurements in a 25-meter-tall Norway spruce forest located in Germany.[18] Most of the foliage is situated in the mid to upper canopy. Daytime air temperature increases as the canopy is approached from above, as expected from boundary layer theory. Within the canopy, temperature has a local maximum near the top of the canopy. Below this height, temperature decreases with greater depth into the canopy so that temperature near the ground is about 1°C cooler than the upper canopy.

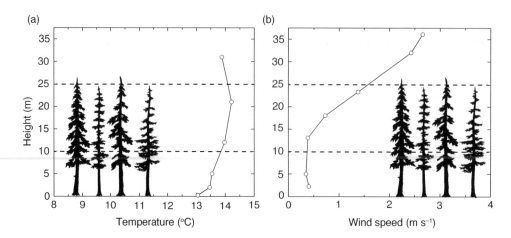

Figure 11.6 Vertical profiles in a Norway spruce forest.
Shown are daytime (a) air temperature and (b) wind speed measured in and above a 25-meter-tall Norway spruce forest in northeastern Bavaria, Germany (September 20–24, 2007). The dashed lines show the height of the main foliage layer. Redrawn from ref. (18).

Figure 11.7 shows similar measurements in an Amazonian rain forest.[19] The main canopy is 30 meters tall, but the tallest emergent trees rise to 40 meters. The upper canopy has the warmest air temperature during the day (Figure 11.7a). Temperature near the ground is nearly 3°C cooler than the upper canopy. At night, temperature decreases with greater depth into the canopy, with a local minimum at about 15 meters height. Temperature in early morning (0630) is fairly invariant with height, but the air in the understory near the ground warms less during the day than does the upper canopy (Figure 11.7b). Air temperature increases to 29°C at a height of 23 meters, but to only 27°C at 1.5 meters. Similar results are seen in other tropical forests. Figure 11.8 compares meteorological conditions in a tropical dry evergreen forest located in Thailand and a nearby clearing.[20] Daytime air temperature is cooler under the forest canopy (by up to 2°C near the ground) compared with the clearing. Nighttime air temperature is similar at both locations because the moist, humid air suppresses radiative cooling in the clearing at night. The air in the clearing warms more during the day than does the forest. The forest soil is considerably cooler than the clearing. The upper forest soil warms by less than 1°C during the day compared with more than 3°C in the clearing.

The Air above Forests

A robust signature of dense forests is the cool daytime temperature under the canopy compared with open locations, especially during the warm season. However, the microclimates of forests within the canopy can differ from the climate influences of forests above the canopy in the atmospheric boundary layer. It is in this layer of the atmosphere extending several hundred meters or more above the ground that forests affect temperature, humidity, and wind through fluxes of energy, moisture, and momentum (Figure 9.1). Measurements of air temperature above forests provide evidence of forest influences on temperature, but also point to the need to differentiate within-canopy and above-canopy meteorology.

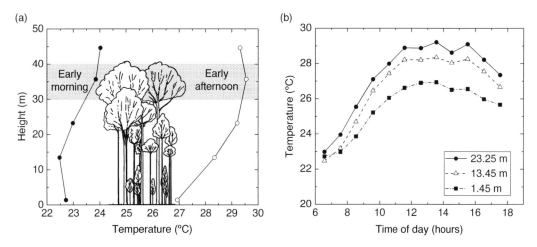

Figure 11.7 Air temperature in a tropical rain forest.

Measurements were made near Manaus, Brazil, during July–August 1984. (a) Vertical profile of average air temperature in early morning (closed circles) and early afternoon (open circles) within and above the canopy. Shading denotes the canopy height. (b) Hourly averages during the day at heights of 1.45, 13.45, and 23.25 meters. Redrawn from ref. (19).

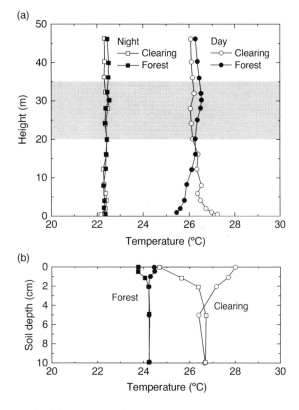

Figure 11.8 Temperature in a tropical dry evergreen forest.

Measurements were made in Thailand (June 20–30, 1970). Shown are profiles of (a) air temperature and (b) soil temperature in the forest (closed symbols) and a nearby clearing (open symbols). Shading denotes the height of the overstory. Temperatures are shown at night (squares) and during the day (circles). Redrawn from ref. (20).

Temperature measurements at paired forest and open sites in North America, the tropical Americas, and eastern Asia reveal the signature of forests on air temperature above the canopy.[21] The forest measurements were obtained using eddy covariance towers extending above the canopy. Nearby surface weather stations located in grassy fields provided the contrasting open land. In the tropics between latitudes 15° S and 20° N, the annual mean air temperature above forests is almost 0.7°C cooler than at open locations. The cooling at the forest sites compared with the open sites occurs from a decrease in daytime air temperature in all months of the year; nighttime temperature is similar at both locations. However, in boreal regions at latitudes greater than 45° N, the air above the forests is nearly 1°C warmer in the annual mean compared with the open sites. The forest warming in northern regions occurs throughout the year due to warmer nighttime temperature relative to open sites; daytime temperature is essentially unchanged. In both tropical and boreal regions, diurnal temperature range decreases over forests, but for different reasons. Forests decrease daily maximum temperature in tropical zones but increase daily minimum temperature in boreal zones. Weaker temperature differences occur between these regions, with latitude 35° N marking the approximate transition in which forests create a cooler local climate in southern latitudes and a warmer climate in northern latitudes.

These results can be understood in terms of the effects of surface albedo, surface roughness, and evapotranspiration on temperature.[21] The low albedo of forests contributes to daytime surface warming. From this alone, forests should be warmer than open sites during the day. In boreal regions, warming from the low forest albedo is offset by the cooling effects of surface roughness and evapotranspiration. At night, when temperature inversions are common, the enhanced mixing associated with tall, rough forests brings heat from aloft down to the forest canopy. In the tropics, high rates of evapotranspiration from forests contribute to their lower daytime temperature, offsetting the warming influence of surface albedo. The difference between the effect of forests on the microclimate under the canopy and their influence on the atmospheric boundary layer above the canopy becomes important when considering macroclimates (Chapter 14).

■

Trees shade the ground, block winds, and create a cooler microclimate under the canopy compared with open areas. The forest canopy moderates under-canopy air temperature and buffers the understory from extreme temperatures, possibly lessening the ecological impacts of higher temperatures in a warmer world. Forest microclimates are now recognized as an indispensable but neglected ecological service of forests, essential for species conservation in light of planetary warming. Further practical implications of forest microclimates are seen in residential landscapes and urban parks, where trees provide shade to cool residences, act as windbreaks to block cold winter winds, and create a cooler urban climate. The unique microclimates of forests are well documented, spanning more than two hundred years of study. The macroclimate influences of forests are only now being identified over the last forty years of research. This research shows that the within-canopy effects of forests on temperature can differ from their above-canopy effects. The influences of forests on macroclimate may not align with their microclimate benefits. These influences arise from the storage of carbon in forest ecosystems, which removes CO_2 from the atmosphere; the hydrology of forests and the water returned to the atmosphere in evapotranspiration; and exchanges of energy and chemicals between forests and the atmosphere. The following chapters consider these other climate influences of forests.

12   Water Yield

The amount of water that runs off the land into a stream is known as water yield. It is the water available for human uses. Evapotranspiration is a loss of water that reduces streamflow. Forests increase annual evapotranspiration and reduce annual streamflow compared with grasslands and other types of vegetation. The science, however, is not precise and our understanding is more qualitative than quantitative. Water yield and the climate services of forests represent conflicting demands for water. Forests cool the surface climate through evapotranspiration and remove CO_2 from the atmosphere during growth, but in doing so they consume water and reduce water yield for human usage. This chapter delves into the science of forests and streamflow. There is a desire for an exactness in science and certainty in predictions, but, as with so much else in the story of forests and climate, we are constrained by inexactness, ambiguity, and uncertainty. The study of the natural world humbles our expectations.

The Economy of Forest Water

The question of whether forests produce rain or, by evaporating considerable quantities of water, consume rainwater was part of the forest–rainfall controversy of the nineteenth century. Supporters of the latter view framed the argument in terms of the public good, that forest evapotranspiration reduces the supply of water to cities and farms. The assertion put forth by the US Geological Survey chief geographer Henry Gannett, given during his rebuttal to Bernhard Fernow's lecture on forest–rainfall influences in 1888, that in arid climates "it is advisable to cut away as rapidly as possible all the forests" so as to increase streamflow exemplifies the latter notion.[1] So, too, did Herbert Wilson, also of the Geological Survey, express a similar sentiment in an 1898 article on forests and water supply.[2] He likened evapotranspiration to a loss of water and asked "of what value is it to enhance the rainfall, if at the expense of water taken from the soil and transpired through the leaves?" Interception of rainfall by leafy forest canopies is a similar loss of precipitation that does not reach the ground, "thus neutralizing their supposed beneficial influences." More than one hundred years later, the debate continues.

The modern debate borrows the language of economics to tell the hydrologic influences of forests in the context of supply-side (forests produce rainfall and increase the overall water supply) and demand-side (forests consume water and reduce the water supplied by streams). Some academics are exuberant in the capability of forests to supply water by promoting rainfall.[3] Others assert that the science supports the demand-side and highlight the gap between forest benefits perceived by the public and the actual science.[4] Still others are reticent and urge caution; they point to the long history of the forest–rainfall controversy.[5] Modern-day climate scientists, too, have shaped the debate. The high evapotranspiration and carbon storage of forests is widely accepted as a means by which forests cool the climate. The notion of forests as global public goods arches across the science of forest–climate influences, but what goods and services should we manage?

The Hydrologic Cycle on Land

Water flows among the oceans, land, and atmosphere, but the balance of water differs between oceans and land (Figure 12.1).[6] More water evaporates annually from the oceans than is replenished by precipitation; there is an annual deficit of water. Over land, however, annual precipitation exceeds evaporation. The surplus of water runs off into streams and rivers where it flows to the oceans to replace the net loss of water. This is the general hydrologic cycle described by Edmond Halley in 1691.[7]

The full hydrologic cycle shown in Figure 12.2 is more complex than the simplified depiction in Figure 12.1.[8] Precipitation in the form of rain or snow provides much of the input of water. However, not all the precipitation reaches the ground. Leaves and branches intercept rain and snow and temporarily store the water. Intercepted water readily evaporates and so never replenishes the soil. The water that is not intercepted falls to the ground through openings in the plant canopy and by dripping down from leaves and

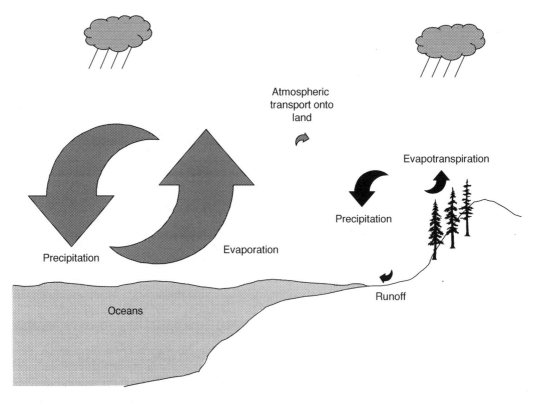

Figure 12.1 The annual global hydrologic cycle.
Arrows are proportional to the size of the fluxes. The deficit of water in oceans, where annual evaporation exceeds precipitation, is replenished by runoff of freshwater from land, where annual precipitation exceeds evapotranspiration. Adapted from ref. (6).

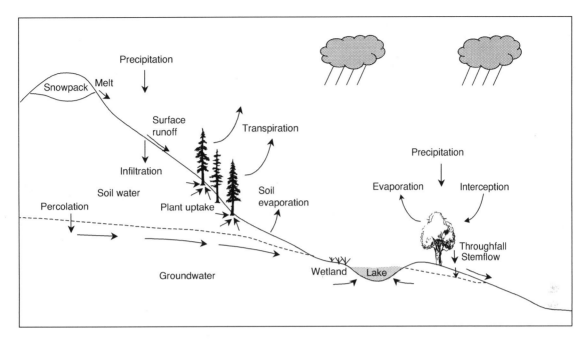

Figure 12.2 Main features of the hydrologic cycle on land.
Shown are the component water fluxes between the land and atmosphere, over the land, and in the soil. Adapted from ref. (8).

branches, collectively known as throughfall. Water also flows down plant stems and tree trunks, reaching the ground as stemflow. Winter storage of precipitation in a snowpack followed by spring snowmelt is an important source of water in seasonally cold climates. Some forests obtain water droplets from fog. These droplets collect on leaves and coalesce into larger drops that fall to the ground. In the coastal redwood forests of northern California, for example, 34 percent of annual water input can originate from fog drip.[9] Similar input of water from fog drip occurs in tropical and temperate montane cloud forests, which are frequently enshrouded in low clouds.[10] This is the basis for the rain-making trees so commonly mentioned during the forest–rainfall controversy.

Liquid water reaches the ground as rainfall where vegetation is absent, as throughfall and stemflow under vegetation, or from snowmelt. Some of this water infiltrates into the soil. When the infiltration capacity of the soil is exceeded, the water runs off over the surface, first in small rills and gullies and then into creeks and streams that feed large rivers. The water that infiltrates into the soil wets the soil and is stored as soil water in an upper layer of soil known as the unsaturated, or vadose, zone. Soil water returns to the atmosphere through evaporation from bare ground and transpiration from plants, collectively called evapotranspiration. Soil water also moves vertically and horizontally due to internal forces in the soil. In most cases, gravity is the greatest force, causing water to flow downwards. Groundwater is the subsurface region below the vadose zone that is saturated with water. The top is defined by the water table, which separates the saturated and unsaturated zones, and the bottom is defined by bedrock. Lateral flow of groundwater recharges lakes and wetlands and provides the base flow to maintain rivers in the absence of rainfall.

The complexity of the hydrologic cycle shown in Figure 12.2 can be reduced to a simplified expression in which the change in water storage is equal to precipitation input minus losses from evapotranspiration and runoff. Over long time periods such as a year, change in storage is negligible so that annual precipitation input is balanced by evapotranspiration and runoff losses. Annual precipitation is either returned to the atmosphere as evapotranspiration or runs off into streams and rivers. The difference between precipitation and evapotranspiration is equal to runoff and is the amount of water flowing in streams. Water yield is defined as the annual streamflow at the outlet to the watershed and is an indicator of surface water supply. An increase in evapotranspiration necessitates a decrease in runoff and vice versa. This simplified water balance equation provides a means to estimate evapotranspiration. If precipitation and streamflow are measured, evapotranspiration can be estimated as the difference. By eliminating transpiration and canopy interception, deforestation increases runoff to streams.

The amount of precipitation that is intercepted depends on rainfall intensity, frequency, and duration. During brief, moderate storms, the canopy storage capacity may not be exceeded, and most of the rainfall is intercepted. In long, intense storms, the storage capacity is quickly exceeded and water drips to the ground. Plant canopies generally intercept 10–20 percent of annual precipitation, but the amount can be larger.[11] The completeness of the canopy cover determines how much water can be intercepted; the denser the foliage, the greater the amount of water stored. The seasonal emergence and senescence of leaves also matters. Deciduous forests intercept less rainfall when leafless than during the growing season when leaves are present. Evergreen forests, on the other hand, have foliage year-round and can intercept a larger portion of annual rainfall.

Evapotranspiration is a complex process that depends on micrometeorological, plant physiological, and soil physical processes. The net radiation available to evaporate water is a chief determinant when soils are moist, as are the dryness of the air and turbulent mixing that carries water vapor away from the evaporating surface. Leaf area index, stomatal physiology, and rooting depth are also important. When plants cover a small portion of the surface, evaporation is the dominant flux. Transpiration becomes more important as plant cover increases. Plants cannot grow if stomata are not open, but when open, water diffuses out of the leaf to the surroundings. Too much transpiration can result in desiccation. Plants differ in how they open stomata to obtain CO_2 while also preventing water loss. Trees have deeper roots than do herbaceous plants such as grasses and crops and so have access to more soil water that can sustain transpiration during the growing season. Soil texture and water content also regulate evapotranspiration. In dry conditions, the amount of water in the soil that can be accessed by plants is a key determinant of evapotranspiration. Sandy soils have large pores and drain quickly; little water is stored to sustain evapotranspiration. Clays have small pores and can store much water, but the water is tightly bound to soil particles and is unavailable for plant usage. Loams are intermediate, draining slower than sands and retaining more water.

The rate of evapotranspiration varies depending on climate, forest type, and other environmental factors. In Figure 12.3, for example, monthly evapotranspiration in the tropical rain forest averages 2.5–3.5 millimeters (mm) per day and has little seasonal variation.[12] Annual water loss is 1,100 mm. The temperate deciduous and ponderosa pine forests have peak monthly rates comparable to the tropical rain forest during the active growing season, but strong seasonal variation yields much less annual water loss (400–500 mm). The seasonally cold temperate deciduous forest has a distinct annual cycle with high evapotranspiration during the warm summer months and low rates in the cold winter season. The annual cycle at the ponderosa pine forest relates to precipitation with markedly less evaporation during the dry season. The boreal forest has a lower maximum rate (1.5–2 mm per day), a shorter growing season, and the least annual water loss (230 mm).

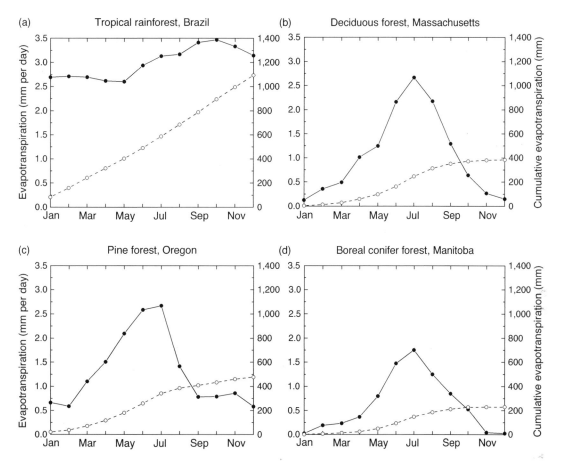

Figure 12.3 Monthly evapotranspiration measured at four forest sites.
Shown are monthly evapotranspiration (solid line, left axis, millimeters per day) and cumulative evapotranspiration (dashed line, right axis, millimeter) for (a) tropical rain forest in Brazil, (b) temperate deciduous forest in Massachusetts, (c) pine forest in Oregon, and (d) boreal conifer forest in Manitoba. Adapted from ref. (12). See Figure 9.5 for the surface energy fluxes.

Paired Watershed Experiments

Paired watershed experiments that compare streamflow in a deforested watershed with that of a nearby control watershed with intact forest cover routinely demonstrate enhanced streamflow with removal of trees. Although comparison of streamflow in watersheds with different amounts of forest cover had occurred previously in France and Switzerland, the first paired watershed experiment was conducted at Wagon Wheel Gap, Colorado, during 1910–26 by scientists at the US Forest Service and the Weather Bureau as an outgrowth of the forest–rainfall controversy.[13] After obtaining baseline measurements over an eight-year reference period, trees were cleared from one watershed and both watersheds were monitored for the next seven years. Numerous similar experiments have since been conducted at locations throughout the world.

Figure 12.4 Effect of deforestation on the water balance.
Shown are summer precipitation, streamflow, and evapotranspiration for two watersheds at the Hubbard Brook Experimental Forest. One watershed (open symbols) was deforested in 1965–66 and vegetation regrowth was suppressed for three years. The other watershed (closed symbols) remained forested. Adapted from ref. (14).

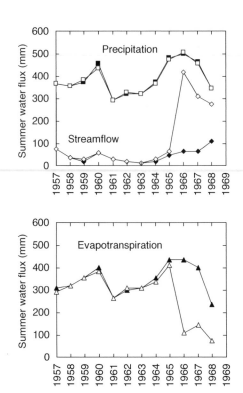

Comparison of two small forested and deforested watersheds at the Hubbard Brook Experimental Forest in New Hampshire demonstrates the increase in streamflow following deforestation (Figure 12.4).[14] Prior to deforestation, both watersheds had similar streamflow and evapotranspiration. One watershed was deforested in late autumn 1965, and regrowth was suppressed during the subsequent three growing seasons. After the loss of trees, streamflow increased and evapotranspiration decreased in the deforested watershed compared with the forested watershed. Subsequently, when trees were allowed to regrow, streamflow decreased and evapotranspiration increased. Likewise, the deforestation resulted in a rapid increase in the concentration of dissolved nutrients and sediments in the stream water. Indeed, water purification is a key hydrologic service provided by forests.[15]

Paired watershed experiments conducted in many different types of forests and in climate regions throughout the world generally show that annual streamflow is less in intact forests, increases following loss of tree cover, and decreases again if the forest regrows.[16] As with much in the study of forests, however, the precise magnitude of the enhancement is highly variable, and sometimes inconsistent, and the effects of forests on seasonal flows and peak flows are less clear. Climate, soils, topography (slope, aspect), forest type, and the amount of forest cover removed affect the magnitude of the streamflow response. Tree species differ in their utilization of water: some have a conservative water-use strategy; others grow fast and transpire large amounts of water. Deciduous and evergreen species differ in their effects on streamflow. Annual interception and evapotranspiration can be higher for evergreen conifer forests than for deciduous forests. This is evident in an experiment at the Coweeta Hydrologic Laboratory in North Carolina. Conversion of mature deciduous forest to pine reduced annual streamflow compared with the control watershed.[17] Forest water use also varies with stand age (e.g., young trees or old-growth forest).

Responses can also vary depending on specific treatments such as deforestation, forest regrowth, conversion of non-forested land to forest, change from one type of forest to another, and various types of tree thinning. An often-neglected aspect of the water yield question is whether forests, where leaf litter forms a forest floor that impedes runoff and with roots and organic matter that make the soil porous, promote infiltration.[18] On the other hand, soil compaction and disturbance during logging can reduce infiltration, leading to more streamflow. In seasonally snow-covered climates, changes in forest cover affect winter snow accumulation and spring snow melt so that changes in streamflow may be related to snow processes rather than to evapotranspiration. Because the enhancement in streamflow declines as forests regrow, the time period at which the streamflow response is measured introduces variability among studies. A further caveat is that the small spatial scale of watershed experiments, often one square kilometer or so, limits the representativeness of the results.

An alternative to paired watershed experiments examines the water balance of catchments that very in climate, vegetation, and soils.[19] These studies do not directly examine the effects of vegetation changes on water yield. Rather, they relate variation in water yield among the catchments to various factors, including vegetation characteristics. In contrast with small watershed experiments, these studies can examine the water balance at large spatial scales of 100–1,000 square kilometers. A common generalization, consistent with less streamflow in forests, is that forests have higher annual rates of evapotranspiration than do grasslands.

Eddy Covariance Studies

Paired watershed studies are taken as an indicator that forests evaporate more water annually than do grasslands or other non-forest vegetation. Measurements of evapotranspiration at eddy covariance sites provide conflicting evidence. Flux tower measurements at pine and hardwood deciduous forests in North Carolina, for example, show that these forests evaporate more water during the growing season and over the full year compared with an adjacent grassland, and this is seen in the cooler surface temperature of the forests.[20] The pine forest has more evapotranspiration than the deciduous forest. Analysis of 28 paired flux tower measurements in North America, Europe, and Australia for forests and nearby cropland, grassland, or shrubland also finds lower summer evapotranspiration at non-forest locations compared with forests, though there is considerable variability among sites.[21] Other analyses of flux tower measurements question the conventional understanding of high rates of evapotranspiration in forests. Measurements at forest, grassland, and cropland sites in Europe show that evapotranspiration is larger at the non-forest sites than at the forests when soils are moist.[22] The forests have a lower surface albedo and absorb more energy than the other sites. However, sensible heat flux, rather than evapotranspiration, is a primary means of surface cooling because the forests have strong stomatal control of transpiration (restricting transpiration water loss) but are well coupled aerodynamically with the atmosphere (enhancing sensible heat loss). With dry soils, such as during a drought or heatwave, evapotranspiration declines sharply in the short-rooted grassland and cropland and sensible heat flux increases, while the deep-rooted trees maintain their evapotranspiration. A global synthesis of flux tower measurements across forest, grassland, cropland, and other vegetation types also shows that forests do not evaporate a larger fraction of annual precipitation compared with grassland or cropland.[23]

The inconsistent findings on forests and evapotranspiration obtained from paired watershed experiments and from eddy covariance flux tower measurements may arise from methodological issues. Watershed experiments do not measure evapotranspiration directly, but instead calculate it as the difference between annual precipitation and runoff. Eddy covariance measurements, too, have uncertainty related to the accuracy of the instruments, the processing of the data, and theoretical assumptions required to infer fluxes from the measurements. Another observational method uses lysimeters to measure evapotranspiration. Lysimeters provide a complete mass balance for a small area of vegetation, from which evapotranspiration can be inferred. One type of instrument, known as a weighing lysimeter, measures the weight of the soil. A large system for a forest might be an area 25 meters by 25 meters with soil 2–3 meters deep in which trees are planted. From additional measurements of precipitation input and drainage loss, evapotranspiration can be calculated as the residual in the water balance equation. Lysimeters provide precise measurements but have small spatial resolution, are expensive to construct and maintain, and therefore are not very common. Analysis of four different lysimeter stations across western Europe finds higher annual evapotranspiration over forest than non-forest vegetation, consistent with paired watershed experiments.[24]

■

Science strives to increase understanding, to provide exactitude; inexactness and ambiguity are seen as failures.[25] The argument about forests and water yield is, on the one hand, a controversy in that the science does not lead to precise predictions. The vagaries of site conditions, forest type, and forest treatments favor broad generalizations with respect to paired watershed experiments rather than quantitative predictions. In general, it is commonly accepted that the presence of trees increases annual evapotranspiration and thereby reduces runoff to streams, compared with deforested watersheds or grasslands. Many studies have demonstrated this using paired watershed experiments, but there are always exceptions to the rule. The exact amount by which loss of trees increases streamflow depends on the vagaries of the forest, the characteristics of the watershed, the precise treatment applied to the experimental watershed, and the length of the study. There is a difference between the generality that forests regulate streamflow and specifics such as the annual amount of water flowing in a river, the extremes of high and low flows, and the frequency of extreme events. Eddy covariance measurements, too, have uncertainty and can give results conflicting with watershed experiments.

Furthermore, the controversy over forests and water yield reflects conflicting views of forests and water. Do forests, with their high rates of evapotranspiration, cool the surface climate and possibly increase rainfall, thereby lessening the deleterious consequences of human-caused climate change? Climate science provides support for this view. Or, as some hydrologists advocate, should forests be managed to increase water yield for human consumption? Consideration of spatial scale further muddies the policy implications because any precipitation benefits of forests occur at large spatial scales covering vast regions of land or entire continents while the water cost of forests is felt at the scale of the watersheds that supply towns and cities. The hydrologic benefits and costs of forests are especially relevant in a warmer world with increased water scarcity. In both stances, we seek solutions to a human-made problem in the forests before us. And as is examined in the next chapter, there is a third way to view forests and water. Forest growth removes considerable amounts of CO_2 from the atmosphere, but forests need water to grow. As with much about forests, there are trade-offs in water for climate protection, carbon storage, and human uses.

13   Carbon Sequestration

The climate benefit of forests that is most frequently recognized is their removal of CO_2 from the atmosphere. Some 80 to 90 percent of the mass of a tree is water. Dry mass is the mass remaining after the material has been dried and the water removed. It is commonly referred to as biomass. About 50 percent of biomass is composed of carbon. Over the course of a tree's lifetime, the accumulated carbon in biomass is carbon removed from the atmosphere. The biomass of a tree increases as the diameter of the tree grows. A single large sugar maple tree with a stem diameter of 50 centimeters has an aboveground biomass in excess of 2,000 kilograms (Figure 10.4). Tree species differ in growth rate, size at maturity, and longevity, but the basic principle of biomass accumulation over the lifetime of a tree is the basis for using forests to reduce the amount of CO_2 in the atmosphere. Decomposition of organic material in the soil emits CO_2 and reduces the net carbon gain by forests. Wildfires, insect outbreaks, logging, and other disturbances also release CO_2 to the atmosphere. The combination of these processes – carbon gain from biomass growth; carbon loss from the soil and from disturbances – makes some forests a sink for atmospheric carbon; the forests have a net gain of carbon annually. Other forests are a source of carbon, in which there is a net loss of carbon to the atmosphere. Forests are, at a global scale, an annual carbon sink, which reduces the accumulation of anthropogenic CO_2 in the atmosphere. Nature-based solutions to mitigate climate change aim to enhance the carbon sink.

Ecosystem Carbon Fluxes

Figure 13.1 depicts the general processes in the cycling of carbon between terrestrial ecosystems and the atmosphere. Plants absorb CO_2 during photosynthesis and use the carbon to grow biomass such as foliage, roots, flowers, seeds, and, for trees, wood. This carbon gain is called gross primary production, and it provides the carbon necessary to grow and maintain living biomass. Not all the carbon gained during photosynthesis remains as biomass. Plants respire CO_2 during the construction of new biomass, to maintain living biomass, and in other physiological functions. This carbon loss is called autotrophic respiration. A general rule is that about 50 percent of the carbon gained in photosynthesis is lost as autotrophic respiration. The net carbon gain, called net primary production, is therefore about one-half of gross primary production.

Some of the carbon that accumulates in biomass returns to the atmosphere when leaves, branches, and wood fall to the ground and decompose. Soil microorganisms consume the debris and, in doing so, release CO_2 to the atmosphere (called heterotrophic respiration). The decay of litter on the ground is one part of soil carbon formation, by which fresh litter decomposes into chemically recalcitrant soil organic matter that

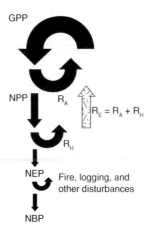

Figure 13.1 Fluxes in the terrestrial carbon cycle.
Carbon is gained during photosynthesis (gross primary production; GPP). Carbon losses include plant respiration (autotrophic respiration; R_A) and decomposition of soil organic matter (heterotrophic respiration; R_H), which together comprise ecosystem respiration (R_E). Net primary production (NPP) is the net carbon uptake by plants (GPP–R_A). Net ecosystem production (NEP) is the net storage of carbon in the absence of fire and other disturbance losses (NEP=NPP–R_H = GPP–R_E). Additional carbon is removed during disturbance so that net biome production (NBP) is the net carbon accumulation.

resists decay. The carbon that accumulates after accounting for heterotrophic respiration is known as net ecosystem production. In forests, long-term carbon storage is materialized in the woody trunks and roots of large trees and in the soil organic matter. Still more carbon can be lost when wildfires burn biomass and directly release carbon to the atmosphere during combustion or leave residual dead organic material that decomposes over time. Other disturbances from strong windstorms and hurricanes that fell trees, insects and pathogens that defoliate or kill trees, or herbivores that consume foliage also disrupt the carbon cycle. Harvesting of wood is another disturbance to the carbon cycle. The carbon remaining after accounting for losses from disturbance is the net biome production.

As in all things ecological, the exact cycling and storage of carbon in a forest is complex. Site variation arising from geographic location, topography, soils, species composition, stand age, and many other factors determine the carbon fluxes at a specific forest. However, Figure 13.2 shows representative carbon fluxes in three different types of mature forests found in the tropical, temperate, and boreal regions.[1] Differences among the three forest types should be interpreted with caution because of the limited number of sites available in the dataset, and the data are not indicative of all forests, or even the average forest, in these regions. Nonetheless, carbon fluxes are generally largest in the tropical broadleaf forests, intermediate in the temperate broadleaf forests, and smallest in the boreal conifer forests – a general geographic pattern found in other studies. Tropical forests have high carbon uptake during photosynthesis and a similarly large net primary production. Boreal conifer forests have low gross and net primary production. Across a variety of ecosystems from tundra to tropical rain forest, annual net primary production increases with warmer climates. Net primary production is generally largest in moist tropical forests with relatively constant temperature throughout the year and high annual rainfall. In very warm climates, excessive temperatures, often combined with low water availability, limit production. Net primary production decreases in colder and drier climates.

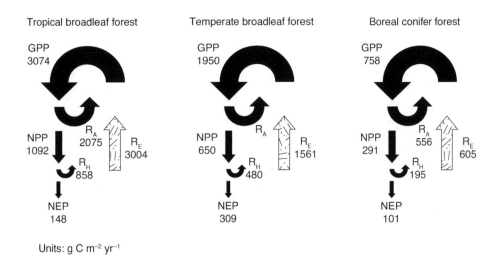

Figure 13.2 Annual carbon fluxes in mature tropical broadleaf, temperate broadleaf, and boreal conifer forests. Units are the mass of carbon (grams; g C) exchanged per unit area (square meters; m^2) per year (yr). GPP – gross primary production; NPP – net primary production; NEP – net ecosystem production; R_A – autotrophic respiration; R_H – heterotrophic respiration; R_E – ecosystem respiration. Fluxes do not necessarily sum because of sampling uncertainties. Adapted from ref. (1).

Respiration returns a considerable amount of carbon to the atmosphere so that the net accumulation of carbon is a small fraction of gross primary production. On average, across many biomes growing in many different climates throughout the world, 82 percent of the annual carbon gained during photosynthesis is lost as respiration.[2] In this dataset, the differences in net ecosystem production among the three forest types are not statistically significant. Net ecosystem production is the small residual difference between two large fluxes (gross primary production and ecosystem respiration) and is therefore subject to considerable uncertainty. A small measurement error in gross primary production, for example, produces a large error in net ecosystem production. Figure 13.3 shows the distribution of carbon throughout the forest ecosystems. Carbon stocks have less systematic differences across the three forest types. Tropical forests have large aboveground and total biomass, followed by temperate broadleaf forests and boreal conifer forests, but other carbon stocks do not statistically differ among the forests in the dataset.

Carbon fluxes vary over the course of a day, primarily in relation to solar radiation, and throughout a year in relation to solar radiation, temperature, soil moisture, and other factors. Forests typically gain carbon during the day when photosynthesis exceeds respiration and lose carbon at night when photosynthesis stops (Figure 9.6). They similarly gain carbon during the growing season and lose carbon in the dormant season. The metabolic signature of terrestrial ecosystems is evident in seasonal variation in the concentration of CO_2 in the atmosphere. Concentrations are lower during the growing season than during the dormant season, when there is a net release of CO_2 in respiration (Figure 8.5). This seasonal variability is most evident in the Northern Hemisphere, where the area of land is largest, and is less evident in the Southern Hemisphere, where ocean influences are large. Changes in the seasonal cycle of atmospheric CO_2 are one indicator of a changing biosphere.[3]

The signature of terrestrial ecosystems on atmospheric CO_2 is also seen in the long-term accumulation of carbon, especially in forests. Figure 13.4 illustrates idealized changes in carbon following forest

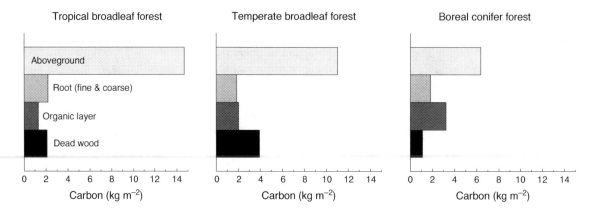

Figure 13.3 Carbon stocks in mature tropical broadleaf, temperate broadleaf, and boreal conifer forests. Units are the mass of carbon (kilograms; kg C) per unit area (square meters; m^2). Aboveground biomass is foliage and wood. Root biomass is fine and coarse roots. Organic layer is the material on the forest floor. Adapted from ref. (1).

Figure 13.4 Idealized changes in ecosystem carbon pools resulting from harvest and regrowth in a temperate forest.
(a) Biomass is initially lost with harvesting and then accumulates as the forest regrows. (b) Carbon is lost to the atmosphere in the early years but is subsequently sequestered as the forest becomes a carbon sink. From ref. (4).

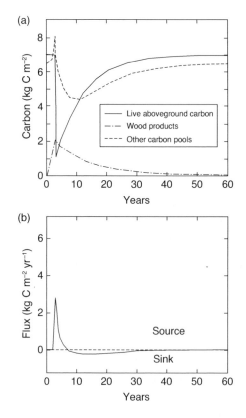

harvesting.[4] In this example, harvesting reduces living biomass from 7 kg C m^{-2} to 1 kg C m^{-2}. Carbon accumulates in living biomass as the forest regrows, recovering to preharvest values by about 40 years. Dead biomass and soil organic carbon (collectively called other carbon pools) initially increase with harvesting as a result of debris left on site. This carbon pool decreases over time as the debris decomposes and then increases as the forest regrows, litterfall increases, and detritus accumulates. In considering the full carbon balance of forests, it is important to account for wood removed from the site and used for wood products. Harvesting transfers the biomass to wood products that decay over time. In this example, the net carbon flux is an immediate release of about 3 kg C m^{-2} yr^{-1} to the atmosphere. The forest remains a source of carbon over the next several years before becoming a small sink.

The Global Carbon Budget

The amount of CO_2 in the atmosphere is the balance of geological processes, biological processes, and human activities (Figure 13.5).[5] All but a small portion of the carbon on Earth is buried in sedimentary rocks. Only about 0.04 percent of the carbon is biologically active, and oceans by far hold most of that carbon (38,000 Pg C; petagram, 1 Pg = 10^{15} g or a thousand million million grams). Considerably less carbon is stored in terrestrial ecosystems. It is estimated that vegetation contains 450–650 Pg C worldwide while soils hold 1,500–2,400 Pg C with an additional 1,700 Pg C frozen in permafrost. The atmosphere held 871 Pg C in 2019 with a CO_2 concentration of 410 parts per million (ppm).[6]

The global carbon cycle is the interaction of two superimposed cycles: the geological carbon cycle, in which carbon cycles among atmosphere, oceans, and continents in response to the chemical weathering of rocks over a period of millions of years; and the biological carbon cycle, in which carbon cycles among the atmosphere and marine and terrestrial ecosystems through biological and physical processes over shorter timescales of days, seasons, years, and decades. Fluxes in the geological carbon cycle are small compared with the biological fluxes. Over a timescale of decades to centuries, changes in ecosystem productivity and ocean biology alter atmospheric CO_2, but over timescales of millions of years, the geological carbon cycle is most important. In the preindustrial era, the carbon cycle is thought to have been in balance with no net carbon gain by the land or ocean (Figure 13.5a). The biosphere was a small annual sink of carbon (1 Pg C yr^{-1}) because photosynthetic uptake exceeded losses from respiration and wildfire. This and additional carbon from weathering (0.4 Pg C yr^{-1}) washed into rivers and lakes, where it returned to the atmosphere in freshwater outgassing (0.3 Pg C yr^{-1}), was buried in sediments (0.55 Pg C yr^{-1}) or was carried into oceans (0.65 Pg C yr^{-1}). The carbon input to oceans balanced a small net loss from air–sea exchange (0.4 Pg C yr^{-1}) and burial on the ocean floor (0.2 Pg C yr^{-1}).

Human activities have significantly impacted the global carbon cycle (Figure 13.5b). Burning oil and coal to generate heat and electricity, combustion of gasoline for transportation, and other industrial processes release CO_2 to the atmosphere. During the ten-year period 2010–2019, these activities emitted 9.4 Pg C yr^{-1}. In addition, human uses of land, primarily from deforestation, emitted another 1.6 Pg C yr^{-1}. Slightly less than half of the total anthropogenic emission remained in the atmosphere (5.1 Pg C yr^{-1}; 46 percent). The rest was taken up by the terrestrial ecosystems (3.4 Pg C yr^{-1}; 31 percent) and oceans (2.5 Pg C yr^{-1}; 23 percent). Without these land and ocean sinks, the amount of carbon in the atmosphere would be considerably larger than it presently is. Indeed, over the industrial era (1850–2019), the land is estimated to have absorbed nearly one-third of the total anthropogenic CO_2 emissions. Whether the terrestrial and ocean sinks for anthropogenic carbon are maintained or decrease in the future – and even whether the terrestrial biosphere becomes a source of carbon rather than a sink – is the subject of considerable scientific research.

(a) Preindustrial carbon cycle (c. 1750)

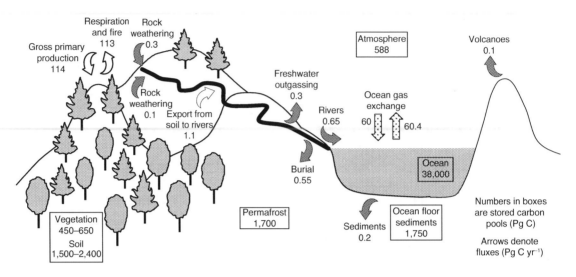

(b) Modern carbon cycle (2010–19)

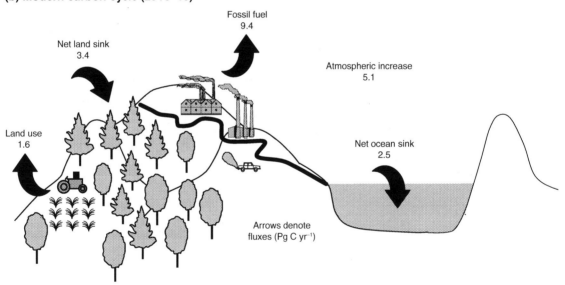

Figure 13.5 The global carbon cycle.
(a) The preindustrial carbon cycle (c. 1750) shows component land fluxes of gross primary production and losses from respiration and fire (white arrows) and ocean gas exchange (stippled arrows). Gray arrows denote fluxes in the geologic cycle. Boxes denote carbon pools. Annual fluxes are given in Pg C yr^{-1}, and pools are Pg C. One petagram (Pg) = 10^{15} g. (b) The modern carbon cycle shows fluxes averaged for 2010–19. Shown are fossil fuel and land use emissions, the net land uptake, the net ocean uptake, and the increase in the atmosphere. Adapted from ref. (5).

The Land Carbon Sink

The influence of the terrestrial biosphere on the amount of CO_2 in the atmosphere manifests from two opposing fluxes: land use emission from human activities and carbon uptake by intact ecosystems. The land use flux shown in Figure 13.5b is the net flux of carbon at the global scale from deforestation and other anthropogenic uses of land such as conversion of forest to cropland or pasture, abandonment of existing farmland followed by forest regrowth, and specific agricultural or forestry management practices.[7] The net flux consists of gross emissions (e.g., loss of carbon with deforestation, harvesting, or soil tillage) and gross removals (e.g., as forests regrow after harvesting or on abandoned farmland). Currently, emissions exceed removal so that there is a source of carbon to the atmosphere from land use. Much of the emissions results from tropical deforestation. The land use emission flux is much smaller than emissions for fossil fuel combustion. However, changes in land use that decrease emissions or enhance removals have a nonnegligible effect on atmospheric CO_2. Indeed, reduced emissions from avoided deforestation, reforestation of cleared lands, and even afforestation of non-forest lands are means to limit the growth of CO_2 in the atmosphere.

Forest growth is the primary driver of the carbon sink, though the exact geographic location, especially the contribution of tropical versus northern forests, is still being debated.[8] Emission from tropical deforestation counters carbon uptake during the growth of intact tropical forests, whereas temperate and boreal forests are carbon sinks. Figure 13.6 shows results of one study.[9] Moist tropical forests contain about 40 percent of the world's live vegetation biomass. These forests store a substantial amount of carbon each year in tree growth, but this is countered by comparable emissions from forest clearing, biomass burning, and forest degradation. Dry tropical and subtropical forests similarly have large uptake and emission fluxes. Together, these ecosystems are mostly carbon neutral. Extratropical temperate and boreal forests combined contain about one-half the live biomass compared with tropical ecosystems. Carbon fluxes in these forests are substantially less than in the tropics, but the low emission from disturbance results in a net carbon sink. Carbon uptake during biomass growth and emission from disturbance are each four times larger in the tropics than the comparable fluxes in the extratropics. That tropical forests have large uptake and emission fluxes highlights the need to reduce tropical deforestation and enhance the removal of carbon in forest growth.

The terrestrial carbon sink term is critically important to atmospheric CO_2 (Figure 13.5b). Without it, the accumulation of CO_2 in the atmosphere would be substantially higher. The fate of the terrestrial sink over the coming years will, to some extent, determine the magnitude of planetary warming. A weakening of the sink with a warmer climate, by which a greater fraction of anthropogenic CO_2 emissions remain in the atmosphere, would accelerate the rate of warming. Projections of the future sink are uncertain. The magnitude of the land sink is expected to increase with greater amounts of atmospheric CO_2, but the fraction of anthropogenic emissions taken up by the land is expected to decrease.[10] The future of the carbon sink is largely understood in terms of two feedbacks related to atmospheric CO_2 and the magnitude of climate change.

The rising concentration of CO_2 in the atmosphere stimulates plant productivity in a process known as CO_2 fertilization. This is a negative feedback on atmospheric CO_2 because elevated CO_2 increases the terrestrial carbon sink. The rate of photosynthesis directly depends on the concentration of CO_2 in the air, and measurements of leaf gas exchange unequivocally show that CO_2 assimilation during photosynthesis

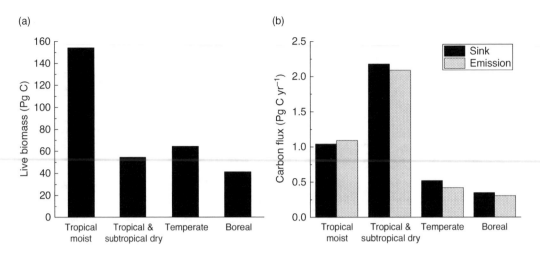

Figure 13.6 Regional carbon stocks and fluxes during the period 2001–19.
(a) Carbon in live biomass for tropical moist forest, tropical and subtropical dry forest and shrubland, temperate forest and shrubland, and boreal forest and shrubland. This excludes dead biomass and soil carbon. (b) Annual carbon sink from biomass growth and emission from disturbance attributed to forest clearing, biomass burning, and forest degradation. Data from ref. (9).

increases as a leaf is exposed to air enriched with CO_2.[11] The stimulation of photosynthesis saturates at some high concentration of CO_2, and the enhancement differs among plants. Plants utilizing the C_3 photosynthetic pathway (e.g. trees, shrubs, and many grasses) respond more and saturate at a higher CO_2 concentrations than do plants utilizing the C_4 photosynthetic pathway (e.g., tropical grasses and some agricultural crops). In addition, stomata close at high CO_2 concentrations so that less water is lost during transpiration for each molecule of CO_2 assimilated during photosynthesis. This is known as a higher water-use efficiency and can increase productivity by alleviating soil moisture stress.

While CO_2 fertilization is well observed at the scale of a leaf, it is less understood at the scale of a whole ecosystem where additional factors determine plant productivity and the long-term accumulation of carbon. Nonetheless, observational, experimental, and modeling evidence shows that the increased concentration of CO_2 over the industrial era has stimulated plant productivity and increased water-use efficiency.[12] One such line of evidence is free-air CO_2 enrichment (FACE) experiments (Figure 10.5). However, the efficacy of CO_2 fertilization in the future is uncertain because plant productivity depends on other factors in addition to atmospheric CO_2 concentration.[13] The amount of nitrogen in the soil, for example, limits productivity in many ecosystems, and the supply of nitrogen may be insufficient to sustain high rates of productivity in the future. Whether the new growth is allocated to leaves and fine roots, which turn over and decay rapidly, or accumulates as wood determines the long-term carbon storage in forests, as does the age of the forest. Nonetheless, CO_2 fertilization is a primary driver for the land carbon sink.

Climate change provides a second feedback with the carbon cycle. Warmer temperatures and longer growing-season length can increase plant productivity in boreal and arctic ecosystems where cold temperatures and short growing seasons limit growth. Satellite measurements of vegetation greenness, analyses of atmospheric CO_2 concentration, and models show that such changes are indeed occurring and contribute to the land carbon sink.[14] Soil water availability controls plant productivity in many regions of the world, and soil water is a main driver of variability in the carbon cycle from one year to the next.[15] In the tropics and other regions with high temperatures, increased evaporative demand in a warming climate may produce soil water stress unless accompanied by greater precipitation. Warming and drying trends are expected to reduce forest productivity in some regions of the world, such as the tropics. Strong, widespread droughts in Amazonia have already increased tree mortality, decreased net primary production, and caused loss of carbon.[16] Drying, wildfires, and deforestation may already be changing some portions of the region to a carbon source.[17]

Warmer temperatures accelerate soil carbon loss through decomposition, and this is a primary reason why the efficacy of the land to sequester anthropogenic CO_2 emissions is expected to decrease in the future. One region especially vulnerable to soil warming is the northern high latitudes of tundra and boreal forest. Some 1,700 Pg C is locked into frozen permafrost soils. The size of this carbon pool exceeds the 700 Pg C cumulatively emitted by human activities since 1850.[18] Warming of high latitudes is likely to accelerate the loss of this carbon to the atmosphere.[19] In general, planetary warming provides positive feedback with the carbon cycle whereby the capacity of the terrestrial biosphere to store carbon decreases with increases in temperature.[20]

Other processes may also weaken the terrestrial carbon sink. High concentrations of ozone in the troposphere damage leaves and decrease photosynthesis.[21] Wildfires emit carbon to the atmosphere in the combustion of biomass, thereby reducing the magnitude of the terrestrial carbon sink. Global fires emitted 2.2 Pg C each year over the period 1997–2016, considerably more than from land use emissions.[22] Insect infestations also kill trees and alter the carbon cycle. The outbreak of mountain pine beetles in western North America, for example, has devastated forests over a wide region of Canada and the United States, resulting in widespread tree mortality and reduced carbon sequestration in forests.[23]

Some processes increase land carbon uptake. Industrial, automotive, and agricultural activities emit nitrogen to the atmosphere that is subsequently deposited onto land. Nitrogen inputs from atmospheric deposition can increase productivity, resulting in enhanced terrestrial carbon storage.[24] This storage is realized primarily in forests as accumulation of either wood in living trees or soil organic matter from increased litterfall. However, nitrogen deposition has other harmful effects that cascade through the environment.[25] Even the amount of aerosols in the atmosphere can affect plant productivity. Aerosols enhance the scattering of solar radiation. Because scattered diffuse radiation penetrates more deeply into forest canopies than does unscattered direct beam radiation, photosynthesis can, with the right conditions, increase.[26]

The Water Cost of Carbon Uptake

A substantial amount of water is required to sustain forest growth. This is seen in the close relationship between photosynthesis and transpiration. Stomata open to allow CO_2 to diffuse into the leaf during photosynthesis (a benefit), but in doing so water diffuses out through the stomatal opening in the transpiration stream (a cost). The physiology of stomata represents a compromise between the two conflicting goals of permitting CO_2 uptake while restricting water loss.[27] Stomata are regulated so as to minimize the water cost of carbon gain, or conversely maximize the marginal carbon gain of water loss. Water-use efficiency is defined as the photosynthetic gain for a given amount of water loss and is quantified by the ratio of photosynthesis to transpiration. At the canopy scale, the ratio of gross primary production to evapotranspiration is a measure of water-use efficiency. A typical estimate for trees is 1–4 grams carbon per kilogram water.[28]

Large water loss accompanies photosynthetic CO_2 uptake. If sufficient soil water is not available to sustain transpiration, stomata close to prevent desiccation. The rate of photosynthesis declines because less CO_2 diffuses into the leaf. Over a long time period, the plant can die from carbon starvation or hydraulic stress.[29] Across a variety of forest, grassland, cropland, and other ecosystems, gross primary production increases by about 3 g C m^{-2} per year for each millimeter increase in annual evapotranspiration.[30] An uptake of 3,000 g C m^{-2} per year requires 1,000 millimeters of water loss as evapotranspiration. Consequently, there is a trade-off between water and carbon.[31] Water-use efficiency is expected to increase with higher CO_2 concentrations.[32] Higher water-use efficiency may be contributing to a greening of vegetation by lessening the water cost of carbon gain, especially in warm, arid environments.

Forests and Oxygen

Associated with the carbon cycle is a cycling of oxygen, also with biological and geochemical components. Photosynthesis releases oxygen to the atmosphere during the assimilation of CO_2, and the supply of oxygen produced by trees has galvanized public attention since Benjamin Franklin, in 1772, extolled it as one reason to preserve forests.[33] Oxygen featured prominently in the writing of many forest advocates in the nineteenth century, and it is seen again today in popular apprehension that clearing tropical forests – the so-called lungs of the planet – is decreasing the amount of oxygen in the atmosphere. Scientific concern about a decrease in oxygen arose with awareness of human impacts on climate in the latter part of the twentieth century, but it was readily understood that the production of oxygen by forests, while important on geologic timescales of hundreds of millions of years, is insignificant in the present-day oxygen budget.[34] The modern analogy with lungs is itself misleading. Stephen Hales drew a similarity between the life-sustaining functions of leaves (photosynthesis) and lungs (breathing) when he told in *Vegetable Staticks* (1727) that leaves "perform in some measure the same office for the support of the vegetable life, that the lungs of animals do, for the support of the animal life."[35] But the modern conception of tropical forests as the lungs of the planet associates photosynthesis more with the production of oxygen than the greater function to assimilate CO_2 into biomass.

Oxygen comprises 21 percent of the atmosphere by volume, with a mass of more than one million petagrams. The large concentration of oxygen is a tell-tale signature of life on Earth, and it does, indeed, represent an accumulation of oxygen produced from photosynthesis over the history of the planet. Photosynthesis consumes CO_2 and produces oxygen, but the respiration that maintains plant life operates in the reverse: oxygen is consumed and CO_2 is released. As trees grow and accumulate biomass, there is a gain of carbon and a concomitant production of oxygen. In contrast, soil microbes and other organisms that decompose organic material consume oxygen during heterotrophic respiration. The production of oxygen by trees and the microbial consumption of oxygen are mostly in balance so that the net amount added to the atmosphere each year is very small. The world's forests do indeed produce an immense amount of oxygen annually during photosynthesis, but the decomposition of organic matter in forest soils consumes a comparable amount. In the short timescale of the human experience, forests have little net impact on atmospheric oxygen. Over geologic timescales, however, some organic matter is buried in swamps, peat bogs, and ocean sediments and does not decompose. In this way, slightly less oxygen is consumed than is produced so that there has been a slow accumulation in the atmosphere. The annual net flux of oxygen from forests to the atmosphere is small, especially compared to the mass of oxygen in the atmosphere, but it is significant over millions of years. Nonetheless, the amount of oxygen in the atmosphere is one indicator of planetary stress. Its concentration is decreasing, though by a minuscule amount each year, due to fossil fuel combustion.[36] The metabolic activity of the biosphere does impart minute variations in atmospheric oxygen, which can be used to infer changes in the global carbon cycle.[37]

■

Terrestrial ecosystems absorb a sizeable portion of the CO_2 emitted to the atmosphere each year by human activities, and carbon storage on land is one of the key climate services of forests. Without the land carbon sink, the amount of CO_2 in the atmosphere would be considerably larger than it currently is, and planetary warming would be greater. The increasing concentration of CO_2 in the atmosphere and the deposition of nitrogen on land enhance the terrestrial carbon sink. Climate change can enhance productivity in some regions where cold temperatures limit tree growth, reduce productivity in warm regions that experience greater drought, and increase the turnover of soil carbon in cold, permafrost soils. Counter to this, deforestation and other human uses of land emit CO_2 to the atmosphere. Nature-based solutions to climate change aim to enhance the terrestrial carbon sink and reduce land-use emissions through avoided deforestation, reforestation of cleared lands, and afforestation of lands not needed for agricultural production. The long-term fate of the sink is uncertain. Overall, however, the terrestrial carbon cycle provides positive feedback whereby a warmer climate decreases the capacity of the terrestrial biosphere to store anthropogenic carbon emissions. While the carbon cycle and other biogeochemical benefits of forests receive much attention, forests affect climate through surface albedo, surface roughness, evapotranspiration, and other biogeophysical processes. These processes can, in some instances, counter the carbon benefits of forests.

14   Forest Macroclimates

The contrasting microclimates of forests and open lands are easy to discern by comparing, for example, air temperature measurements taken in nearby locations. The climate influence of forests over large regions – the forest macroclimate – is much harder to establish by direct observations. The correct way to do so, explained Auguste Mathieu in his 1878 study of forest influences, is to compare measurements in a region that was first forested and then deforested.[1] This was impossible for Mathieu to do, and he, like other scientists of his era, settled for a comparison of locally forested and non-forested lands. Nearly 50 years later, the modern purveyor of forest microclimates, Rudolf Geiger, noted that the study of forest influences on large-scale climate "excludes the possibility of testing it out by means of experiment."[2]

Careful statistical examination of climate data can sometimes reveal an ecological influence. Geographic comparison of forested and non-forested regions was, in fact, used by Mathieu's contemporaries, both to prove and disprove forest influences, though mostly without satisfaction. The modern climate scientist has tools that were not available to Mathieu and others. Geographically extensive networks of eddy covariance flux towers provide a comprehensive means to sample many locations in different regions of the world, and advances in remote sensing aid comparison of land surface temperature in different biomes and different regions. More often, however, scientific understanding of how forests and other ecosystems affect climate comes from numerical models of Earth's climate with their atmosphere, ocean, and biosphere components. Climate models, and the new generation of Earth system models with their advanced representation of planetary physics, chemistry, and biology, provide a theoretically robust and mathematically sophisticated but tractable means to study forests and climate in ways that prior generations of scholars could not have even envisioned (Figure 10.8). Paired climate simulations, one serving as a control to compare against another simulation with altered vegetation, demonstrate an ecological influence on climate (Figure 10.9). As Geiger acknowledged, the problem can be addressed "theoretically."[3] Mathematical models of climate are a key theoretical tool in the climate scientist's toolkit. They have identified the processes by which forests influence climate at the local scale; atmospheric transport that propagates that influence to remote, nonlocal scales; and the changes in climate caused by past and future changes in forest cover (Figure 14.1).

A Desert World

Imagine a world without forests. It is hard to conceive of the untold consequences for biodiversity, water resources, human cultures, and life itself. A world devoid of trees defies imagination. Climate scientists have, however, conducted just such a thought experiment and have contrasted the climate of a forested

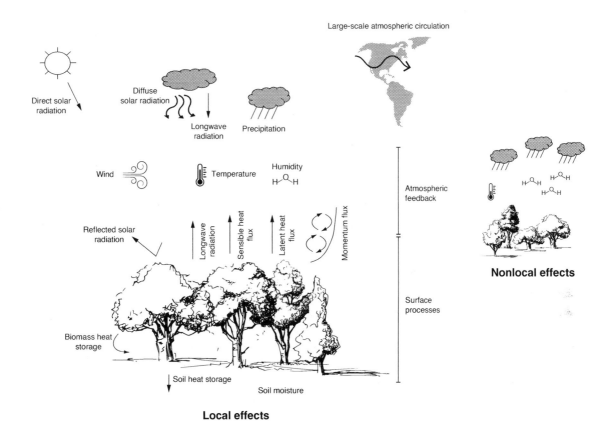

Figure 14.1 Spatial scales at which forests interact with climate.
Shown are local surface processes and their interaction with the atmospheric boundary layer (left). Large-scale atmospheric circulation transports local influences to nonlocal locations to make, for example, the climate cooler, rainier, and more humid (right).

planet with a deforested planet. William Ferrel, the great American meteorologist, posed just this question in 1889 when he considered what would happen to the rain "if a whole continent had mostly a hard and barren surface" or "if the continent mostly without forest should become covered by dense forests."[4] Now, this experiment can be performed using paired climate simulations that, as an extreme, simulate a planet completely covered with well-watered vegetation and contrast the simulated climate with that of a dry planet completely devoid of vegetation. The first such experiment published in 1982 demonstrated the importance of evapotranspiration for global climate.[5] The vegetated planet with high rates of evapotranspiration produces a climate that is cooler and has more rainfall and altered atmospheric circulation compared with a barren planet without evapotranspiration. The study, the authors concluded, "confirms the long-held idea that the surface vegetation, which produces the evapotranspiration, is an important factor in the earth's climate."[6]

Subsequent studies adopted similar methodology to quantify the maximum effect of vegetation on global climate and altered additional features of the land beyond just evapotranspiration. An early study contrasted a desert planet in the absence of vegetation and a green planet where all non-glaciated land is covered by forest.[7] The desert planet has no vegetation, is brightly reflective, presents a smooth surface to the atmosphere, and has low soil water storage capacity. The green planet is covered by trees with high leaf area index. The green leaves absorb much of the solar radiation, and the tall trees present a rough surface. Soil water storage is high because of the porous and organic soil. Annual land evapotranspiration in the green planet more than triples compared with the desert planet while precipitation nearly doubles. The annual mean surface air temperature decreases by more than 1°C averaged over all land area and by several degrees depending on geographic location and time of year. These simulations highlight the importance of evapotranspiration in creating a cool and moist planet. A later study with a more advanced model highlighted the complexity of land–atmosphere coupling.[8] The study contrasted a maximally forested world with a grassland world and demonstrated a prominent influence of the albedo, or reflectivity, of the land in addition to evapotranspiration. In this model, forests warm much of the planet, most prominently in northern high latitudes where trees decrease the surface albedo when snow is on the ground. Tropical forests, in contrast, cool temperature. The presence of trees increases annual precipitation worldwide.

Forest–Climate Processes

Studies of desert and green worlds or maximally forested and deforested worlds are part of a vast modern-day literature in climate science that demonstrates the influence of forests. The interactions between forests and the atmosphere are broadly categorized into physical and chemical processes. Biogeophysics is the study of physical interactions among the biosphere, geosphere, and atmosphere. It considers the exchanges of heat, moisture, and momentum between the land and atmosphere and the meteorological, hydrological, and ecological processes regulating these exchanges. These fluxes affect the dynamics of the atmospheric boundary layer, including wind, temperature, humidity, precipitation, and clouds, which, in turn, alters surface processes in complex ways. Atmospheric radiative transfer, for example, is impacted by the temperature, water vapor, and clouds in the column of air above the land. A warm, dry boundary layer, which commonly happens when there is little evapotranspiration, may reduce cloud cover, thereby further warming the land by allowing more solar radiation to reach the surface. Downwelling longwave radiation from the atmosphere onto the surface increases because of the warmer air, but a decrease in atmospheric water vapor and clouds also reduces longwave radiation. Additional feedbacks arise from changes in large-scale atmospheric circulations.

Biogeophysical processes are largely understood through the surface energy balance, its coupling to the atmospheric boundary layer, and the associated hydrologic cycle. Processes important for climate include the frictional drag of vegetation and other surface elements on wind; absorption of radiation by plant canopies and at the ground; the partitioning of net radiation at the surface into fluxes of sensible heat and latent heat (evapotranspiration); physiological processes such as stomatal conductance that regulate these fluxes; heat transfer in soil; and the hydrologic cycle, including interception, evaporation, transpiration, infiltration, runoff, soil water, snow, and groundwater. These processes are included in climate models using advanced representations of the ecology, hydrology, and meteorology of the land surface, including plants and terrestrial ecosystems.[9]

Figure 14.2 illustrates key forest–climate feedbacks.[10] One of the primary influences is from surface albedo (Figure 14.2a). Forests have a lower albedo than pastures or croplands, even more so when snow is on the ground. Deforestation increases albedo, and reforestation and afforestation have the reverse effect. An increase in albedo decreases the absorption of solar radiation at the surface, decreases net radiation, and

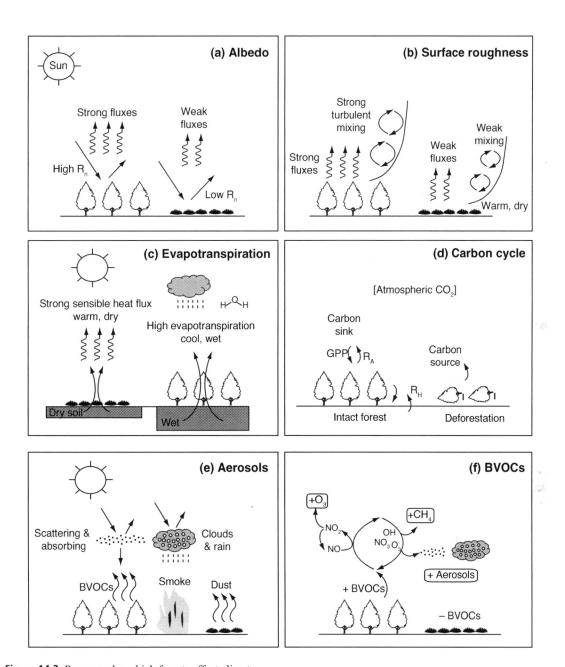

Figure 14.2 Processes by which forests affect climate.
Shown are (a) surface albedo and net radiation (R_n); (b) surface roughness and turbulent mixing; (c) evapotranspiration, precipitation, and surface heating; (d) cycling of carbon through gross primary production (GPP), autotrophic respiration (R_A), litterfall, and heterotrophic respiration (R_H); (e) aerosols, radiation, and clouds; (f) biogenic volatile organic compounds (BVOCs), atmospheric chemistry, and secondary organic aerosols. Adapted from ref. (10).

cools the surface. Less energy returns to the atmosphere as sensible and latent heat, and the decreased surface heating of the atmospheric boundary layer can, in some instances, reduce cloud cover and precipitation. In cold, snowy climates such as the boreal forest, tall trees protrude above the snowpack and reduce surface albedo. Vegetation masking of the high albedo of snow creates a warmer climate than in the absence of trees.

Another important aspect of land–atmosphere coupling is surface roughness (Figure 14.2b). Trees are taller than grasses and crops and are aerodynamically rougher. Rough surfaces create more turbulence and have higher sensible and latent heat fluxes than smooth surfaces, all other factors being equal. Deforestation decreases surface roughness and alters turbulent heat fluxes. During the day, the lower roughness reduces turbulent heat exchange with the atmosphere, resulting in a warm surface and a warm, dry atmospheric boundary layer. However, at night or in winter in seasonally cold climates, the overlying air can be warmer than the surface. The enhanced atmospheric mixing in tall forests brings heat downward from aloft and contributes to surface warming. In this situation, deforestation cools the surface by reducing the mixing of warm air from above.

Evapotranspiration, and the associated flux of latent heat, affects climate in many ways. Forests can have high evapotranspiration rates (latent heat flux) compared with grasslands because of their large leaf area and deep roots (Figure 14.2c). A decrease in leaf area reduces the surface area for transpiration and for the interception of rainfall. Evapotranspiration from the plant canopy decreases, but soil evaporation may increase. A high latent heat flux cools the surface and moistens the boundary layer. Additional feedbacks arise from the effects of clouds and water vapor on atmospheric radiation. Clouds reduce the transmission of solar radiation through the atmosphere to the surface and increase the proportion of the radiation that is diffuse in contrast with direct beam. Water vapor is a powerful greenhouse gas that reemits longwave radiation back onto the surface. A decrease in forest cover or leaf area that reduces evapotranspiration warms the surface, decreases atmospheric water vapor and clouds, and may reduce precipitation, as is particularly evident with tropical deforestation. Wet soil can likewise sustain a high latent heat flux and create a cool, moist atmospheric boundary layer – conditions that can feed back to increase precipitation. In contrast, dry soil decreases latent heat flux and amplifies droughts and heat waves.

In addition to biogeophysical processes, other forest–climate feedbacks relate to biogeochemistry. Biogeochemistry is the study of element cycling among the biosphere, geosphere, and atmosphere. Biogeochemical influences on climate manifest through effects on atmospheric radiation and are considered radiative processes. The albedo of the land is also a radiative process. Surface roughness and evapotranspiration affect climate through non-radiative processes. The distinction between radiative and non-radiative effects is important because there is a strong theoretical relationship between radiative processes and planetary temperature. Climate scientists often speak of the radiative forcing of surface albedo, CO_2, and other greenhouse gases. Radiative forcing is a measure of the change in the atmospheric radiation balance and the resulting warming and cooling of the planet. A positive radiative forcing increases atmospheric heating; a negative radiative forcing reduces atmospheric heating. Surface roughness and evapotranspiration, in contrast, do not conform to the concept of radiative forcing.

The most prominent biogeochemical cycle is the carbon cycle, which regulates the concentration of CO_2 in the atmosphere. Forest ecosystems exert a negative radiative forcing (i.e., a biogeochemical cooling) by removing some of the anthropogenic CO_2 emissions (Figure 14.2d). Carbon emission from deforestation counters the terrestrial carbon sink, and managing the land-use carbon flux is an important component of climate change mitigation policies. In addition to CO_2, the atmospheric concentrations of methane (CH_4) and nitrous oxide (N_2O), two other important greenhouse gases, are regulated by terrestrial ecosystems.

Additional chemical exchanges include mineral dust, biomass-burning aerosols, biogenic volatile organic compounds, and other gases. These chemical emissions alter atmospheric composition. In doing so, they provide positive and negative radiative forcings. Terrestrial ecosystems regulate the emission of aerosols to the atmosphere (Figure 14.2e). Biomass burning during wildfires injects black carbon (soot) and other chemicals into the atmosphere. Barren surfaces can have high entrainment of mineral dust. Many trees emit a class of chemicals known as biogenic volatile organic compounds (BVOCs), mostly isoprene and monoterpenes, and these emissions produce ozone (O_3), increase the amount of methane in the atmosphere, and form aerosols (Figure 14.2f). The effects of aerosols on climate are complex. They can lower temperature by scattering solar radiation to space, warm climate by absorbing solar radiation, increase cloud brightness, and suppress rainfall.

The Climate Influences of Forests

Comparison of a maximally forested world with a grassland world is a common climate modeling protocol to demonstrate forest influences on climate. Forests and grasslands differ in albedo, surface roughness, evapotranspiration, and in other ways, and isolating the importance of a specific mechanism is difficult. A more advanced method is to individually vary key features of forests and grasslands to identify the specific ways in which forests influence climate and which process is most important.[11] In this approach, one simulation depicts a world with the maximum extent of forests, and another simulation replaces forests with grasslands. This provides the net climate effect of all processes taken together. Three more simulations individually consider the albedo, roughness, and evapotranspiration differences between forests and grasslands. The latter process represents various model parameters including leaf area, rooting depth, canopy water-holding capacity, and stomatal conductance. For example, trees are more efficient at transpiring water than are grasses because of their deeper roots and larger leaf area.

Global-scale replacement of forests by grasslands decreases global annual mean temperature by 1°C, but the effect varies with region (Figure 14.3a).[12] Tropical latitudes between 20° S and 20° N warm by up to 1°C in the annual mean, while latitudes north of 30° N cool. The cooling is largest in high latitudes, by more than 3°C north of 50° N. This is the net outcome of cooling from higher albedo and warming from lower evapotranspiration and surface roughness (Figure 14.3b). Forests have a lower albedo than grasslands so that deforestation reduces the solar radiation absorbed at the surface and cools the surface climate. The albedo difference is most prominent when snow is on the ground, during which there is a stark contrast between bright snow-covered surfaces and darker trees. Consequently, the albedo-induced cooling is strongest in northern temperate and boreal regions of the Northern Hemisphere north of 50° N (4–6°C in the annual mean) and smallest in the tropics (1°C). Evapotranspiration and surface roughness counter this cooling. The conversion from forest to grassland decreases evapotranspiration and warms the surface climate. The evapotranspiration-mediated warming is largest (1°C in the annual mean) in the tropics, occurs year-round, and is larger during the dry season (2–5°C) than at other times of the year. Temperate regions of North America and Europe have less evapotranspiration warming in the annual mean, but the response varies seasonally. There is little warming in winter, when evapotranspiration is weak, but 2–5°C warming in summer, when the decrease in evapotranspiration with deforestation is large. Conversion from forest to

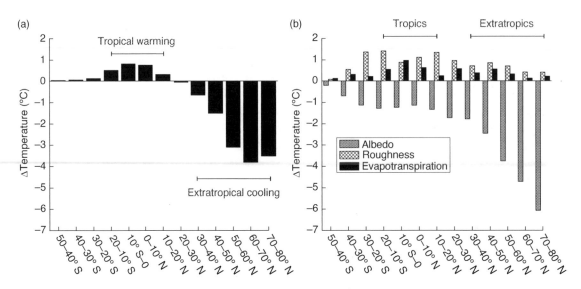

Figure 14.3 Change in temperature from climate simulations contrasting a maximally forested world with one in which grasses replace the forests.

Shown is the difference in annual mean temperature (grassland minus forest) over deforested areas for 10° latitude bands spanning 50° S to 80° N. (a) Net temperature change from all processes. (b) Temperature change from surface albedo, surface roughness, and evapotranspiration efficiency individually. Adapted from ref. (12).

grassland decreases surface roughness. The change in roughness increases temperature by about 1°C over most land areas and is largest in the tropics. In northern high-latitude regions, the albedo effect dominates the temperature response and causes the net cooling with deforestation. The net effect of deforestation in the tropics is warming because evapotranspiration efficiency and roughness are the dominant influences. The response in temperate latitudes is intermediate between the tropical and boreal responses because of large and opposing effects of albedo (cooling) and evapotranspiration and roughness (warming).

Subsequent modeling studies have confirmed the importance of albedo, evapotranspiration, and surface roughness in affecting climate.[13] The general understanding is that forests warm the surface climate because of their low albedo compared with grasslands and other herbaceous land cover, cool the climate because of their high latent heat flux arising from their evapotranspiration efficiency, and additionally cool the climate because their rough surface enhances turbulent heat transport away from the surface. The net outcome of these processes varies among tropical, temperate, and boreal forests.[14] There is a general consensus that tropical deforestation creates a warmer climate with less precipitation. Outside of the tropics, the effects of temperate and boreal deforestation on temperature and rainfall are less robust across models, and the effects also vary seasonally. In general, however, boreal deforestation cools the northern high-latitude climate in the annual mean, as seen in numerous model simulations.

Macroclimate sets the broad context within which to understand forest–climate influences, and model simulations that deforest the entire world and separately tropical, temperate, and boreal forests identify the climate effects of specific forests.[15] Tropical rain forests, for example, operate in a warmer, moister climate than do temperate or boreal forests. Evaporation of the plentiful rainfall provides strong evaporative cooling that, along with surface roughness, counters a smaller albedo warming. Climate model simulations that deforest the Amazon or the entire pantropical rain forest routinely find a warmer climate and less rainfall in the region of deforestation, though the magnitude varies among models. Boreal forests, in contrast, grow in snowy, seasonally cold climates with moderate summer temperatures. Annual evapotranspiration is low. A prominent climate signal is that the low albedo of forests when snow is on the ground provides a strong warming. Boreal deforestation, in contrast with tropical deforestation, cools the climate because of the large increase in albedo. Surface roughness has an additional influence on climate. The aerodynamically rough forest canopy aids surface cooling in the summer but contributes to surface warming during the winter.[16] In winter, when the surface is cooler than the overlying air, heat is transferred from the warm air aloft to the surface because of temperature inversions.

The switch between tropical warming and boreal cooling as a result of deforestation occurs in the midlatitudes.[17] In these latitudes, however, the climate forcing of temperate forests is more uncertain, and the climate response to deforestation varies seasonally. The low albedo of forests, especially during winter in snowy climates, contributes to warming, while high rates of evapotranspiration during summer contribute to cooling. The opposing effects are comparable in magnitude so that the net change in temperature from deforestation is less certain. Models generally agree on the albedo-induced winter cooling because of deforestation but differ in the summer when evapotranspiration is the dominant mechanism.[18] Models simulate either summer warming or cooling as a result of deforestation, mostly due to differences among models in evapotranspiration. Evapotranspiration is the modeled outcome of complex interactions among meteorological factors (e.g., net radiation, atmospheric humidity), the physiology of stomata, the amount of leaves in the canopy, surface roughness, and soil water. Moreover, temperate forests encompass a broad variety of forests across North America, Europe, and Asia growing in conditions ranging from mesic to arid and spanning warm climates to cold northern climates (Figure 8.3). The high evaporative cooling of forests is most prominent on mesic sites. Drier soils limit evapotranspiration.

Changes in the atmospheric boundary layer as a result of surface processes feed back to determine the sign and magnitude of the climate response to altered forest cover. A decrease in evapotranspiration, for example, can warm and dry the lower levels of the atmosphere, leading to a decrease in cloud cover.[19] The reduction in clouds reinforces the surface warming by allowing more solar radiation to reach the surface. Observational and theoretical analyses show that forests may increase cloud cover, which may contribute to forest cooling, especially in midlatitudes.[20] Water vapor is also a powerful greenhouse gas. Although the primary response to decreased evapotranspiration is to warm the surface, the reduction in atmospheric water vapor also cools the surface by reducing greenhouse warming.[21] There is a trade-off between local warming from reduced evapotranspiration and larger-scale cooling from reduced atmospheric water vapor.[22] Feedbacks with the ocean that change ocean circulation or sea ice extent also affect the climate response to deforestation.[23] For example, an increase in sea ice as a result of the cooler climate with boreal deforestation reinforces the cold temperatures.

The Spatial Scale Problem

Models of Earth's climate are an essential tool to understand the effects of forests on climate. The simulations typically deforest large regions of the world or conversely increase forest cover over large treeless regions. As such, they present idealized experiments of continental-scale forest change, whereas the actual change in forest cover is much more local and occurs at a smaller scale.[24] Much of the historical controversy over forests and climate was a problem of spatial scale and of local versus large-scale influences on climate. Today's science, too, is conflicted with a misunderstanding of scale. This is especially evident when comparing model simulations with observations.

Observationally based assessments of the climate influence of forests compare temperatures measured in undisturbed forests and nearby non-forest sites (e.g., open fields) as a proxy for the response to deforestation, assuming the same background climate at both locations. One analysis compared air temperature measured above forests at 33 eddy covariance flux towers in the United States and Canada with that at surface weather stations located in open grass fields.[25] The average distance between paired forest and fields was 28 kilometers, and the average elevation difference was 59 meters. The study found that forest influences on air temperature vary seasonally, differ between day and night, and depend on the background climate. Forests warm air temperature in the annual mean compared with open land, with greater warming north of 45° N and less warming southward to 35° N. In a more extensive geographic analysis, forests were found to cool air temperature at tropical latitudes and warm temperature at northern latitudes.[26] The annual mean air temperature at forests in the tropics and subtropics between 15° S and 20° N was 0.67°C cooler than in open locations. Conversely, in boreal latitudes north of 45° N, the forests were 0.95°C warmer than the open fields. Weaker temperature change occurred between these regions, with the transition between the two regimes occurring at approximately 35° N. South of this latitude, the cooling influence of forests was seen in a decrease in daily maximum air temperature in all months of the year. The warming at northern locations occurred from an increase in the daily minimum temperature throughout the year.

Other studies have examined remotely sensed land surface temperature. In general, forests present a cooler surface than open land during the day and are somewhat warmer than open land at night, but the effects vary by season and with geographic region.[27] There is an overall pattern of annual mean cooling in tropical forests, lesser cooling in temperate forests, and warming in some of or all the boreal forest. A cooler daytime temperature is prominent throughout the year in tropical forests and during warm seasons in temperate and boreal forests. In winter, the daytime temperature of forests is warmer than open land north of latitude 40° N, where snow is present in winter. Warmer nighttime temperature compared with open land occurs throughout the year in all but tropical forests.

Still other studies derive surface temperature from fluxes measured at eddy covariance flux towers. These studies find, for example, that temperate forests in the United States are generally cooler than grasslands throughout the year, with larger cooling during the summer growing season.[28] Care must be taken in interpreting surface temperature because it is not the same as air temperature.[29] Satellite-derived radiometric surface temperature is the temperature at which the land surface emits longwave radiation, and the flux-derived surface temperature is the temperature that balances the surface energy budget. Satellite measurements are further limited to clear sky conditions and so exclude the influence of clouds. Nonetheless, the temperature can be decomposed into component processes to identify the mechanisms driving the change in temperature. For example, the higher nighttime temperature at forests compared with open fields relates to the roughness length of forests, which enhances turbulent heat transfer from the overlying air to the surface when there is a temperature inversion.[30] The cooler daytime temperature relates to the rough forest surface that transports heat away from the surface during the day, as well as ecophysiological differences that cool temperature through evapotranspiration and non-radiative processes.[31] These offset the warming from the low albedo of forests.

Although surface temperature measurements provide valuable insight into the influence of forests on climate, they do so at a different spatial scale than that used in climate models. Global climate models typically have a spatial resolution of approximately 100 kilometers in latitude and longitude, and they model a column of air extending several tens of kilometers in height from the ground to the top of the atmosphere. An increase or a decrease in forest cover in a model grid cell has local effects on air temperature, humidity, wind speed, and clouds within the atmospheric column mediated through surface fluxes and their interaction with the atmospheric boundary layer (Figure 14.1). These feedbacks with the atmosphere themselves alter land–atmosphere coupling and the surface response to changes in forest cover.[32] Furthermore, it is now understood that these changes can be transported by atmospheric circulations to remote regions far away from the local changes in forest cover. Climate model simulations of large-scale changes in forest cover include both local and nonlocal influences. The observations, on the other hand, measure only local effects. Much of the discrepancy between models and observations relates to this problem of scale.[33] Observationally based studies of forest influences on climate capture the local changes in surface fluxes, which are attributed mostly to non-radiative influences, but not the large-scale atmospheric feedbacks seen in climate models. In models, the nonlocal effects of albedo dominate the response to deforestation, which is especially important in northern temperate and boreal forests and can spread to other regions. In contrast, the local effects in these forests are mediated through turbulent fluxes as affected by surface roughness and the amount of foliage. The surface albedo influence is largely excluded at the local scale, but not at the large scale.[34]

Atmospheric Circulations

One of the central questions in the historical forest–climate controversy was whether forest influences are local or transmitted to other regions through atmospheric circulations. The Americans Hugh Williamson and Thomas Jefferson proposed a mechanism for nonlocal influences when, toward the end of the eighteenth century, they hypothesized a circulation between warm open lands and surrounding cooler forests.[35] The nineteenth-century scholars Antoine-César Becquerel, Ernst Ebermayer, Julius Hann, and Josef Lorenz further advanced the premise of mesoscale circulations created by the temperature contrast between forests and fields (Figure 5.1). Heterogeneity in the landscape created by interspersing fields among forests was thought to induce precipitation because of these circulations (Figure 5.2). Today's science confirms the existence of mesoscale circulations and also an influence of forests on the large-scale circulation of the atmosphere.

Mesoscale circulations develop because of differential surface energy fluxes and atmospheric heating across a heterogeneous landscape.[36] Dry fields lack sufficient water for sustained evapotranspiration. As a consequence, they have large sensible heat flux and less latent heat flux, and the overlying air is warm and dry. In contrast, wet forests have large latent heat fluxes, and the air is cool and moist. The contrast in surface fluxes needed to create mesoscale circulations is particularly evident in semiarid climates where wet sites are interspersed in a dry landscape. Wet forests among dry grasslands can, for example, create mesoscale circulations on summer days.[37] Surface winds flow from the cooler forest to the warmer grassland while upper-level winds flow in the opposite direction. This is accompanied by upward movement of air over the warm, dry grassland and descent on the forest. Local differences in albedo can also generate mesoscale circulations.[38] In dry landscapes, for example, the low albedo of forests produces strong surface heating. Air flows toward the warm forests and rises at the forest edge.[39] By creating gradients in surface heating, a heterogeneous mixture of vegetation can induce precipitation. Landscapes that are a mixture of wet forest and dry farmland or that intersperse irrigated cropland and forests in dry, short grass prairie may favor cloud formation and convective precipitation in contrast with a homogenous landscape.[40]

Large-scale atmospheric circulations can also be altered by forests because of their influence on planetary energetics. The Hadley circulation operates in both hemispheres and redistributes energy between the tropics and midlatitudes. Air rises over the equator and flows poleward, descending in the extratropics at about 30° latitude, and returning equatorward as surface winds (Figure 8.2). The warm, moist air rising in the tropics creates heavy rainfall in the region known as the intertropical convergence zone (ITCZ). Afforestation of northern midlatitudes between 30° and 60° N warms the Northern Hemisphere and in doing so changes the Hadley circulation.[41] The greater forest cover decreases surface albedo, increases the absorption of solar radiation, and increases temperature. The additional energy absorbed in the Northern Hemisphere necessitates a change in heat transport between the hemispheres. The Hadley circulation shifts northward, moving the ITCZ and bands of precipitation northward (Figure 14.4).[42] The extent of the circulation change scales directly with the spatial extent of afforestation.[43] Conversely, deforestation in northern middle and high latitudes shifts the ITCZ southward and decreases precipitation in monsoon regions of the Northern Hemisphere.[44]

Afforesting the northern midlatitudes raises the possibility of remote teleconnections whereby a change in forest cover in one region affects precipitation in another region. Other studies have also considered teleconnections arising from tropical deforestation. Some studies find that deforesting the Amazon rain forest can affect the extratropical climate.[45] The warmer, drier climate created by deforestation can, for

Figure 14.4 Changes in the Hadley circulation with midlatitude afforestation.
Shown are (a) present-day grasslands and (b) forest expansion. More forest cover increases energy absorption in the Northern Hemisphere. The Hadley circulation shifts northward to transport more heat across the equator, causing the intertropical convergence zone (ITCZ) to also migrate northwards. Adapted from ref. (42).

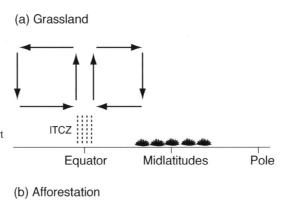

example, alter the Hadley circulation, with consequences for precipitation in the extratropics. The altered Hadley circulation can also affect atmospheric transport from North Africa across the Atlantic to South America.[46] Deforesting boreal, temperate, and tropical regions can affect the climate of other regions.[47] Forest loss in some ecoregions of the United States, either from deforestation or tree mortality, changes atmospheric heating and can alter atmospheric circulation over the United States, thereby affecting climate in faraway regions of the country.[48] Dieback of southwestern forests, for example, may influence eastern regions. Afforestation and vegetation greening in China may have remote influences on the Arctic climate.[49] Atmospheric teleconnections are complex to diagnose and require careful statistical analyses.[50] Many more studies are required to confirm their existence, but a growing body of evidence suggests an influence of forests on the large-scale circulation of the atmosphere.

Droughts and Heat Waves

Deforestation in temperate regions of North America, Europe, and Asia may exacerbate the temperature of hot days.[51] The reduction in evapotranspiration and increase in sensible heat flux that accompanies deforestation creates a warm, dry atmospheric boundary layer that feeds back to further the warming. Likewise, deforestation increases the occurrence of hot, dry summers,[52] and tropical deforestation exacerbates local warming to produce unsafe outdoor temperatures.[53] Restoring forest cover could, to the contrary, protect against extreme heat through evaporative cooling. The type of vegetation, through differences in evapotranspiration, can mitigate heat waves in complex ways based on plant water-use strategy. A key distinction is between shallow-rooted herbaceous plants and deep-rooted trees. Measurements at forests and herbaceous sites (predominantly grassland) in Europe that experienced heat waves show that evapotranspiration is initially less at the forests than at the non-forest sites when the soils are moist.[54] The low evapotranspiration of the forests occurs because they have strong stomatal control of transpiration. With dry soils, however, evapotranspiration declines sharply in the short-rooted grassland, while the forests maintain their evapotranspiration. The grasses, with their exploitive water-use strategy and shallow roots, experience soil moisture stress more rapidly than the trees, which are more conservative and have deep roots. The decline in evaporative cooling contributes to the hot temperatures during the heat wave. Similar results can be seen in satellite measurements of daytime surface temperature over open lands compared to forests during drought.[55]

The feedback between soil moisture and evapotranspiration to accentuate droughts and heat waves is well established.[56] Dry soils reduce evapotranspiration, leading to a warmer surface, warming and drying of the atmospheric boundary layer, and less clouds and rain. Although droughts and heat waves generally originate in large-scale atmospheric circulation, land–atmosphere interactions mediated through soil moisture and evapotranspiration can reinforce the intensity. Dry soils accentuate droughts and heat waves across Europe.[57] Dry soils likely accentuated drought in the North American Great Plains during the 1930s Dust Bowl. Sea surface temperatures in the tropics regulate the occurrence of drought across North America.[58] However, the Great Plains is a region of strong coupling between soil moisture and precipitation.[59] Dry soils during the Dust Bowl likely reinforced the low precipitation by reducing evapotranspiration.[60] Crop failure and vegetation dieback during the drought further amplified the hot, dry conditions.[61] Wind erosion and increased dust emissions with the loss of vegetation still further reduced rainfall and decreased soil moisture to prolong the drought.[62]

The Physiological Forcing of Climate

Forest influences on climate do not require a change in land cover. Changes in the physiology of stomata, for example, alter transpiration, with consequences for climate and water availability. Elevated concentrations of CO_2 in the atmosphere increase photosynthesis (commonly referred to as CO_2 fertilization) and concomitantly reduce stomatal opening so that less water is lost during transpiration per unit gain of CO_2. By reducing transpiration, decreased stomatal conductance reinforces the surface warming arising from higher amounts of CO_2. A groundbreaking, early application of the terrestrial biosphere models coupled to climate models identified this mechanism and contrasted the physiological effects of increased CO_2 (through stomata) with the radiative warming from CO_2.[63] Numerous subsequent studies that compared the climates with low and elevated CO_2 have confirmed the finding.[64] Leaf area can increase with higher CO_2 because of greater productivity, which can further change climate. More leaves increase the surface area for transpiration and interception loss but reduce soil evaporation and decrease surface albedo. The decrease in albedo reinforces the stomata-mediated warming, while the increase in transpiring leaf area partially offsets the reduction in transpiration. The direct physiological effects of stomatal closure and the indirect effects from greater leaf area produce a net warming that contributes to planetary warming.[65] Stomatal closure with high CO_2 can increase the runoff of water to rivers.[66] Physiological changes in stomatal conductance can also alter tropical precipitation through complex changes in atmospheric circulations.[67]

Leaves impart a discernible signal on air temperature. The seasonal emergence of leaves in the Northern Hemisphere decreases the rate of springtime air temperature warm-up because of increased transpiration after leaf emergence.[68] A similar process may be cooling climate because of a greening of the biosphere. Satellite measurements show an increase in leaf area over many regions of the planet during the past few decades due to CO_2 fertilization, climate change, and land use.[69] The effect on surface air temperature of the greening of the biosphere is a balance among evaporative cooling, albedo warming, changes in atmospheric water vapor and cloud cover that affect solar and longwave radiation, and changes in atmospheric circulation. The net outcome of vegetation greening during the growing season is a cooling of air temperature over much of the planet where leaf area has increased, which has lessened the rate of planetary warming over the past few decades.[70] Other studies confirm widespread cooling,[71] or find warming in cold temperate and boreal regions because of a decrease in surface albedo.[72] The difference among studies may relate to the time of year when the greening occurs.[73] Albedo effects differ between springtime, when vegetation masking of snow albedo is prominent, and the summer growing season. Earlier spring leaf emergence in northern temperate and boreal regions can, for example, warm climate.[74]

Atmospheric CO$_2$

In addition to their biogeophysical influences, forests remove CO_2 from the atmosphere. Deforestation, conversely, releases CO_2 to the atmosphere. Climate simulations show the necessity of considering both the biogeophysical effects arising from albedo, evapotranspiration, and surface roughness and the biogeochemical effects related to the carbon cycle when examining the climate response to past and future land cover change.[75] Forests cool climate by reducing the concentration of CO_2 in the atmosphere, whereas deforestation contributes to planetary warming by emitting CO_2 to the atmosphere. Because CO_2 is well mixed in the atmosphere, the biogeochemical signal of deforestation or reforestation is distributed globally. Biogeophysical effects are more regional in the area where forests are cut or planted and can amplify or reduce the biogeochemical CO_2 signal. The net outcome of biogeophysical and biogeochemical processes is uncertain, and whether there is a net warming or cooling varies among models.

Understanding the net effect of forests on climate requires quantifying the sometimes-opposing biogeophysical and biogeochemical influences and how they vary geographically in different climate regions and forest conversion contexts. The high productivity of tropical forests reduces the accumulation of anthropogenic CO_2 emissions in the atmosphere, augmenting the biogeophysical cooling of the forests. Tropical deforestation, consequently, warms the climate through biogeophysical and biogeochemical processes. Avoided deforestation or reforestation have the counter effect and lessen planetary warming. Outside of the tropics, in temperate and boreal forests, the net outcome is less clear. Of particular concern in boreal forests is that the albedo warming counters the benefits of carbon storage.[76] Some studies have combined the biogeophysical and biogeochemical influences into a single metric of climate service value. These analyses generally find high climate mitigation value for tropical forests, less for temperate forests, and still much less, and even little value, for boreal forests.[77]

Atmospheric Chemistry

Forest–atmosphere interactions can dampen or amplify anthropogenic climate change. However, an integrated assessment of forest influences entails an evaluation beyond albedo, evapotranspiration, surface roughness, and CO_2. An emerging research frontier is to link the biogeophysical and carbon cycle influences of forests with a full depiction of biogeochemical feedbacks mediated through atmospheric chemistry. Forests, in addition to being carbon sinks, are sources for aerosol particles from chemical emissions and biomass burning. Aerosols affect the radiative balance of the atmosphere by absorbing and scattering radiation and by altering the formation, optical properties, and longevity of clouds (Figure 14.2e). Fires, for example, influence climate though emissions of long-lived greenhouse gases, organic and black carbon aerosols, and short-lived reactive gases. The net effect is the balance of these biogeochemical emissions and also biogeophysical effects from changes in surface albedo and energy fluxes. The aerosols contained in smoke also affect clouds and precipitation, and in some regions (e.g., the Amazon) may decrease precipitation in positive feedback whereby biomass burning promotes drought and greater susceptibility to fire.

One emerging area of study concerns the many chemical gases emitted by forests, which affect climate through what is known as short-lived climate forcings. Trees emit numerous organic chemicals collectively known as biogenic volatile organic compounds (BVOCs) in trace amounts, many of which are commonly recognized as scents and odors (e.g., pines, cedars, eucalyptus). Chief among these are isoprene and monoterpenes. The concentration of BVOCs in the atmosphere is low (a few parts per trillion to several parts per billion) and their lifetime is short (minutes to hours), but they undergo chemical transformations that affect the concentration of greenhouse gases and aerosols (Figure 14.2f). Tropospheric ozone is a greenhouse gas that warms climate, and BVOCs increase its concentration. Methane is a powerful greenhouse gas, and BVOCs increase its concentration by decreasing the naturally occurring chemical reactions in the atmosphere that consume it. BVOCs also produce secondary organic aerosols. These aerosols directly scatter solar radiation, thereby cooling climate, and they also indirectly affect climate by making clouds more reflective. By altering BVOC emissions, changes in forest cover can affect climate.

Tropical deforestation, for example, reduces isoprene emissions by replacing high-emitting trees with lower-emitting cropland and pastureland. Over the twentieth century, there has been a reduction in global isoprene emissions because of deforestation.[78]

The chemistry of these reactions is complex, and the precise climate outcome is uncertain. However, studies find that loss of forests, either as observed over the historical era or through idealized global deforestation, substantially reduces the emission of isoprene and monoterpenes.[79] As a result, the concentration of ozone, methane, and secondary organic aerosols in the atmosphere decreases with deforestation. The decreases in ozone and methane are negative radiative forcings that cool climate, while the reduction in aerosols is a positive radiative forcing that warms climate. The net effect is less certain, with studies finding either a net cooling from deforestation or a net warming depending on what processes are considered and the precise details of the calculations.

Figure 14.5 shows the results from one modeling study in which the forests of the world are removed.[80] Global deforestation reduces emissions of isoprene and monoterpenes with the result that the production of secondary organic aerosols also decreases, providing a positive radiative forcing (planetary warming).

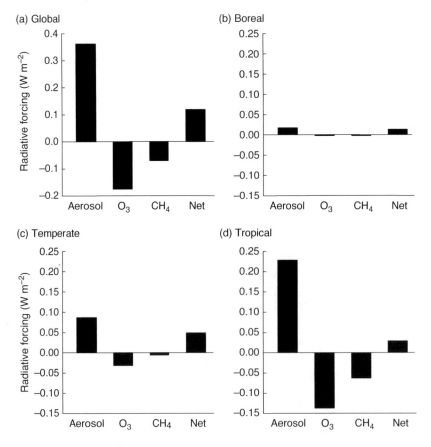

Figure 14.5 Climate influence of BVOCs as measured by the radiative forcing arising from global and regional deforestation.
Shown are the individual effects from secondary organic aerosols, ozone (O_3), methane (CH_4), and the net effect for (a) global, (b) boreal, (c) temperate, and (d) tropical deforestation. Redrawn from ref. (80).

Counter to this, the reduced BVOC emissions with deforestation cause ozone concentrations to decrease and also shorten the lifetime of methane in the atmosphere. Both of these outcomes provide a negative radiative forcing (planetary cooling). The combined effect is the balance between planetary warming because of fewer aerosols and planetary cooling from less ozone and methane. In this study, the aerosol warming outweighs the ozone and methane cooling, producing a net warming due to deforestation. The warming from chemistry–climate interactions augments the warming from the CO_2 emissions that accompany deforestation and counters the associated increase in surface albedo. The influence on climate varies geographically among boreal, temperate, and tropical forests (Figure 14.5b–d). The climate influence of aerosols is largest in tropical forests but is offset by ozone and methane so that the net effect is small. The climate effects of BVOCs in boreal forests are small in this study, and the major outcome of boreal deforestation is planetary cooling from the increased surface albedo.

These estimates are likely to be revised with further studies, but they show another way in which forests affect climate. An additional complexity relates to atmospheric radiation and plant productivity. By scattering solar radiation, the aerosols produced from BVOCs increase the fraction of radiation that is diffuse rather than direct beam.[81] This increases plant productivity because diffuse radiation penetrates deeper into the canopy and illuminates leaves that would otherwise be shaded. Nonetheless, our current knowledge suggests that altered emission of BVOCs from deforestation has a nonnegligible effect on climate that must be considered in addition to the biogeophysical and carbon cycle effects of deforestation. Conversely, planting trees for carbon sequestration increases BVOC emissions, which may further aid in mitigating planetary warming.

∎

The mechanisms by which forests influence climate are well known, but the specific response to changes in forest cover is less well understood and varies regionally with background climate (e.g., tropical, temperate, boreal), the extent of forest change and type of land cover conversion (e.g., forest clearing to grow crops), and time of year (e.g., winter versus summer), and differs between day and night. Forests remove CO_2 from the atmosphere, thereby lessening planetary warming (high confidence).[82] Forests also affect climate through biogeophysical exchanges of energy, water, and momentum with the atmosphere, which can warm or cool the surface climate depending on geographic location, time of year, and time of day. There is a distinct latitudinal pattern from tropical to temperate to boreal forests, with different degrees of influence on temperature and different underlying biogeophysical mechanisms. In general, forests cool the daytime surface climate during the growing season through evapotranspiration and other non-radiative processes, and they warm the nighttime surface climate (high confidence).[82] Outside of the growing season, forests are generally warmer than open fields, especially in locations where snow is present (high confidence).[82] Further changes in climate occur through emissions of biogenic volatile organic compounds (low confidence).[83]

Much of this understanding comes from models of Earth's climate that combine the atmosphere, oceans, sea ice, and terrestrial biosphere. Models are imperfect and do not capture all the processes or the complexity of forests, climate, and their interactions, but they do provide a comprehensive picture of the local and nonlocal effects of forests. Observational studies from satellites or eddy covariance flux towers provide only local effects, and not the feedbacks with the atmosphere. Together, however, they provide a clearer understanding of forests and their influence on climate at spatial scales spanning local to continental. The next chapter examines specific case studies of forests and climate and how deforestation, or conversely increased forest cover, affects climate.

15 Case Studies

Many of the concepts expressed during the forest–climate controversy are central to the science of forests and climate today. Fundamental to this is that evapotranspiration cools the surface climate and provides water vapor that condenses and falls back onto the land in a recycling of precipitation. In the tropics, the presence of forests aids in lessening the tropical heat and in promoting rainfall through positive feedback between forest cover and rain. Georges-Louis Leclerc, Comte de Buffon, Alexander von Humboldt, Hugh Williamson, and others described this feedback, and conversely, the lack of rain in barren deserts. The premise, too, of irreversible climate change, voiced often in the decline of ancient civilizations as a result of deforestation, was common. In today's climate science, the Amazonian rain forest is a case study for forest–climate coupling, and one in which deforestation is a possible tipping point that irreversibly changes the climate. Buffon, Antoine-César Becquerel, and others speculated that planting trees in the Sahara would increase rainfall. A green Sahara has, in fact, occurred in the past and is known to have increased rainfall at the time. This, too, provides another climate tipping point. The boreal forest presents a further example of positive feedback by which the northward advancement of the treeline in a warmer climate decreases surface albedo and reinforces the warming; the southward retreat of the treeline with a cooler climate conversely exacerbates the cooling.

One central theme courses throughout the history of the forest–climate question: that humankind can manage the world's forests to purposely engineer climate. To this end, some forests are more useful than others. Buffon voiced this idea expressively in saying that the temperature can be set to a desired level by clearing forests where warmth is needed and planting trees to lessen the tropical and desert heat. Alexandre Moreau de Jonnès, the gold prize winner of the 1825 competition sponsored by the Royal Academy of Brussels, likewise wrote that tropical forests are more effective at cooling the climate than are temperate forests because of the lesser importance of evapotranspiration away from the tropics. And in a thought prescient of the modern science, the French botanist Bernardin de Saint-Pierre proposed that tropical forests cool the climate while northern forests warm the climate. Yet even then, scholars such as Becquerel and the French forester Jules Clavé warned against generalities and cautioned that the full suite of forest influences, some counteracting others, needs to be examined. The latitudinal dependence of forest influences on temperature, seen in both observations and models, confirms that not all forests are equal in their climate benefits. New processes not envisaged by eighteenth- and nineteenth-century scholars, such as carbon storage, biomass-burning aerosols, biogenic volatile organic compounds, and biogenic aerosols both further the cause for forest conservation for climate and also muddle the net climate benefits of forests. The influence of forests on climate, and their effectiveness in mitigating climate warming, varies regionally. This chapter examines forests in specific regions of the world (Figure 15.1).

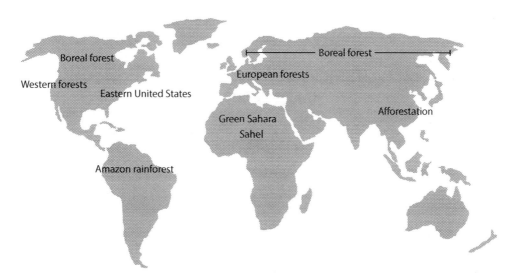

Figure 15.1 Regions where changes in forest cover are altering climate.
Deforesting the Amazon, the greening and browning of the boreal forest, and the green Sahara have been identified as possible climate tipping points. Loss or gain of tree cover in the temperate forests in western North American, eastern United States, Europe, and China also change climate. The Sahel region of Africa is another region in which human uses of the land influence climate.

Amazonian Rain Forest

Concern about the deforestation of the Amazon basin of South America provided the impetus to renew the study of forest–climate influences in 1970.[1] One of the first applications of the newly developed models of the land surface, including plant canopies, coupled to atmospheric models was to study tropical deforestation. Climate model simulations contrasted the climate produced with an intact Amazon rain forest with one in which the forest was changed to degraded pasture.[2] The change from trees to grasses decreases roughness length, leaf area index, and vegetated fraction and alters stomatal conductance. In the simulations, soil texture was also changed to finer soil with more clay to reduce water-holding capacity, and soil color was made lighter to increase soil albedo. Removal of the forests was found to reduce evapotranspiration and produce a warmer climate with less rainfall over the deforested region. Numerous subsequent studies have confirmed these initial findings.[3] Most studies of Amazonian deforestation find that complete transformation of forest to pasture results in a warmer and drier climate. A warmer, drier climate upon deforestation is also found throughout other areas of the tropics, though the magnitude of change varies with region.

A key question is whether precipitation actually decreases with deforestation, as the models predict. Central to this is the recycling of water, by which moisture from evapotranspiration precipitates back onto the land as rain. Although air from the Atlantic Ocean provides much of the water that falls as rain across the Amazon basin, evapotranspiration provides 25–35 percent of the rainwater.[4] Transpiration is the source of rainwater in some portions of the Amazon,[5] and also drives the transition from dry to wet seasons across the southern Amazon.[6] The rising stream of vapor from foliage – the "torrents of vapor" – was witnessed by Humboldt and informed his views of forests and climate. Today's science has likened the transpiration streaming upward from an individual tree in the Amazon to a geyser, though much more elegant in its biological intricacies, and the total amount of water rising from the hundreds of billions of trees is greater than the flow of the Amazon River.[7] Evapotranspiration is similarly an important source of rainwater for the rain forests of the Congo basin in Africa.[8] Observationally based analyses find that air that passes over tropical forests produces more downwind rainfall than air that moves over sparse vegetation.[9] Despite strong theoretical and observational evidence for forest influences on precipitation, there is great difficulty in finding a signature of deforestation in the observational record.[10] Climate model studies of Amazonian deforestation are idealized and deforest the entire basin, whereas only about 20 percent of the rain forest has been cleared.[11] The present-day amount of deforestation may be too small to produce a detectable reduction in precipitation in the observational record.[12]

A further difficulty is that the effects of deforestation are more complex than represented in large-scale, global climate models, which have a spatial resolution of about 100 kilometers or larger. The actual deforestation is localized and patchy. Small-scale, heterogeneous deforestation affects atmospheric processes at the mesoscale, in some cases even enhancing rainfall.[13] Mesoscale atmospheric circulations generated by the contrast between forests and pastures (primarily the increased sensible heating over warm pastures in contrast with cooler forests) can increase cloud cover and produce conditions favorable for precipitation. There may, therefore, be a critical spatial threshold of deforestation in which small patches of forest clearing can increase rainfall but larger deforestation decreases rainfall. During the dry season (June–September) in the southern Amazon, for example, widespread deforestation has increased rainfall downwind of cleared regions, but decreased rainfall in the upwind areas. This reflects a regime shift from small-scale clearing (a few kilometers in size) that creates thermally driven local circulations resulting from spatial variation in surface heating between forests and pastures to large-scale deforestation (a few hundreds of kilometers) that reduces surface roughness and atmospheric transport of moisture (Figure 15.2).[14] The effect of deforestation on the development of shallow clouds in turn regulates light, temperature, and water vapor and therefore fluxes of carbon and evapotranspiration.[15]

The difference in evapotranspiration between forests and pastures is a critical determinant of the climate response to deforestation. Evapotranspiration decreases with conversion from forest to pasture because of the lower surface roughness of grasses, because trees have deep roots that sustain transpiration during the dry season, because interception loss decreases, and because the higher albedo decreases net radiation. Numerous observational and modeling studies across the Amazon have confirmed these findings.[16] Less evaporative cooling is also seen in satellite measurements of land surface temperature, which increases in deforested areas.[17]

The climate effect of tropical deforestation, and especially Amazonian deforestation, is one of the most intensely studied examples of forest–climate coupling. The research highlights two aspects of the science that also tested nineteenth-century scholars. The first is the challenge of spatial scale and the difficulty in relating local-scale observations to the large-scale climate. Measurements of the meteorology of forests and pastures demonstrate differences in microclimate whereas global climate models simulate larger spatial scales and large-scale deforestation. That there may be a scale dependence to the effects of deforestation on rainfall – an increase with small-scale deforestation; a decrease with large-scale deforestation – is exactly the issue raised by German and Austrian meteorologists in the mid-nineteenth century.[18] Mesoscale models simulate a regional domain at much higher spatial resolution. They can more realistically depict the fine scale pattern of deforestation but lack the dynamic connection back to global climate.

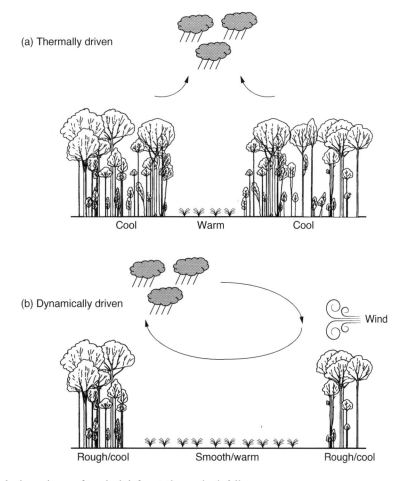

Figure 15.2 Scale dependence of tropical deforestation and rainfall.
(a) Small-scale deforestation enhances clouds and rain over the cleared area through local circulations driven by the thermal contrast between cool forests and warm pastures. (b) Large-scale deforestation decreases surface roughness such that clouds and rain are suppressed over the upwind region and enhanced downwind. Redrawn from ref. (14).

The second point concerns how to quantify forest influence on climate. Nineteenth-century scholars knew that the most direct evidence would be found in long-term observation of a region that underwent a change in forest cover, but as this was impractical, some sought to correlate variation in rainfall with changes in forest cover over a large geographic region. A recent review of the effects of tropical forests on precipitation spoke of exactly the same difficulty: "The most straightforward way to try and identify a link between deforestation and rainfall is either to determine whether rainfall has changed after deforestation over a region has started, or to look at whether spatial patterns in the surface correlate with rainfall patterns. Both of these approaches, however, have significant limitations."[19] The first approach is limited in that the signal of forest clearing is often small compared to the year-to-year variability in precipitation in the region of deforestation. The second approach has the challenge of separating cause and effect because high rainfall also leads to greater forest cover. Scientists of today have a third method – climate models – not available to earlier scientists, but models, too, are imperfect, with oftentimes considerable uncertainty in the simulated climate. Some processes are poorly represented in the models, and some that are thought to be important are not included because of difficulty in formulating the necessary mathematical equations.

Biogeochemical feedbacks arising from biomass burning, biogenic volatile organic compounds (BVOCs), and aerosols in the Amazon are one example of a missing process.[20] Biomass burning injects black carbon (soot) into the atmosphere. Pollen, spores, bacteria, plant debris, and other organic materials provide additional aerosols from primary biogenic particles. The BVOCs emitted by forests create secondary organic aerosols. Aerosols affect climate through complex interactions with radiation, clouds, and precipitation. Aerosols scatter and absorb solar radiation, thereby altering the radiative balance of the atmosphere, and they also affect cloud microphysics by serving as nuclei for the condensation of cloud water. The exact outcome in terms of surface warming or cooling is uncertain, and there are indirect effects arising for the physiological response to changes in direct and diffuse radiation. The biogeochemical coupling between the forests and the atmosphere through aerosols also affects the hydrologic cycle. Observations and models show that biomass burning during the dry season in the Amazon inhibits clouds and rainfall.[21] By scattering and absorbing radiation, the aerosols alter the heating of the atmosphere, thereby changing the conditions for the formation of clouds and precipitation. Smoke particles also affect cloud condensation nuclei and the number and size of cloud droplets.

A defining concept during the nineteenth-century forest–rainfall controversy was that evaporation from oceans provides much of the water vapor that condenses and falls onto the land as rain. That forests could influence this cycling of water was unimaginable to many meteorologists of the day. The Amazon rain forest presents a counterview, and the extensive forest across the basin has been called the "green ocean" for its similarity to maritime clouds and rain during the wet season.[22] A further point of contention was that of reinforcing feedbacks, whereby forests enhance the rains necessary for their existence. Buffon, Williamson, and Humboldt spoke of the influence of tropical forests on rain in a language that is readily familiar to scientists of today.[23] Others, notably the Prussian meteorologist Heinrich Dove, envisioned the possibility of irreversible climate change from deforestation, in which loss of forests decreases rainfall to the point that the forests cannot recover.[24]

Today's science speaks of the loss of the Amazon rain forest, either from clearing or climate change, as a so-called tipping point that irrevocably alters the climate, hydrology, and ecology of the region.[25] A tipping point occurs when small incremental changes reach a critical threshold to create a larger, destabilizing change with cascading effects that may be irreversible; the system switches from one state to another. In the Amazon, fewer trees mean less evapotranspiration, less rainfall, and warmer temperatures (Figure 15.3). Increased wildfires further the loss of trees and the decline in rain. If the Amazon is sufficiently deforested, the concern is that the region may switch to a dry climate with savanna vegetation or dry seasonal forests.[26] Loss of the vast carbon stores contained in tropical forests would further exacerbate the warm, dry climate. One estimate is that the threshold could be 20–25 percent deforestation.[27] There is, indeed, evidence of positive forest–rainfall feedback by which forest cover extends the geographic region in which tropical rain forests grow by increasing rainfall, and that conversely accentuates forest loss with declining rain.[28] Of further concern is that the Amazonian carbon sink is declining or even changing to a source because of climate stresses and deforestation.[29] The warmer and drier climate resulting from deforestation can itself increase carbon loss in remaining intact forests.[30] Some evidence suggests that the Amazon is approaching a critical threshold of forest dieback.[31] Embedded in the notion of a tipping point is that tropical forests create a climate that favors the existence of forests. A similar

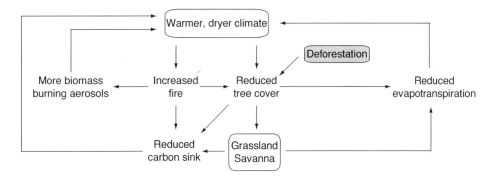

Figure 15.3 Processes by which the Amazon rain forest may be a climate tipping point.
Ecological, hydrological, and meteorological feedbacks arising from a warmer, drier climate lead to forest dieback and a loss of forest cover that reinforces the warm, dry climate and may convert the region to a savanna climate. Deforestation initiates the same feedback processes, and large-scale deforestation of the region may likewise create a savanna climate.

understanding is seen in the Gaia conceptualization of Earth, in which life makes and maintains the conditions suitable for a livable planet.[32] The geologic record abounds with many examples of the profound influence plants have had on the history of the planet.[33]

Boreal Forests

If tropical forests present a textbook example for the beneficial climate services of forests, boreal forests offer a more complex case study. In Alaska and northern regions of Canada, Europe, and Russia, the winters are bitterly cold, the summers are cool and short, and the vegetation is a mix of evergreen and deciduous trees that form boreal forests (Figure 8.3). Needleleaf evergreen conifers – spruce, pine, fir, and larch – are abundant. In the extreme cold of Siberia, deciduous larch trees that drop their needles in winter are common. Broadleaf deciduous trees include aspen, birch, poplar, alder, and willow. In the cold, snowy climate, the growing season is short and annual productivity and evapotranspiration are low compared with those of tropical forests.

A central characteristic of boreal forests is the low albedo of forests compared with that of open land, especially during winter when snow is on the ground. The overall surface albedo is the combined reflection of all plant material (leaves, branches, and stems) and the underlying ground. With low leaf area, the albedo is largely that of the underlying ground, but as leaf area increases, albedo responds more to the optical properties of the leaves. Measurements for 22 North American boreal forest stands initiated by wildfires spanning a 150-year chronosequence show the dependence of albedo on forest age (Figure 15.4).[34] The general trend is a decline in albedo with forest recovery from burning. Summer albedo immediately following fire is low (about 0.05) because of charring but increases to about 0.12 for a 30-year period as a deciduous canopy develops, and then decreases to about 0.08 as the deciduous trees are replaced with mature spruce forest.

The effect of forests on albedo is especially evident when snow is on the ground. Fresh snow can reflect 80–90 percent of incoming solar radiation. Foliage reflects less sunlight, and canopies of evergreen needles

Figure 15.4 Surface albedo in relation to stand age in boreal forests. Shown are daily mean albedo during (a) summer and (b) winter in boreal forests of Alaska and western Canada in relation to time since disturbance. The linear line for summer is for stands older than 12 years. Reproduced from ref. (34).

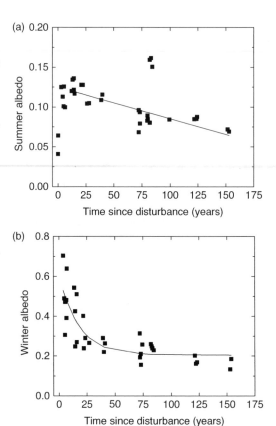

effectively mask the underlying snow. Across the sites shown in Figure 15.4b, winter albedo has a pronounced decline from about 0.7 for young forests with a sparse canopy to about 0.2 for mature pine and spruce forests with a dense canopy. Other studies find similar decreases in winter and summer albedo as boreal forests age and the canopy closes.[35] Forest masking of snow is also seen in Figure 9.7. In this figure, the albedo of cropland is much higher than that of the forests in January. Species composition also affects surface albedo. Broadleaf deciduous forests have a higher albedo compared with needleleaf evergreen conifers, seen also in Figure 9.7 during late spring to autumn. Climate model simulations show that the boreal forest warms climate, primarily because of its low albedo in winter and spring.[36] Sea ice melts with the warmer climate, and polar oceans warm. Warming persists into summer despite smaller differences in albedo between forest and treeless land because of the warmer oceans and reduced sea ice. When compared among various global biomes, the boreal forest has the greatest effect on annual mean temperature as a result of large changes in albedo.[37]

The geographic location of the forest–tundra ecotone has played a role in past climates. Loss of forests provided positive feedback for glacial inception and glacial climates.[38] The climate of the Northern Hemisphere was much colder at the onset of the last glaciation 115,000 years ago because of reduced solar radiation and lower atmospheric CO_2. High-latitude forests died back and were replaced with tundra and grassland-like vegetation. The increased albedo of the forestless landscape reinforced the cold climate.

The location of the treeline separating forest and tundra changed over the past 18,000 years with the transition from glacial to interglacial. As climate warmed and the glaciers retreated northwards, the treeline migrated northwards. The decrease in albedo caused by northward expansion of forest accentuated the warming.[39]

The influence of boreal forests on climate is more complex than just albedo. The forests contain large stores of carbon. This is a beneficial climate service that cools the planet by removing CO_2 from the atmosphere; deforestation releases this carbon to the atmosphere. The effect of boreal forests on climate is the balance between the warming from albedo and the cooling from carbon storage. The net effect of these two opposing processes is unclear, but the low forest albedo decreases the carbon benefits of the forests.[40]

The various types of boreal vegetation also differ in evapotranspiration.[41] Summertime evapotranspiration is larger for deciduous forests compared with conifer forests, and this difference in evapotranspiration affects climate. Climate model simulations show that expansion of broadleaf deciduous trees in the northern high latitudes not only decreases albedo, but also increases evapotranspiration.[42] Water vapor in the atmosphere is a powerful greenhouse gas. Increased water vapor warms the climate and initiates positive feedback whereby warmer temperature melts sea ice, which decreases ocean albedo and increases evaporation from the ocean, producing still greater warming.

Another feedback between boreal forests and climate arises from BVOCs and aerosols. Conifer forests are strong emitters of BVOCs, and observations of the air over Finnish boreal forests highlight the influence of the forests on clouds. Atmospheric measurements find a high amount of secondary organic aerosols above the forests originating from BVOCs.[43] Measurements further show that secondary organic aerosol formation together with water vapor from forest evapotranspiration causes aerosols and clouds to increase as air passes over the forests.[44] The aerosols cool temperatures over the forests by brightening the clouds so that less solar radiation reaches the surface, seen in both atmospheric measurements above the forests[45] and more generally in modeling studies.[46] Emission of BVOCs initiates feedbacks that affect climate change. Boreal forests provide negative climate feedback whereby higher forest productivity in a warmer, CO_2-enriched climate increases BVOC emissions and the production of secondary organic aerosols, which in turn lessens the climate warming.[47] This feedback among forests, aerosols, and climate, in which warmer temperatures enhance BVOC emissions, aerosol formation, and cloud reflectivity has been observed.[48] Further cooling is provided by carbon storage in the forests, which may be enhanced because the aerosols boost productivity by increasing the fraction of solar radiation that is diffuse compared to direct beam.[49] Deforestation or conversion of forests to broadleaf deciduous trees, with less BVOC emissions, would alter this negative climate feedback.

Boreal forests, like tropical forests, require an integrated biogeophysical and biogeochemical understanding of their climate influences (Figure 15.5).[50] The forest–albedo feedback is positive feedback whereby a gain in forest cover leads to climate warming. Evapotranspiration provides additional feedback that can both cool and warm the climate. An increase in evapotranspiration with forest expansion cools the surface climate during summer but warms the climate by increasing the amount of atmospheric water vapor. Countering these is the forest–carbon feedback that removes CO_2 from the atmosphere. The forest–aerosol feedback in conifer forests is likewise negative feedback that lessens planetary warming. Analysis of a Scots pine forest in southern Finland highlights the conflicting effects on climate.[51] Conifer forests cool the climate by serving as a sink for carbon, further the cooling by increasing secondary organic aerosols in the air above the forests but warm the climate by decreasing albedo. The aerosols offset the albedo warming so that there is a net cooling from carbon sequestration in this fast-growing forest. The climate effects of

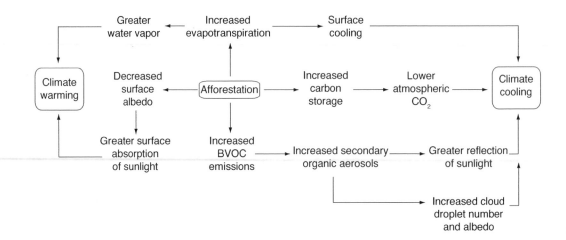

Figure 15.5 Climate feedbacks in boreal forests.
Shown are biogeophysical (albedo, evapotranspiration) and biogeochemical (carbon, aerosols) processes by which increased tree cover in the boreal forest affects climate. Revised from ref. (50).

secondary organic aerosols need to be included in addition to carbon storage and surface albedo when considering the management of Finnish forests to mitigate climate change.[52]

Still more impacts on climate arise from wildfires. Burning and postfire recovery of forests alter surface energy fluxes.[53] Most prominently, the albedo of the land changes following burning. Summer albedo can decrease because of charring and blackening of the surface, but the more notable effect is that winter albedo increases with loss of trees that exposes snow on the ground and then decreases as the burned forest recovers (Figure 15.4). Carbon and other greenhouse gases are released to the atmosphere during combustion, and ozone is also produced. Black carbon aerosols pollute the air. Some of the soot is deposited onto snow and ice, decreasing surface albedo. The climate effects of these various factors can be seen in a study of fire in a black spruce forest in Alaska (Figure 15.6).[54] In the first year, the post-fire increase in surface albedo provides a negative radiative forcing that cools climate. Emission of CO_2 and methane during burning is a positive radiative forcing that warms climate. Ozone, black carbon deposition, and biomass-burning aerosols are an additional positive radiative forcing. The net outcome is a positive radiative forcing. However, the effects of ozone, black carbon deposition, and aerosols are short lived. Over an 80-year postfire forest recovery, their contribution to radiative forcing is negligible. Carbon uptake by the regrowing forest, as well as oxidation of methane in the atmosphere, diminishes the radiative forcing from these greenhouse gases. Instead, the dominant term is from surface albedo, yielding a net negative radiative forcing when averaged over 80 years.

Changes in the fire regime reshape the landscape and the age and composition of forests, thereby altering feedbacks with climate. In North American spruce forest, for example, deciduous aspen and birch trees may initially colonize a burned site before they give way to a mature spruce forest after several decades. There is a clear temporal trend in surface energy fluxes related to the immediate fire, regrowth by deciduous trees, and recovery of a mature spruce forest. Increased forest burning, as expected with a warmer climate, will give rise to younger forests, which can cool the climate by increasing surface albedo.[55] However, a reduction in snow cover with climate warming lessens the importance of the forest–albedo feedback.[56] The effect of burning on surface albedo varies geographically, so these results are not generalizable across

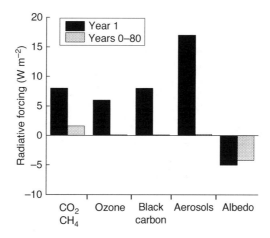

Figure 15.6 Radiative forcing associated with wildfire in interior Alaska. Shown are the radiative forcing from CO_2 and methane (CH_4), ozone, black carbon, aerosols, and surface albedo. Radiative forcing is shown for the first year and averaged over 80 years. Data from ref. (54).

the circumpolar boreal forest. High-intensity crown fires that kill most trees are common in the North American boreal forest and have a different impact on albedo than do low-intensity surface fires, which are common in Eurasia.[57] The deciduous larch forests of eastern Siberia, for example, have a different postfire recovery than in North America.[58]

The boreal forest has been identified as a region of possible tipping point.[59] Forest dieback in a warmer climate, along with logging, wildfire, and insect outbreaks, lessen the capacity of the forests to store carbon – possibly even shifting them from a carbon sink to a source – thereby accelerating the warming.[60] The forest–albedo feedback and loss of the forest–aerosol feedback could be additional destabilizing mechanisms. Inherent in this is the notion of coupled climate–vegetation dynamics in which the geographic extent of the boreal forest both affects and is affected by climate. Ecologists have historically thought that climate determines the geography and functioning of the boreal forest.[61] It is likely, however, that the boreal forest creates a climate favorable for its growth. For example, the present-day northern and southern geographic boundaries of the boreal forest correlate with the position of the July 13°C and 18°C isotherms, respectively, across much of the boreal forest. In climate model simulations that remove the entire circumpolar boreal forest, temperatures cool and the isotherms shift far to the south beyond the present-day biogeography of the forest.[62] This suggests that the cooling caused by deforestation is sufficient to prevent forest regrowth in much of the presently forested area. Loss of the forest may initiate irreversible feedback in which the forest does not recover and the treeline moves progressively southward.

Temperate Forests

Temperate forests encompass a broad variety of broadleaf deciduous and needleleaf evergreen trees in the midlatitudes situated between tropical and boreal forests (Figure 8.3). Growing in a diverse range of climates from warm to cold and mesic to arid, they further complicate the climate services of forests from albedo, evapotranspiration, and carbon storage. Large areas of Europe and the United States have been deforested over many centuries of human inhabitation. Other regions have been reforested following farm abandonment, and still more regions are proposed as locations for large-scale reforestation and afforestation to mitigate greenhouse gas emissions. Disturbance from wildfires and insects threatens temperate forests throughout much of the world.

The outcome of historical deforestation is obtained from climate models. The most recent generation of models shows that land-cover change in temperate regions of North America and Eurasia since the mid-1800s, during which time the area of forests decreased and croplands and pastures increased, has reduced winter temperature and increased summer daytime temperature in those regions.[63] The winter cooling occurs from increased albedo with deforestation. Although the winter cooling is robust across various models, larger differences occur among models in their summer temperature, related in part to differences in evapotranspiration and the partitioning of net radiation into sensible and latent heat fluxes.[64] A general understanding from models is that temperate forests are transitional between the annual mean warming of tropical deforestation and the cooling of boreal deforestation.[65] A key distinction is between northern temperate forests growing in seasonally snowy climates and southern forests in warmer climates. Analyses of local air temperature measured over forests and open fields and also satellite measurements of land surface temperature show a similar latitudinal change, though the exact temperature patterns differ from models for reasons discussed in Chapter 14.[66]

Regional studies of European forests further highlight the difficulty in ascertaining forest influences and the net balance of albedo, evapotranspiration, and surface roughness. Comparison of a non-forested, grassland Europe and a maximally forested Europe shows that forests warm winter and spring temperatures over much of the continent by decreasing surface albedo.[67] The effect of forests on summer temperature differs depending on the particular model (either widespread cooling or widespread warming). The contrast in evapotranspiration between grassland and forest is one cause of the model discrepancies, as are atmospheric processes such as cloud feedbacks. Climate benefits of forests may be found in their effects on clouds and precipitation. Observations show an increase in clouds over forested regions of western Europe.[68] Rainfall measurements reveal that converting agricultural land to forest increases precipitation in Europe, both locally (during winter) in the forested location and downwind (during summer).[69] The physical mechanism for this may be that forests create more turbulence and slow the movement of moisture-laden air masses. Furthermore, higher evapotranspiration over forests contributes moisture that then precipitates downwind of the forests. The authors of the study caution, however, that their analysis is based on spatial correlations between precipitation and forest cover rather than temporal change in forests. This reasoning for an influence of forests, as well as the cautionary caveat, is remarkably similar to the arguments for and against forest influences during the late-nineteenth-century forest–rainfall controversy.

Deforestation in the United States has also been studied with models. A key issue is what type of land cover replaces the forests. Productive crops with a summer growing season and fall harvest can cool the climate by increasing surface albedo and by transpiring substantial quantities of water.[70] Observations suggest that intensification of agriculture in the Midwest has decreased temperature.[71] In contrast, portions of northeastern United States have been reforested following farm abandonment, which warms winter temperatures.[72] Analyses of paired eddy covariance flux towers located in forests and grasslands in the eastern United States give insights to the physical processes controlling surface temperature.[73] In the forests studied, the forests are generally cooler than grasslands throughout the year, with larger cooling during the summer growing season. Warming from the lower albedo of forests is offset by ecophysiological differences (e.g., leaf area, stomatal conductance) and aerodynamics (surface roughness) that cool temperature. The high evapotranspiration of forests contributes to the summer cooling. The large sensible heat flux from forests, which are taller and have more turbulent mixing compared with grassland, also contributes to cooling. Other studies also point to the importance of surface roughness in cooling forest temperatures during the summer.[74] Further observations from paired flux towers (open versus forest) in North America find an increase in summertime net radiation, evapotranspiration, and sensible heat at forests compared with open sites.[75] Carbon sequestration contributes to the climate benefits of US forests. In eastern forests, loss of carbon and reduced productivity following deforestation outweigh the albedo cooling for a net warming effect from forest loss, suggesting a high priority to protect these forests, but not so in the snowy climates of the Rocky Mountain western forests.[76]

Large-scale ecological disturbances such as wildfire and insect defoliation alter climate by disrupting ecosystem functions. Forest dieback in western North America illustrates the complexities and nuances of forest influences. The mountain pine beetle epidemic in western North America has killed tens of millions of hectares of forests in Canada and the United States.[77] The loss of trees reduces carbon uptake by forests and alters biogeochemical cycles.[78] Still more consequences are seen in energy fluxes and the hydrologic cycle. Surface albedo increases, most noticeably in winter and spring when snow is on the ground.[79] Satellite-based measurements show that the canopy dieback decreases summertime evapotranspiration, which warms the surface climate because energy that previously fueled the evaporation of water instead heats the land.[80] Decreased evapotranspiration should increase water yield, which is the amount of water running off to streams. Catchment measurements have provided variable results, with some studies finding little change in water yield after tree mortality.[81] In fact, however, multiple meteorological, hydrological, and ecological processes come into play that both increase and decrease water yield (Figure 15.7).[82] The loss of needles reduces transpiration, and defoliated canopies also intercept less rain and snow. The open canopy allows more snow to reach the ground and accumulate in the snowpack. However, more solar radiation penetrates to the ground, resulting in earlier and larger peak snowmelt and increased springtime streamflow. These processes act to increase water yield. Counter to this, greater soil evaporation and snow sublimation (direct conversion from frozen water to vapor) can increase in the sunnier and winder

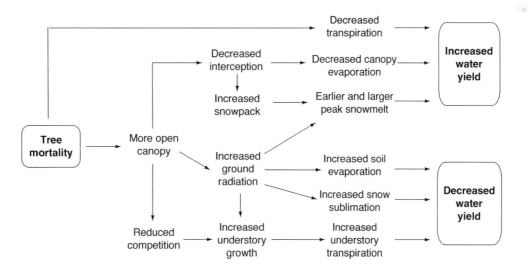

Figure 15.7 Processes affecting water yield.
Pine beetles that kill trees and open the canopy can increase or decrease water yield by altering evapotranspiration and snowmelt. Adapted from ref. (82).

understory. Transpiration can increase because reduced competition with overstory trees and more sunlight at the forest floor promote dense understory growth. These processes decrease water yield by increasing evapotranspiration losses. The complexities preclude simple generalities. Even predicting whether a change in forest cover increases or delays snowmelt is exceedingly difficult, despite the well-known fact that forests affect the accumulation and melting of snow.[83]

Wildfire activity has also increased in the western United States as a result of climate change.[84] In addition to the carbon cycle and biogeophysical effects of forest loss, burning releases aerosols into the atmosphere. Black carbon (soot) deposited on snow, as well as charred plant debris, blackens the snow, decreasing its albedo and causing more solar radiation to be absorbed. Greater solar heating of the snowpack, combined with loss of shading from the tree cover, accelerates spring snowmelt by several days.[85] The smoke and airborne particles from wildfires can alter cloud properties. More numerous and smaller cloud droplets are found in smoke-filled clouds over the western United States compared with those in clean air, which should, consequently, cause them to reflect more sunlight and produce less rain.[86] Forest fire emissions affect not only air quality, but can also influence weather itself such that the accuracy of weather forecasts is degraded if the emissions are not accounted for in the models.[87]

China has undertaken a massive program to protect and restore forests through afforestation and by reducing timber harvests.[88] The greening of China is evident in satellite measurements of vegetation greenness and leaf area.[89] Remote sensing satellite measurements show that afforestation decreases daytime surface temperature because of enhanced evapotranspiration.[90] Modeling studies find that large-scale afforestation and vegetation greening cool temperature, though they differ in the magnitude, region, and seasonality of the temperature change.[91] Southern regions of China may experience winter warming from afforestation. Precipitation may also be impacted, increasing in some regions and unchanged in others depending on the atmospheric controls of precipitation.[92] The higher evapotranspiration accompanying afforestation is likely to decrease runoff to streams and reduce water availability, and some regions may already be at vegetation capacity for sustainable water yield.[93] A further outcome is that emissions of BVOCs have likely increased because of afforestation, with consequences for air quality.[94]

Northern Africa

Northern Africa from the equator to the Sahara is a region of decreasing rainfall and increasing aridity progressively northward (Figure 15.8).[95] Tropical evergreen forests grow in the south where rainfall is plentiful, giving way to woody savanna, grassland, dry shrubland, and desert as the climate becomes increasingly more arid. It is a region of widely studied vegetation influences on climate, from the past through to the present. The dryness of the desert and lack of vegetation have long been considered interrelated. Buffon, in *Des époques de la nature* (1778), thought that planting trees would lessen the desert heat and bring rain.[96] In *Ansichten der Natur* (1808), Humboldt advanced a theory in which the barren desert landscape arises because the sparse vegetation contributes to the low rainfall.[97] George Perkins Marsh's 1864 call to action to prevent deforestation invoked the imagery of the dry, barren desert landscape, as did many other nineteenth-century forest enthusiasts.[98] In his 1853 study of forests and climate, Becquerel speculated that a forested North Africa, by lessening the desert heat, would alter the climate of Europe.[99] Of particular interest to the modern science is the Sahel – the region along the southern boundary of the Sahara – which exhibits a close coupling between vegetation and rain. Degradation of

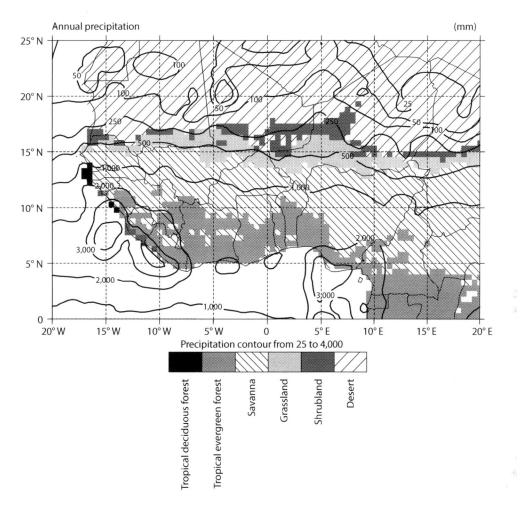

Figure 15.8 Land cover and rainfall in northern Africa.
Shown are vegetation types in a region of western Africa between the equator and latitude 25° N and from longitude 20° W to 20° E. Vegetation types are tropical deciduous and evergreen forest, savanna, grassland, shrubland, and desert. Overlain on the vegetation distribution are contours of annual rainfall in millimeters (mm). Precipitation decreases from in excess of 3,000 mm near the equation to 25 mm in the Sahara. The Sahel is the region between latitudes 13° N and 20° N and longitudes 15° W and 20° E. Reproduced from ref. (95).

vegetation during drought feeds back to accentuate the drought. Conversely, greening of the Sahara, such as has happened in past wetter climates, reinforces the rain. Throughout the region, there is a possibility of two alternative stable states: a wet–green ecoclimate with much rain and abundant vegetation; or a dry–desert ecoclimate lacking in rain and with sparse vegetation. For these reasons, the region is thought of as another possible climate tipping point.[100]

The initial concept for vegetation–rainfall feedback in northern Africa proposed by Jule Charney and colleagues was based on the higher albedo of exposed desert soil compared with vegetation.[101] The albedo of the land increases with the loss of vegetation, which initiates a series of atmospheric changes that decrease rainfall (Figure 15.9).[102] There is less net radiation at the surface to heat the atmosphere. Less

Figure 15.9 Positive feedback between
vegetation and rain in the Sahel.
A decrease in vegetation increases
surface albedo and causes changes in the
atmosphere that reduce rainfall.
Modified from ref. (102).

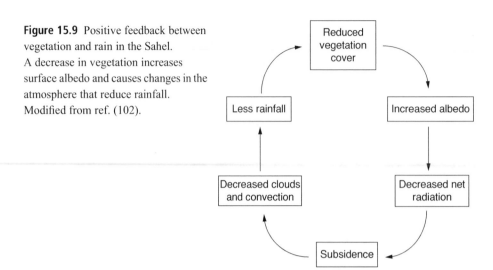

surface heating promotes subsidence of air aloft. Subsidence decreases cloud formation and convection, leading to less rainfall. The drier soil further reduces vegetation cover in positive feedback. Counter to this, the reduction in clouds causes more solar radiation to reach the ground. However, the downward flux of longwave radiation decreases with the drier atmosphere so that the net radiation at the surface decreases. Evapotranspiration was subsequently recognized as an additional feedback.[103] Charney's theory of vegetation–rainfall coupling is notable in that Charney, a preeminent leader in advancing dynamic meteorology and numerical weather prediction, provided a much needed dynamical framework for the reemerging science of forest–climate influences in the 1970s. Charney called the science "biogeophysics" to emphasize biological control of energy exchanges with the atmosphere. It was an inability to conceptualize surface forcing of the atmosphere that had led to the heated argumentation one hundred years earlier.

Further studies built upon and refined the initial concept. A more complete depiction of land–atmosphere coupling includes changes in addition to surface albedo. Sparse vegetation not only increases albedo, but also decreases surface roughness and leaf area. Soil degradation reduces soil water-holding capacity. Numerous climate model simulations that replace dryland vegetation with degraded or desert vegetation find decreased precipitation in the Sahel arising from increased albedo, reduced evapotranspiration, and changes in atmospheric circulation.[104] Conversely, reforesting degraded lands or afforesting drylands can have the reverse impact.[105] Additional modeling studies in which the growth and survival of vegetation depends on climate, so that the vegetation changes in response to climate, reveal a dynamically coupled system.[106] Desert expansion in a dry climate feeds back to reinforce the dryness and further the desert land. Conversely, a vegetated landscape in response to a wet climate augments the wetness and enables still more expansive vegetation. The region may have two alternative stable states of either wet or dry climate. While models provide a strong theoretical case for vegetation–rainfall interactions, they represent idealized changes in land cover. As with tropical deforestation, the actual land degradation occurs on a much smaller spatial scale so that there is a discrepancy in the scale at which observations of change can be obtained and the scales at which the region is modeled.

The West African Sahel experienced a severe drought during the latter part of the twentieth century. Charney framed his theory as a human-caused explanation of the drought in which overgrazing and land degradation, by reducing vegetation cover and increasing surface albedo, alter precipitation. A more complete understanding now shows that the Sahel is a region of closely coupled interactions among ocean, atmosphere, and land.[107] Precipitation in the region is characterized by a summer monsoon driven by the thermal contrast between land and ocean, and drought is linked to anomalies in tropical sea surface temperatures.[108] However, while sea surface temperatures drive decadal precipitation variability, vegetation and soil moisture enhance the variability and the severity of drought.[109] Decreased rainfall leads to drier soils and reduced vegetation cover, which in turn leads to higher surface albedo and reduced evapotranspiration. This weakens atmospheric circulation by reducing the energy and water flux in the atmosphere, resulting in less rainfall.

Charney's mechanism by which surface albedo controls precipitation is central to understanding the wetter Saharan climate of the past. About 6,000 years ago, the climate of North Africa was much wetter than today.[110] Changes in Earth's orbital geometry increased summer solar radiation, heated the land, and strengthened the summer monsoon. Paleobotanical data indicate grasses and shrubs covered much of North Africa, including areas that are presently desert, as a result of the wetter climate.[111] By altering surface albedo and evapotranspiration, the green Sahara amplified the precipitation increase.[112] The period 6,000 years ago was part of a longer time beginning several thousand years earlier when the climate of North Africa was wetter than today. Then, the climate abruptly became drier, the vegetation died back, and the land became desert. A reduction in summer insolation weakened the summer monsoon and loss of vegetation amplified the drying. At some critical threshold, the system collapsed and switched to the dry desert of today.[113] The abrupt change over a period as short as a few centuries is further evidence of two different stable states: a wet climate with vegetation or a dry climate with desert. This finding applies specifically for the western Sahara, but also more broadly to West Africa spanning tropical forests to the desert.[114]

Interest in purposely modifying the Sahara to increase precipitation, first voiced by Buffon more than two centuries ago, continues today. Covering sandy desert areas along the Mediterranean coast of Africa with asphalt so that the albedo of the land is reduced was proposed in the 1960s as a means to promote rainfall.[115] A modern revival of this idea advocated planting vegetation in the dry landscape to generate mesoscale circulations that increase precipitation.[116] Many subsequent studies further propose planting trees to increase rain, restore degraded lands, and remove CO_2 from the atmosphere.[117] Other studies find that placing large-scale solar farms across the Sahara to harness solar energy has the additional benefit of increasing rainfall by reducing surface albedo.[118] Expansion of vegetation in response to the wetter climate reinforces the rainfall benefits of the solar panels. Wind turbines produce a similar rainfall boost by increasing surface roughness.[119]

Studies of a pine forest in the semiarid landscape of Israel point to the complexities of afforesting the desert.[120] Even with the dry environment, the forest gains carbon at a high rate comparable to European pine forests. This biogeochemical climate benefit is offset, however, by the low albedo of the forest. In the semiarid climate, the clear sky allows strong solar heating of the land, and the forest absorbs more solar radiation than the surrounding sparsely vegetated shrubland. Despite more net radiation and low evaporative cooling, the temperature of the forest is several degrees cooler, on average, than the shrubland. Large sensible heat fluxes carry the heat away from the forest canopy into the overlying air. The forest, with its low tree density and open canopy, has high aerodynamic coupling with the atmosphere.[121] The forest achieves a cool surface by having a small aerodynamic resistance to heat transfer. In this water-limited environment, the turbulent transfer of heat, not evapotranspiration, is the primary mechanism to cool the forest. The large heating of the atmosphere can, in the right conditions, generate secondary circulations of updrafts and downdrafts over the forest and shrubland.[122] A further consequence of forests in the region is that they reduce water yield (i.e., precipitation minus evapotranspiration) because they evaporate more water than native shrublands.[123]

∎

A central tenet of the forest–climate question is that forests can be managed to control climate. The same idea arises in today's advocacy of forests as providing a natural solution to climate change. Yet the issues that muddied the historical debate are still evident today. Forest influences vary regionally, at different times of the year, even differing between day and night, and all forests are not equal in their climate benefits. The climate influences of forests defy simple generalizations and need to be considered in terms of the full suite of biogeophysical and biogeochemical processes. Conflicting opinions arise when only some processes are considered. The next chapter examines forests as solutions to climate change in light of the many nuances of forest type, background climate, and what processes to consider.

16   Climate-Smart Forests

Limiting planetary warming to 2°C or less requires reducing emissions of CO_2 to the atmosphere, and even removing CO_2 to lower the atmospheric concentration. Principal among the climate services of forests is their carbon storage. This is the basis upon which forest advocates call to protect, restore, and manage forests to mitigate the harmful effects of climate change. Large areas of new forests must be planted to offset anthropogenic CO_2 emissions, and the way in which land is used for agriculture must also be transformed.[1] The possibility of reforesting the world to alleviate the climate problem is intoxicating, and its appeal, much like that of the nineteenth-century call to plant trees to ensure rainfall, has garnered the interest of scientists, governments, and the public. Yet the climate services of forests extend beyond just carbon sequestration. As the preceding chapters have shown, forests influence climate through many biogeophysical and biogeochemical mechanisms, often in ways that augment the carbon benefits and sometimes to the contrary. To the carbon benefits of forests must be added their influence on temperature and precipitation through albedo, surface roughness, evapotranspiration, and biogenic aerosols. If the potential of forests to lessen planetary warming over the coming century is to be realized, their many influences on climate must be synthesized into an integrated understanding. In this, tropical forests are readily recognized as beneficial for climate. Temperate and boreal forests, on the other hand, have confounding influences that lessen their utility for climate security.

Nature-Based Climate Solutions

Nature-based, or natural, climate solutions use forests, wetlands, grasslands, and agricultural lands to alleviate the climate problem with better ecosystem conservation, restoration, and management.[2] The solutions are framed in terms of enhancing terrestrial productivity or the retention of carbon in soils to remove CO_2 from the atmosphere or through avoided emissions by reduced deforestation and improved land management. Because carbon accumulates in ecosystems over time, nature-based solutions are assessed for specified time periods (e.g., the CO_2 benefits by 2050). Natural climate solutions are one way to limit planetary warming below specified targets, and forests provide a critical means to achieve CO_2 mitigation goals. Additional solutions use bioenergy plantations as an alternative to fossil fuels.

Forest solutions include avoided deforestation, restoration of degraded forests, reforestation of cleared lands, afforestation of non-forest lands, management of natural forests, and management of forest plantations. Reforestation and avoided deforestation routinely rank high for their mitigation potential. Conserving, restoring, and managing the world's natural ecosystems and agricultural lands can provide one-third of the CO_2 mitigation needed to keep planetary warming below 2°C, and forests account for two-thirds of the attainable mitigation.[3] In the tropics, natural climate solutions can mitigate over half of the national emissions in the majority of countries and completely offset national emissions in one-quarter of tropical countries.[4] Much of the gain is from reforestation and avoided deforestation. In the United States, one-fifth of net annual CO_2 emissions (2016) can be offset with ecosystems.[5] Reforestation has the single largest maximum mitigation potential, followed by forest management. These findings have led to the conclusion that "regreening the planet through conservation, restoration, and improved land management is a necessary step for our transition to a carbon neutral global economy and a stable climate."[6]

Natural climate solutions value forests for their CO_2 mitigation potential, but some analyses acknowledge the contrary warming influence of forest albedo. Pity, then, the spruce, pine, fir, larch, and other trees that collectively form the boreal forest. The circumpolar boreal forest spans the northern lands of North America, Europe, and Russia with a global area comparable to tropical forest (Figure 8.3).[7] It is central to the ecology, economy, and cultural identity of those lands, but the low albedo of the forest warms climate and lessens its efficacy as a solution to climate change. The boreal forest was thrust onto the public consciousness on December 8, 1992, when the *USA Today* newspaper ran a graphic with the headline "Northern forests warm the Earth" (Figure 16.1).[8] Further text explained that cutting down the forest cools the climate by reflecting more solar radiation and would make the summer temperature of Washington, DC, comparable to Boston. Thirty years of research has found new ways that boreal forests influence climate through carbon storage, atmospheric water vapor, and biogenic aerosols, but the net outcome on climate – whether the forests warm or cool climate – is still unknown. As a result, boreal forests are excluded from

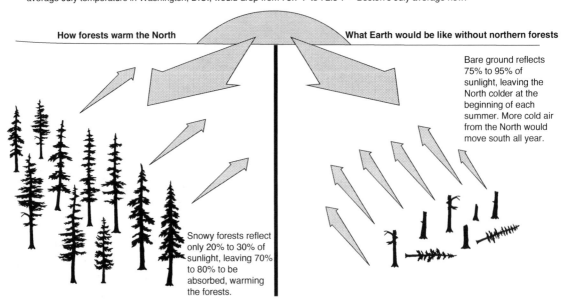

Figure 16.1 Boreal forests warm the planet.
A graphic was published in the December 8, 1992, edition of *USA Today* showing the climate warming caused by northern forests. See ref. (8) for the published graphic.

consideration of nature-based climate solutions.[9] Reforestation of cleared lands or protection from deforestation are not worthy of consideration for these forests. Other analyses find that Canadian forests are less helpful to fight climate change when their effects on albedo are considered.[10] Temperate conifer forests, too, have a lower albedo than deciduous forests, and nature-based climate solutions in the United States lessen the carbon benefit of conifer forests to account for unwanted albedo-induced warming.[11]

Reforesting the Planet

Human transformation of the land for agriculture, forest products, and other uses extends back several thousand years.[12] Over the past millennium, from 850 to 2015 CE, the area of primary land undisturbed by humans is estimated to have decreased from 125 million square kilometers (km^2) to 50 million km^2 as forests were cleared and the prairie was broken for agricultural uses (Figure 16.2).[13] Forest area decreased by 10 million km^2 over this time, which is approximately equal to the land area of the United States – from 47 million km^2 in 850 to 37 million km^2 in 2015. The area of cropland increased from 1.7 million km^2 to 4.3 million km^2 between 850 and 1800 and further increased to 15.9 million km^2 in 2015. Pastureland increased from 3.3 million km^2 (850) to 9.2 million km^2 (1800) and to 32.8 million km^2 (2015). It is estimated that more than 15 billion trees are cut down annually and that humans have decreased the number of trees by 46 percent worldwide.[14]

Future human activities will continue to alter the face of the planet as more land is appropriated to grow crops, raise livestock, and extract forest products. The notion of forests as natural climate solutions will further change humankind's relationship with forests. One study optimistically estimated that 9 million km^2 of land could be potentially reforested for CO_2 sequestration.[15] The shared socioeconomic pathways (SSPs) introduced in Chapter 8 provide plausible trajectories of the future. They describe not only scenarios of greenhouse gas emissions; they also describe the underlying population trends, economic growth, social and cultural change, technological advances, energy consumption, and land use that drive the emissions.[16] The SSPs depict changes in land use arising from demand for food, timber, and bioenergy, and they account for possible regulations, changes in agricultural productivity, concern for environmental impacts, and

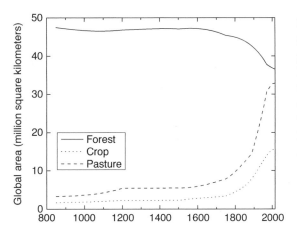

Figure 16.2 Global area of forests, crops, and pastures, 850–2015 CE. Forest area has declined throughout the historical era as the area of land used to grow crops or graze livestock has increased. Adapted from ref. (13).

Figure 16.3 Change in global forest area over the twenty-first century. Shown are four shared socioeconomic pathways (SSPs) and the target radiative forcing, denoted as SSPx-y, for 2005–2100. The change in area is relative to 2005. For data source, see ref. (18).

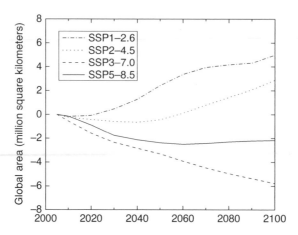

future agricultural and forestry markets. In addition, they consider the mitigation provided by replacement of fossil fuels with clean energy alternatives, advances in energy conservation and efficiency, and use of bioenergy to reduce greenhouse gas emissions. Changes in the land-use sector to reduce emissions include improvements in agricultural practices and utilizing the carbon storage potential of forests through reforestation, afforestation, and reduced deforestation. As a result, the SSPs provide narratives of changes in the area of land in forest, crop, and pasture through the twenty-first century.[17]

The various SSPs provide a broad range of possible futures in terms of land use over the twenty-first century, as shown in Figure 16.3 for forested land.[18] Two SSPs continue the historical trend to deforest the planet. Pathway SSP3 has large global population growth, low improvements in agricultural productivity, little concern for environmental protection, and reliance on meat-rich diets. Growing demand for land to grow crops and raise livestock increases the area of agricultural land and reduces the area of forests compared with 2005. Without concern for climate change mitigation in SSP3-7.0, with a radiative forcing of 7 W m^{-2}, atmospheric CO_2 reaches 867 parts per million (ppm) in 2100 (Figure 8.7a). The loss of 5.8 million km^2 of forest by 2100 is the largest among the SSPs. It is an area equal to more than one-half the total loss of forests over the historical era, and it exceeds the land area of the European Union. SSP5 describes high economic growth driven by high reliance on fossil fuels, with a meat-rich diet and without strong land-use regulation. Crop yield improves so that cropland area increases though the middle of the century to meet demand but thereafter decreases. Pastureland declines throughout the century. Tropical deforestation continues but decreases over time. Without mitigation considerations, SSP5-8.5 has a CO_2 concentration of 1,135 ppm by 2100, the largest of the SSPs (Figure 8.7a). Forest loss peaks in 2060, with a net decrease in forest area of 2.2 million km^2 by the end of the century.

In contrast, SSP1-2.6 and SSP2-4.5 are climate stabilization scenarios with reduced greenhouse gas emission, less planetary warming, and increase in forest area compared with SSP3-7.0 and SSP5-8.5. Scenario SSP1 describes a future with large emphasis on sustainability. Improvements in agricultural technologies that increase crop yields and changes in diet and food consumption mean that some agricultural lands are abandoned. Land use is highly regulated with, for example, reduced tropical deforestation. Scenario SSP1-2.6 is a mitigation pathway that further relies on bioenergy, carbon storage in ecosystems, avoided deforestation, and restoration of degraded forests to reduce greenhouse gas emissions. It has low CO_2 concentration in 2100 (less than 450 ppm; Figure 8.7a). The area of land used for agricultural purposes decreases, especially pastureland, and is lower than in the other SSPs. Forest area increases by 5 million km^2 at the end of the century. In SSP2, social, economic, and technological trends continue as they have in the past. Deforestation continues, but at a slower rate than in the present day. Climate change mitigation is delayed until the middle of the century. Scenario SSP2-4.5 is one of climate stabilization, but at higher CO_2 concentration than SSP1-2.6. Forest loss continues through the middle of the century, followed by an increase in forest area through to 2100; there is a net gain of 2.9 million km^2 by the end of the century. Limiting planetary warming to 1.5°C would require additional transformations in land use – up to 9.5 million km^2 of new forests by 2050.[19]

Biogeophysical Climate Influences

Nature-based climate solutions are framed in terms of CO_2 emissions. The intent is to enhance carbon storage in terrestrial ecosystems (negative emissions) or reduce land use emissions (avoided emissions), and the metric is the CO_2 flux. A financial cost (e.g., 100 US dollars per metric ton) is assigned to CO_2 emissions to assess the feasibility and cost-effectiveness of various solutions.[20] The SSPs shown in Figure 16.3 to mitigate planetary warming likewise consider only CO_2 emissions and other greenhouse gases. These solutions do not generally consider the biogeophysical effects of forests on albedo, surface roughness, and evapotranspiration and their consequences for temperature, precipitation, cloud cover, and atmospheric circulation. Yet the changes in land use required to mitigate anthropogenic CO_2 emissions affect climate in many ways. Where and how forests, croplands, pasturelands, and other vegetation are grown and managed can lessen or exacerbate warming depending on the region and the land cover transition.[21] For example, afforestation reduces atmospheric CO_2 and lessens planetary warming compared with no intervention, but is countered by biogeophysical warming in northern temperate and boreal forests.[22] In this framework, tropical forests provide greater reduction in planetary warming than do northern temperate and boreal forests.

The challenge with obtaining the net climate outcome of forests is in quantifying and combining the various influences. Atmospheric CO_2 affects the radiative balance of the atmosphere and has a direct relationship with planetary temperature. It is a gas that is well mixed in the atmosphere so that its influence is felt at the global scale. The carbon emission or sequestration of a particular forest stand does not alter the local CO_2 concentration and temperature, but rather contributes to the global concentration and planetary warming. The effect of CO_2 on planetary temperature is measured by its radiative forcing. Radiative forcing is the change in Earth's energy balance resulting from, in this case, an increase in atmospheric CO_2 concentration and is formally assessed by the change in net (downward minus upward) radiation (solar plus longwave) at the boundary between the troposphere and the stratosphere. Surface albedo also affects the radiative balance, and some studies have combined the albedo and carbon influences of forests into either a common radiative forcing metric or the change in CO_2 concentration that provides the equivalent albedo-caused radiative forcing.[23] Carbon uptake by forests provides a negative radiative forcing that reduces atmospheric CO_2 and cools the planet. The low albedo of forests is a positive radiative forcing that warms the planet.

Other biogeophysical processes such as surface roughness and evapotranspiration affect the turbulent exchange of heat and moisture with the atmosphere. Some analyses have combined the radiative forcing of

CO_2 with surface energy fluxes (net radiation, evapotranspiration) into an index of climate regulation value.[24] The difficulty is that biogeophysical influences on climate are nonradiative and cannot be directly combined with the radiative forcing of CO_2.[25] Additionally, biogeophysical effects manifest at local to regional scales while CO_2 is felt at the global scale. Biogeophysical effects also occur over short timescales. For example, the changes in albedo and evapotranspiration with reforestation happen rapidly, within the first few years. Carbon sequestration occurs over several decades or more as the forest regrows. Consequently, radiative forcing or other such climate service metrics must be calculated for a specific time interval, and the results can change depending on the time window. This is evident in the radiative forcing of boreal wildfires shown in Figure 15.6. The biogeochemical radiative forcing is large during the first year, when there is an initial pulse of CO_2, ozone, black carbon, and aerosols to the atmosphere during combustion, but much less over an 80-year window. Instead, surface albedo provides the dominant radiative forcing over the longer time frame.

Another approach is to examine changes in land surface temperature arising from land cover conversion. Land surface temperature integrates albedo, surface roughness, evapotranspiration, and other surface features and is observable from satellites. Many studies have used differences in surface temperature between forests and grasslands as an indicator of the climate effects of forests.[26] An extension of this approach is to transform the biogeophysical effects on surface temperature into a CO_2 equivalent for comparison with the carbon benefits of forests. If, for example, reforestation cools the surface, what is the change in atmospheric CO_2 that would produce an equivalent decrease in temperature? This methodology puts the CO_2 and biogeophysical influences into the same scientific units of measurement and allows them to be compared or combined. Carbon services are assessed based on the carbon content of different ecosystems and the conversion, for example, of grassland to forest. Biogeophysical services are assessed based on land surface temperature obtained from satellites. This methodology only accounts for local temperature, not the remote influences of forests. Nonetheless, it is an observationally based measure of climate services and provides an instructive comparison of different regions.

Figure 16.4 provides an example of such a calculation.[27] The study calculated the effectiveness of two forest policies: planting trees on current grassland, shrubland, and cropland (termed forestation); and avoided deforestation of existing forests (termed conservation). The timeframe was 80 years (to 2100). The results again point to the importance of tropical rain forests. The climate benefit of a tropical forest is, on average, almost four times greater than the same area of temperate and boreal forest. The greatest contribution arises from the large carbon storage of tropical forests, but biogeophysical cooling augments the biogeochemical cooling of carbon uptake. In temperate and boreal forests, biogeophysical influences are less and can even conflict with carbon benefits depending on location and time of year. At an annual timescale, biogeophysical warming reduces the benefits of carbon cooling in more than one-fifth of continental temperate regions and more than one-third of boreal regions. North of latitude 56° N, biogeophysical processes cause warming in the annual mean. In summer, however, biogeophysical processes produce local cooling in temperate and boreal forests. Overall, the analysis points to the need to include biogeophysical processes, not just carbon storage, when assessing the climate services of forests. Accounting for biogeophysical effects increases the climate value of tropical forests and further incentivizes their conservation.[28] Forestation and conservation solutions that only account for CO_2 emissions undervalue the benefits of tropical forests and overvalue temperate and boreal forests. Planting trees in the latter regions may risk warming the climate because of their biogeophysical influences.

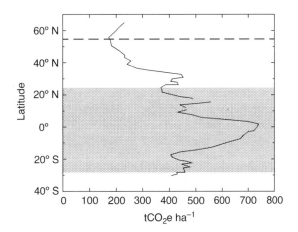

Figure 16.4 Climate effect of forestation and forest conservation.
The mitigation potential of forests at the end of the century was measured in terms of CO_2 uptake and the CO_2 equivalent of the local biogeophysical effect. Shown is the annual response in terms of metric tons of CO_2 equivalent per hectare. The dashed line at 56.1° N is the latitude north of which the biogeophysical effect is counter to the carbon benefit. The shaded region from latitude 22.9° N to 29.2° S is where strong biogeophysical and carbon effects act together. Adapted from ref. (27).

What Is a Climate-Smart Forest?

It is not just deforestation or replanting cleared forests that matters for climate. Two-thirds to three-quarters of the world's forests are managed for timber, wood products, or other reasons.[29] Forest harvesting, the species of trees grown, and other silviculture practices impact climate in similar ways as clearing forests. The practice of forestry itself needs to be rethought so as to lessen impacts on climate while still providing the goods and services that improve the physical environment, protect wildlife, and enrich human societies.

Climate-smart forestry is sustainable forest management that enhances the resilience and provisioning of goods and services by forests in the face of a changing climate.[30] An exact definition is lacking, but it can be described as forest management "to protect and enhance the potential of forests to adapt to and mitigate climate change. The aim is to sustain ecosystem integrity and functions and to ensure the continuous delivery of ecosystem goods and services, while minimizing the impact of climate-induced changes."[31] Climate-smart forestry strives to reduce the adverse impacts of climate change on forests so as to maintain healthy forests while also mitigating planetary warming. Mitigation, in this context, means to increase carbon sequestration in forest biomass and harvested wood products, as well as avoid emissions through substitution of wood for materials and energy. Like nature-based climate solutions, climate-smart forestry recognizes the extraordinary climate value of forests and provides the management practices to achieve the desired outcome; but likewise, it focuses on carbon storage to the exclusion of biogeophysical processes. Forests need to be managed for these climate influences in addition to carbon storage. Maximizing climate benefits entails many purposeful interventions to forests, will require resolving conflicting demands on forest uses, and may give rise to forests quite unlike natural forests.

Consider, for instance, carbon sequestration. The notion of increasing the productivity of forests to remove CO_2 from the atmosphere is readily appealing. It harnesses nature to solve a human-caused problem, and it lends prominence to the science of forest ecology. It also raises problematic questions. The prioritization of carbon uptake likely values some tree species and some types of forests over others. There are two contrasting life strategies to ensure survival.[32] Some trees can be fast growing, but they also tend to be short lived. These species are favored where a disturbance has created a large opening in the forest and are referred to as early successional or pioneer species. They dominate the first few decades of forest recovery. Biomass growth is rapid, but the lifetime accumulation of carbon may be relatively small. Other species use a strategy of slow growth over a long lifetime. These are the species that come to dominate old-growth forests as the early successional species die out. The carbon benefits of slow growing, long-lived trees will only be achieved over many decades. Another way to store carbon and offset anthropogenic emissions is to grow trees for bioenergy, which is touted as a clean energy alternative to fossil fuels. Bioenergy plantations are likely to be industrial forests – monocultures of non-native species with short rotation cycles of young, small, fast-growing trees to draw CO_2 from the air, quite unlike natural forests. Monoculture plantations, in contrast to biodiverse natural forests, will likely contribute to loss of native flora and fauna, and they are vulnerable to disease and pests.

It is not just productivity that determines carbon accumulation. Carbon dioxide is returned to the atmosphere during respiration by trees and soil microorganisms. Forests can be managed to increase photosynthesis by selecting the right species, opening the canopy to provide more light, applying nutrients, and with other means, but any increase in photosynthesis will be accompanied by an increase in respiration loss that lessens the photosynthetic carbon gain. The key issue is the rate at which the forest accumulates carbon in the trees themselves but also in the soil, and the permanence time of the carbon. To this end, there is a debate about the most effective means to maximize carbon storage: protect intact forests, restore degraded forests to a natural state, reforest cleared lands, establish forest plantations, or manage secondary forests.[33]

The long-term fate of the carbon taken up by forests does not simply involve tracking changes in woody biomass and soil carbon. Fully accounting for the carbon offsets provided by forests is difficult and assessing the carbon benefits of forests requires rigorous carbon accounting.[34] Wood products are extracted from many forests. These products emit CO_2 over their lifetime – immediately if the slash left on site is burned, over a few years for paper products, and several decades or longer for timber used in construction. Production substitution such as use of wood rather than steel in building construction poses an additional accounting complexity in that the avoided emissions of the substituted products must be quantified. Bioenergy plantations require further consideration. They are assumed to be carbon neutral because the same amount of carbon is released in combustion as is removed from the atmosphere during biomass growth. However, a carbon debt is occurred when, for example, an intact tropical forest is replaced by an oil palm plantation. This debt must be overcome before a net reduction in CO_2 is achieved.[35] In carbon accounting, the biological carbon sink measured in the storage of carbon in trees and soil is different from the net carbon sink after accounting for wood products, product substitution, and bioenergy. Some of the disagreement over natural versus managed forests to enhance carbon sequestration stems from how to account for wood products.

Bioenergy plantations, reforestation, and other natural climate solutions may harm the environment in that sufficient supplies of water and nitrogen are needed to sustain plant productivity and because of other unintended consequences.[36] A key concern is the water cost of carbon gain. One estimate is that the new forests needed to achieve a sufficient carbon sink to limit planetary warming to 2°C (3.3 petagrams carbon per year in 2100) would encompass 9.7 million square kilometers of land and use 1,040 cubic kilometers of water per year.[37] On the other hand, forests provide many environmental and socioeconomic benefits. Well-crafted forest-based climate solutions can enable adaptation to climate change and facilitate carbon storage while also providing co-benefits for many ecosystem goods and services.[38] This is the intent of climate-smart forestry. For example, restoring deforested lands and protecting intact forests can both lessen species extinctions and sequester carbon, thereby achieving both biodiversity conservation and climate mitigation goals.[39] Tropical and subtropical forests again receive high priority because of their enormous carbon storage and their location in regions of high biodiversity. To this, forest microclimates buffer the impacts of climate change on forest organisms.[40] Intentionally planting trees in tropical pasturelands, even in small contiguous patches, can provide meaningful cooling benefits to both agricultural workers and livestock, thereby lessening exposure to excessive heat.[41]

To the focus of climate-smart forestry on CO_2 mitigation can be added other characteristics of forests that affect climate. Chief among these is surface albedo. Managing the land to increase albedo is one example of purposeful geoengineering to counteract climate warming.[42] Enhancing the reflectance of leaves through crop selection or genetic modification can increase the albedo of the land. Forest management also affects albedo. The openness of the forest canopy, the amount of leaves, the type of species, and the albedo of the ground all affect the albedo of the land. In northern temperate and boreal forests, albedo generally decreases as forests age and the canopy closes or with greater abundance of conifer species (Figure 15.4). This is most prominent in winter when snow is on the ground. These forests can be managed for albedo in addition to timber and carbon.

Traditional forest management emphasizes timber production and strives to optimize the rotation cycle between wood harvests to maximize timber yield. Climate-smart forestry adds the value of carbon storage to forest management, which may necessitate longer rotations to store sufficient carbon. A longer period between logging allows trees to grow larger and store more carbon, and it also allows more carbon to accumulate in the soil. Carbon storage increases in older forests, but this climate benefit conflicts with the higher albedo of younger forests. Valuing carbon storage lengthens the rotation so that carbon accumulates in older forests, but valuing albedo favors younger forests. Economic models that value timber, the climate benefits of carbon storage, and the climate cost of a low albedo provide an optimal age to log forests. When the albedo of forests is included in the economic analysis, the optimal rotation length in northern temperate and boreal forests is shortened so as to decrease forest age and increase the proportion of young forests in the landscape.[43] Broadleaf deciduous trees have a higher albedo compared with conifers so that albedo increases as they occupy a greater proportion of the forest landscape. Conversion from conifers to broadleaf species allows for longer rotation cycles when albedo is considered.[44] Mixed broadleaf and conifer forests maintained through short rotation lengths also optimize timber production, carbon storage, and albedo.[45] For this reason, managing the boreal forest to increase the proportion of broadleaf deciduous species has been proposed as a strategy to counter climate change by increasing the albedo of the land while maintaining the terrestrial carbon sink.[46]

Managing forests to increase albedo may have other climate consequences. The shift to shorter rotation cycles with more frequent harvests creates a younger forest landscape that increases albedo, but it also changes forest composition and structure. Boreal conifer trees emit biogenic volatile organic compounds (BVOCs), which form secondary organic aerosols that are thought to cool climate. Many temperate and boreal broadleaf species are low emitters of BVOCs, which, if the aerosol cooling is proved to be significant, could negate a beneficial influence of conifers to mitigate climate warming. Surface roughness must be considered as well. The aerodynamically rough surface presented by tall forests contributes to the efficient transfer of heat and moisture with the atmosphere. Surface roughness generally increases with the height of the canopy, but the exact relationship depends on canopy closure, the amount of leaves, and their vertical distribution through the canopy.[47] Shorter, more open canopies as a result of increased harvesting will alter surface roughness and turbulent heat exchange with the atmosphere. Indeed, some studies point to reduced logging of old forests, not more frequent harvesting, to counter climate warming. Tall forests across the United States have a cooler surface temperature than shorter forests, especially during the growing season.[48] Older pine forests in temperate and boreal North America are cooler than younger forests.[49] Although albedo decreased, greater evapotranspiration and sensible heat flux with the older forests caused the surface cooling.

There is a trade-off between the water needed to support forest growth and runoff to streams.[50] Foresting vast tracts of land for carbon sequestration will increase evapotranspiration and reduce streamflow. Species conversion, such as replacement of conifer forests with broadleaf deciduous trees to increase albedo, may also impact water yield. Paired watershed experiments and eddy covariance flux towers show that, in general, temperate conifers evaporate more water annually than do broadleaf trees, in part because they retain foliage throughout the year.[51] Tree species differ in water-use strategies. Some follow an exploitive strategy in which transpiration proceeds at high rates, unchecked by stomatal closure. Other species use water more conservatively. These strategies relate to plant traits that regulate leaf gas exchange and plant hydraulic transport, and they influence land–atmosphere feedback during drought.[52] High photosynthetic rate and high hydraulic transport combine to produce high transpiration rates and delete soil moisture rapidly. During drought, the resulting decline in evapotranspiration can reinforce the drought through feedbacks with the atmosphere. Species can also differ in rooting profiles in the soil. A conservative water-use strategy combined with deep roots may sustain forest evapotranspiration and provide a cool surface temperature compared with shallow-rooted herbaceous vegetation during heatwaves and drought.[53] Increased water-use efficiency with elevated CO_2 concentration can produce carbon gain with little additional water cost and will likely change the calculus of forestation and water yield.[54]

European Forests: Case Studies

Studies of Europe's temperate and boreal forests point to the complexity of managing forests to mitigate CO_2 emission and climate change. Over the past two and a half centuries, the area of forests and the type of forests has changed considerably.[55] The forested area decreased from 1750 to 1850, when forests, mostly broadleaf trees, were cleared for agricultural land. Thereafter, abandoned farmland was reforested, oftentimes with conifers. Forest gains exceeded losses so that there was a net increase in forest land between 1750 and 2010. More than 85 percent of the forests are managed for timber, which has released CO_2 to the atmosphere with wood harvesting. Forest management also favors conifers over broadleaf trees so that there has been an overall increase in the area of conifer forests and a loss of broadleaf forests. Conversion of European forests between 1750 and 2010 from predominantly broadleaf to conifer contributed to regional warming by decreasing albedo and evapotranspiration. In contrast, broadleaf trees help to lessen hot extreme temperatures during summer compared with conifer forests, likely related to their higher albedo and larger evaporative cooling, and shifting to more broadleaf trees across Europe has been suggested to mitigate warming.[56]

The conflicting climate change mitigation potentials of forests means that using forests to meet specific climate objectives necessitates different types of European forests than today.[57] One objective may be to maximize the forest carbon sink by storing carbon in forests and wood products and substituting wood for materials and energy. This policy reduces the area of deciduous forests and increases conifer forests compared to no intervention (209,000 km^2 net gain in conifers) and doubles the area of forests that are unmanaged. However, the benefit of lowered atmospheric CO_2 concentration is reduced by a decrease in albedo. In contrast, a policy to reduce surface air temperature requires conversion to predominantly broadleaf trees (493,000 km^2 net decrease in conifers) and that almost one-half of Europe's forests be managed in coppice. The temperature benefit is felt mostly over Scandinavia and in winter and spring, with much less cooling across the rest of Europe. This approach has the benefit of reducing atmospheric CO_2 (though not as much as in maximizing the forest sink) but would reduce wood harvest compared with business-as-usual.

Mitigating climate change is one component in the management of Finland's forests, where the climate services of forests impact management decisions related to species composition and rotation length. Economic analyses that value forests for high timber production, large carbon storage, and high albedo find that albedo is a factor in determining the age at which to log forests comprised of Norway spruce.[58] The decrease in albedo as the forest ages diminishes the value of older forests to store carbon in trees and soil. The economic optimization depends on site conditions. A fertile site may have high growth rates with high carbon storage, thereby favoring longer rotation lengths between harvests. Forests in cold, snowy locations may grow more slowly, have less carbon benefits, and have a greater impact on albedo. Climate change itself alters the cost-benefit analysis by changing the conditions in which forests grow.

Other analyses suggest that unmanaged, old-growth stands may be the best strategy to enhance the climate benefits of Norway spruce forests.[59] Norway spruce is commonly grown in even-aged plantations in northern Europe. Trees are planted at the same time, commercially thinned during stand development, and clear cut upon reaching a desired volume of wood, whereupon the rotation cycle is repeated. A contrasting management is to grow uneven-aged forests by selectively cutting only large, canopy-dominant trees, infilling the canopy gaps with the growth of subordinate trees, and allowing natural regeneration. An analysis for central Finland shows that no management is, in fact, the best option to maximize climate benefits over multiple cycles of forest management by producing old-growth forests. Significant carbon storage in old-growth forests offsets the lower albedo to produce net climate cooling. However, no income is generated because logging is prohibited. In contrast, the opposing effects of carbon storage and albedo cancel each other in the managed forests. Uneven-aged management produces the highest albedo, but the lowest carbon storage. Even-aged forests store more carbon but also have a lower albedo compared with uneven-aged management.

Coniferous Norway spruce forests are common in Finland. Monocultures are favored in forest management, but mixed conifer and broadleaf forests might provide the most climate benefits while maintaining timber production. Mixed forests of Norway spruce and silver birch, for example, exert a larger cooling influence than does a Norway spruce monoculture.[60] There is a trade-off between carbon storage and albedo between the two species. Birch has a higher albedo than spruce, but spruce is longer-lived and provides greater carbon sequestration benefits. Birch has a higher growth rate than spruce during the early years of stand development and gives way to spruce as the stand ages. The optimal management for timber production and climate protection favors a mixture of species achieved through frequent thinning and a short rotation length to control species composition. Mixed conifer and broadleaf forests offer benefits not only for climate change mitigation, but also for biodiversity, recreation, and resilience to wildfire, pests, and disease.[61]

Secondary organic aerosols produced from BVOCs are another beneficial climate service of boreal conifers. When the cooling of biogenic secondary organic aerosols are included in the calculation of climate benefits, a strategy of longer rotations with less harvesting emerges as the most beneficial forest management.[62] The older forests allow for large carbon storage in trees and soil, and the greater production of biogenic aerosols in older stands counters the warming from lower albedo. An additional consideration is the avoided emissions due to substitution of wood for products and energy. Forest management that harvests wood frequently has a net warming effect. A high harvest rate increases albedo as a result of younger stands. It also reduces CO_2 emissions by utilizing harvested wood for products and energy. However, these climate benefits are offset by reduced carbon sequestration in the young forests and less cooling from aerosols. Comparison between Norway spruce and silver birch forests grown on fertile soils illustrates the trade-off in climate services between these species. Both have a similar net climate cooling benefit but differ in their albedo and aerosol influences. Birch has a higher albedo than spruce, but this is compensated by its lower BVOC emissions and smaller aerosol production. Spruce has a low albedo but also high production of aerosols.

Similar considerations of albedo pertain to managing the forests of Norway. Forest management that replaces spruce and pine forests with birch to increase the abundance of broadleaf trees cools surface temperature by increasing albedo and evapotranspiration.[63]

Surface cooling with conversion from conifer to broadleaf deciduous forest is seen throughout boreal North America, Europe, and Russia.[64] Large-scale planting of Norway spruce to replace birch has, however, been proposed to store carbon and mitigate anthropogenic greenhouse gas emissions. Such a strategy does, in fact, produce climate benefits by reducing atmospheric CO_2 in 2100 compared with regrowing deciduous forests, though the benefit decreases by about 12 percent when the difference in albedo between the forests is considered.[65]

These studies of forest management in Finland and Norway, as well as the other studies across Europe, point to the importance of integrative assessment of forest influences on climate that includes biogeophysical influences and biogenic aerosols in addition to carbon storage. A changing climate impacts the utility of forest management decisions. Managing Finland's forests for their climate services plays out on the stage of a changing climate that will threaten forests with windstorms, drought, wildfire, insects, and diseases.[66] Moreover, the threat posed by the low albedo of forests diminishes in a warmer, less snowy climate while the benefits of the secondary organic aerosols produced by conifers may increase as a climate change mitigation option in a warmer world.

■

There is much potential for forests to help mitigate climate change. Through carbon storage, albedo modification, and other ecological processes that influence climate, protecting, restoring, and managing the world's forests is one of many necessary steps to lessen climate change. Furthermore, forest microclimates may provide essential habitats and refuges for wildlife and outdoor workers in a warmer world. Better understanding of the processes by which forests influence climate, at both the microscale and the macroscale, and of the effectiveness of forests as natural climate solutions is required to adequately determine what makes a forest climate-smart. Central to this determination is an assessment of the net climate influences from the myriad biogeochemical and biogeophysical processes by which forests affect climate. The many different scientific tools of in situ observation, modeling, and remote sensing must also be brought together into an integrated assessment of the climate services of forests. Climate solutions that measure only the carbon content of forests undervalue the climate services of tropical forests, where biogeophysical influences augment their carbon benefits. Less well understood are northern temperate and boreal forests. Policies that consider only carbon storage erroneously value the climate benefits of these forests and may even be counterproductive. More careful uses of these forests tailored to local and regional site conditions – where productivity is high to maximize carbon storage and where biogeophysical influences engender local cooling – may be required to maximize their benefits.

Valuing forests for climate benefits picks winners and losers in forest conservation. Tropical forests are widely seen as providing a positive climate benefit, but less clear are other forests. Since the low albedo of boreal forests, especially coniferous spruce, pine, and fir trees, was first identified, boreal conifers have been implicated in warming climate. Some studies of nature-based climate solutions have excluded them from protection because of the possibility they warm, rather than cool, climate. Other studies have suggested replanting conifer forests with broadleaf deciduous trees because of their higher albedo, or similarly logging conifer forests more frequently to increase albedo. Yet there is also the likelihood that the volatile organic compounds emitted by boreal conifers produce secondary organic aerosols that cool climate and offset their albedo warming. We experience these compounds in the aromatic scents of pine and spruce forests. We bring these smells into our homes in air fresheners. Is it the chemical emissions of northern conifers that will convert them from planetary menace to world savior? The climate benefits of forests are complex. Planting trees has benefits, but tree planting is not an easy solution to the peril of climate change. One thing is certain: destroying the world's forests makes mitigating climate change more difficult.

The health of forests over the coming century, too, must be considered. Climate change is already impacting forests. What does the future hold for the major forest biomes of the world – the rain forests of the Amazon and other tropical regions; the temperate forests of North America, Europe, and Asia; and the boreal forests of Alaska, Canada, northern Europe, and northern Russia – that are so central to the carbon balance, hydrology, and climate of the planet? A changing climate with associated risks of droughts, heatwaves, wildfires, insects, and diseases could limit the potential climate services of forests. As the next chapter shows, the future of forests is unclear.

17   Forests of the Future

The notion of planting trees for their climate services – storing carbon, cooling surface climate, enhancing rainfall, providing aerosols that reflect solar radiation, creating favorable microclimate refuges, or other climate benefits – is not small scale or immediate. It requires vast tracts of healthy and thriving forests and setting aside the land to grow those forests over time periods of 50 to 100 years or longer. Achieving the climate benefits of forests requires a permanent forest presence over many decades. Climate will change during that time, and a forest planted today may not thrive in the climate of tomorrow. The forests of the future will grow in a climate different from the present one and likely in regions of the world that differ from those of today. They will be stressed by climate change, increased wildfires, disease, and insects. Forest growth, too, is not one-directional. Recurring disturbances from wildfires, droughts, insects, and windstorms continually reset forests back to young stages of development. An old-growth forest that has accumulated enormous stores of carbon in its trees and soil becomes a young, regenerating forest. Asking forests to solve the climate problem requires a long-term commitment to and investment in forests and their health.

Biogeography and Climate

The geographic distribution of natural vegetation across the planet closely corresponds with climate, and, indeed, many climate zones are named after vegetation. Temperature and precipitation are two principal determinants of plant geography. Temperature is important because sufficient warmth, but not excessive heat, is a prerequisite for the biochemical reactions that support life. Water is important because 80–90 percent of the mass of a plant is water. Figure 17.1 shows the distribution of the major types of vegetation in relation to temperature and precipitation.[1] Two environmental gradients are evident. One is a latitudinal gradient from tropical forest to temperate forest to boreal forest to arctic tundra that reflects increasingly cold climates. Precipitation forms a second axis of vegetation differentiation. In the tropics, vegetation changes from rain forest to seasonal forest to savanna to desert as annual precipitation decreases. In temperate regions, forest gives way to woodland, shrubland, grassland, and desert as precipitation decreases.

Evapotranspiration also relates to plant geography. Evapotranspiration is an integrated measure of temperature and precipitation. Temperature indicates how much energy is available to evaporate water. Precipitation determines the amount of water that can evaporate. Tropical rain forest climates, for example, are warm throughout the year, have a large positive net radiation balance, and have a high rate of potential evaporation. High annual precipitation is required to maintain a positive water balance. In contrast, boreal

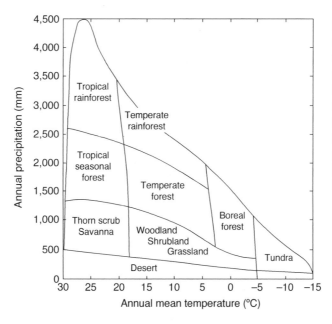

Figure 17.1 Generalized relationships among vegetation biomes and climate.
Shown is the distribution of biomes in relation to annual mean temperature (°C) and annual precipitation (millimeters).
Reproduced from ref. (1).

forests have a shorter warm season, receive less radiation, and have lower evaporative potential. Less precipitation is needed to meet evaporative demand. In general, less water is evaporated in cool climates than in warm climates, and less precipitation is required to support growth. Conversely, more precipitation is needed to support growth in hot climates with high evapotranspiration loss.

The close correspondence between climate and plant geography can also be seen in the zonation of vegetation on mountains. Naturalists at the end of the eighteenth century had recognized this dependence and depicted it in meticulous illustrations of vegetation and climate along mountain slopes.[2] One such study was published by Alexander von Humboldt in 1807 in his *Essai sur la géographie des plantes*.[3] He described his travels in Central and South America, during which he documented the vegetation and climate of the region. While ascending Chimborazo, a volcano in Ecuador, Humboldt observed the change in vegetation from tropical forest at the base to the barren landscape above treeline. In a now famous color illustration entitled *Tableau physique des Andes et pays voisins*, his profile drawing of vegetation in the Andes showed which plants grow at different elevations. Accompanying the visualization was an extensive table of the climate at each elevation. It is this illustration that inspired Frederic Church's *Heart of the Andes* (Figure 2.3). More than two hundred years later, Chimborazo and Humboldt's *Tableau physique* are still being studied for their insights to vegetation and climate change.[4]

The influential ecologist Robert Whittaker provided a more modern depiction of vegetation gradients in his study of forest communities in the Smoky Mountains of Tennessee. In this region, elevation ranges from 460 meters along the bottomlands to 2,000 meters at the summit of the highest peaks. Annual precipitation increases from less than 1,500 millimeters in the lower valleys to more than 2,000 millimeters at high elevations. The abundance of tree species changes markedly in the mountains, segregated along moisture and elevation gradients (Figure 17.2).[5] Vegetation at lower and middle elevations is typically mixed deciduous cove forest and eastern hemlock forest on moist sites. Oak forests, with some hickory, grow on moderately wet slopes at low elevation. A variety of oak forests grow on moderately dry sites, giving way to oak heaths on dry sites and pine forests and heaths on xeric slopes. Red spruce and Fraser fir forests prevail at high

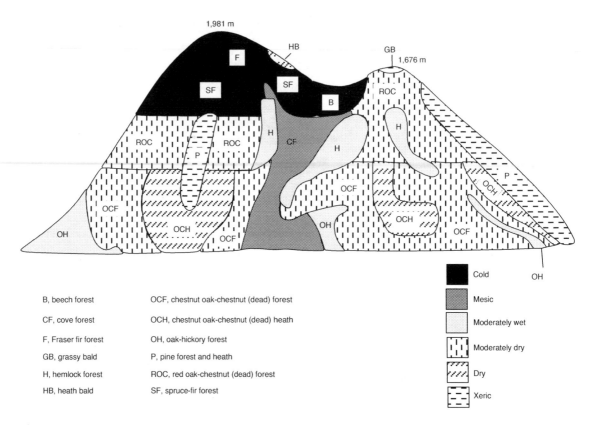

Figure 17.2 Topographic distribution of vegetation on a west-facing slope in the Great Smoky Mountains circa 1940s and 1950s.
The type of forest found in the mountains varies depending on elevation and moisture. Reproduced from ref. (5).

elevations, with beech and oak forests on favorable sites. Patches of grassy balds form on the summits of mountains. Shrub communities dominated by evergreen ericaceous shrubs form heath balds along dry ridges.

Because of the strong relationship between climate and the type of vegetation that grows in a region, simple rules can be developed to describe the vegetation of a region. Annual temperature, precipitation, and evapotranspiration are three variables commonly used in ecological classifications of bioclimatic regions.[6] Biogeography models that employ these rules can be used to study what type of vegetation might occur in a changing climate. A key limitation, however, is that the models are equilibrium models; they assume that the geographic distribution of vegetation is in equilibrium with climate. The models do not consider how long it takes the vegetation to be replaced by another type. Instead, the models describe the potential vegetation that could exist under a given set of climate conditions. As discussed in the following sections, the actual vegetation might differ because of inertia in the vegetation.

Past Forests

Forests have a long history on Earth. Extensive forests had arisen by the Middle Devonian period 393–383 million years ago (Mya).[7] Fossils unearthed at Gilboa, New York, and nearby indicate the presence of conifer-like trees with wood, leaves, and highly advanced roots.[8] These early trees gave way to the forests of the late Carboniferous period 323–299 Mya that formed today's coal deposits.[9] Conifers, a group of gymnosperms that produces seed cones, appeared 300 Mya and diversified during the Jurassic period into the Cupressaceae (cypress) family 171–157 Mya and the Taxaceae (yew) family at a similar time.[10] (The modern cypress family includes redwoods, junipers, cypress, and thuja.) The Pinaceae (pine) family, which today is the largest group of conifers including pines, spruces, firs, hemlocks, cedars, and larches, similarly dates to 188–155 Mya.[11] *Ginkgo biloba*, another gymnosperm, has been essentially unchanged for 200 million years.[12] Angiosperm (flowering) trees are more recent, appearing in the Cretaceous period with well-established forest communities by 100 Mya.[13]

In studying what forests of the future might look like, understanding past forests is key to predicting future forests. In the hundreds of millions of years of forests on Earth, atmospheric CO_2 has varied from less than 300 parts per million (ppm) to more than 1,500 ppm and in climates spanning an ice-free hothouse world to the glacial climates of the ice ages.[14] Forests provide much evidence for where different taxa have grown in the past and how the landscape has changed over time. Fossilized leaves and stumps record vastly different forests tens and hundreds of millions of years ago. Pollen preserved in lake and bog sediments show how vegetation responded following the last glacial maximum. Paleobotanical studies of fossils and pollen reveal a richly dynamic biosphere in which taxa migrate across continents, appearing in regions when the climate is favorable and disappearing when the climate is unfavorable. The forests that we know today are not static in time. They are an expression of ecological and climatic conditions over the past decades, centuries, and millennia.

Over timescales of tens and hundreds of million years, the forests of the world have undergone many changes as climate has varied. Fifteen million years ago, mesic tropical rain forests dominated interior regions of New South Wales, Australia, before the climate dried and the landscape became the arid shrublands, grasslands, and deserts that are present today.[15] Some three million years ago, the forests of Germany were much different from those of today. Fossilized wood, leaves, and pollen recovered at Willershausen, located near Göttingen, Germany, indicate a taxonomically diverse assemblage of trees, many of which are more similar to species found in present-day warm temperate forests of eastern Asia.[16] The region supported a rich forest of broadleaf deciduous trees. Thirty-four species have been identified including *Acer* (maple), *Betula* (birch), *Carpinus* (hornbeam), *Carya* (hickory), *Fagus* (beech), *Fraxinus* (ash), *Quercus* (oak), *Tilia* (linden), and *Ulmus* (elm). Several types of conifers have also been found, likely growing on exposed sites or poor soils. Excavated plant materials have been identified from *Abies* (fir), *Cedrus* (cedar), *Picea* (spruce), *Pinus* (pine), *Pseudotsuga* (Douglas-fir), *Tsuga* (hemlock), *Cathaya* (a genus of the Pinaceae, or pine, family), *Sequoia* (redwood), the Taxaceae (yew) family, and the Cupressaceae (cypress) family.

An extraordinary example of past forests is the polar forests that existed some 40 to 50 million years ago in a climate that was much warmer and wetter than the cold, dry polar desert climate of today. Lush forests with annual productivity comparable to modern-day forests of the Pacific Northwest in North America grew in the High Arctic of northern Canada and Greenland at latitudes extending to 80° N.[17] Deciduous *Metasequoia* (dawn redwood) was prevalent in these unique forests, as were other conifer taxa such as cypress, larch, spruce, pine, and hemlock. Many broadleaf deciduous taxa, whose modern-day species comprise temperate deciduous forests, formed a diverse understory. A distinguishing feature of the high latitudes where these forests grew is the extreme light environment with continuous winter darkness for several months, similar continuous summer light, and short spring and autumn seasons, which challenged the physiology of the trees.[18] Tropical-like polar forests were also found along the Antarctic coast at this time.[19] A 90-million-year-old temperate rain forest has been uncovered close to the South Pole at a paleolatitude of 82° S.[20]

Fossil pollen records forest recovery from the last ice age. Some 18,000 years ago at the height of the last ice age, glaciers covered much of the Northern Hemisphere high latitudes. Over the next several thousand years, summer solar radiation in the Northern Hemisphere increased, atmospheric CO_2 concentration increased, climate warmed, and the glaciers retreated. Vegetation adapted to cold climates shifted northwards to be replaced by warm vegetation types. Pollen sequences extracted from lake and bog sediments provide a record of the type of vegetation found in a locale. Figure 17.3, for example, shows the vegetation history at Anderson Pond, Tennessee, over the past 16,000 years.[21] The pollen of fir, spruce, and pine – species typical of boreal forests – was most abundant at the oldest dates. However, the forest was not representative of the modern boreal forest. Oak, ash, hickory, hornbeam, and other species typical of temperate deciduous forests were also present. Over time, as climate warmed, boreal species declined in abundance and were replaced by temperate deciduous species. Between 12,500 and 9,500 years ago, oak and other taxa found in present-day mesophytic forests became dominant. By about 8,000 to 6,000 years ago, a forest similar to those of today, dominated by oak with ash and hickory also present, had formed.

From pollen sequences obtained at numerous locations, maps of taxa at various points in time can be produced. Vegetation throughout eastern North America underwent similar changes as Anderson Pond over the past 18,000 years.[22] East of the Appalachian Mountains 12,000 years ago, there was a north–south zonation to vegetation with spruce-dominated boreal forest and pine-dominated mixed forest in the north,

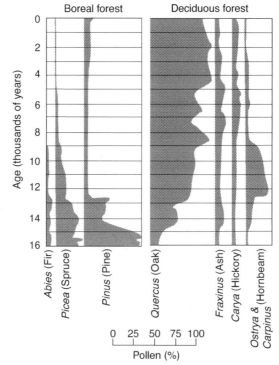

Figure 17.3 Pollen abundance at Anderson Pond, Tennessee, over the past 16,000 years.

Shown is the pollen abundance, expressed as a percentage, of various taxa (horizontal axis) in relation to age before present (vertical axis). Taxa are divided into those common to boreal forest and those of eastern deciduous forest. Adapted from ref. (21).

oak-dominated deciduous forest to the south, and pine-dominated forests in Florida. However, the most significant feature is a large region in which the vegetation was not similar to any modern vegetation. This consisted primarily of spruce-dominated woodland with high amounts of sedge and deciduous trees such as ash and elm. The northern movement of vegetation is the most obvious trend over the last 12,000 years. This movement was gradual before 12,000 years ago, rapid from 12,000 to 9,000 years ago, and then again gradual. Modern-like vegetation did not begin to develop until after 9,000 years ago. At this time, spruce-dominated boreal forest was less common than today and tundra was restricted to east of the retreating ice sheet. The deciduous forest consisted of prairie–aspen parkland in the west and northwest and mesophytic taxa in the south. Southern pine species were largely restricted to Florida until about 9,000 years ago, when southeast pine forests similar to modern forests spread northward. At the same time, prairie vegetation similar to modern prairie formed in the west and the modern deciduous forest formed in the east. Tundra and forest tundra, largely absent prior to 9,000 years ago, became more extensive. Modern-day vegetation patterns were well established by 6,000 years ago. The change in vegetation reflected the overall warmer and wetter climate of eastern North America with deglaciation.

A key aspect of the response of vegetation to climate change is whether it is in equilibrium with climate. If so, changes in climate are matched by changes in plant geography. Alternatively, lags may be introduced if the rate of climate change exceeds the timescale at which plants migrate. Most tree species have poor dispersal, with migration rates on the order of 100–1,000 meters per year or less.[23] Eight thousand years were required for oaks to migrate from their glacial refuge in the southern United States to their current northern range limit (Figure 17.4).[24] This suggests that the geographic distribution of tree species was not in equilibrium with climate at certain periods. The rate of migration was not controlled by climate, but instead depended on the availability of seeds and the ability of seedlings to become established. The result was a temporary absence of certain species for hundreds or thousands of years from regions in which they were potentially able to grow. Alternatively, migration may be accomplished primarily by rare long

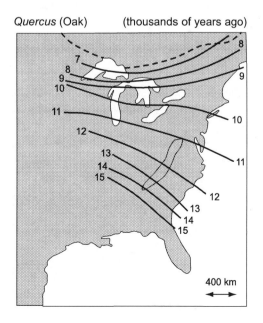

Figure 17.4 Migration map for oak in eastern North America.
Contour lines show the time (thousands of years ago) of the first appearance of oak pollen in sediments. The dashed line shows the present northern range limit of oak. Adapted from ref. (24).

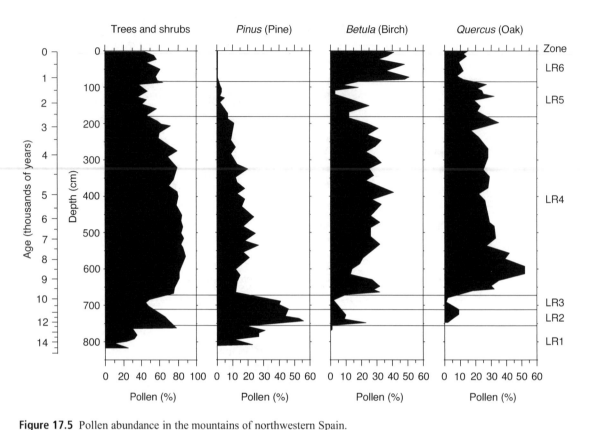

Figure 17.5 Pollen abundance in the mountains of northwestern Spain.
Pollen was collected at Laguna de la Roya located at an elevation of 1,608 meters. Shown are the depth of the sediment core (centimeters) and the corresponding age spanning the past 14,000 years. Pollen abundance is given for all trees and shrubs and separately for pine, birch, and oak. Adapted from ref. (26).

dispersal that creates outlier populations in advance of the main migrating front. These outlier trees would be too sparse to be detected in the pollen record but would serve as foci hundreds of kilometers in advance of the migrating front for rapid colonization as climate became more favorable. In this view, plants migrate fast enough to track climate change and migrate faster than seen in the pollen record.

The vegetation of Europe underwent similar change since the last glacial maximum occurred in Europe.[25] Figure 17.5 shows the pollen found at a site located in the northwestern Iberian Peninsula.[26] In this example, the sediment core extends to a depth of over 8 meters and spans the past 14,000 years. The region is thought to have been covered with ice during the last glacial maximum. The pollen records a dynamic landscape in which forests and woodlands were extensive when the climate was wet and retreated during cooler, drier conditions. Six pollen zones are evident in the sediment core. The deepest zone (LR1; 14,559–12,425 years ago) reveals the presence of pine in the landscape around the lake at that time, possibly at low elevations, but the contribution of trees to total pollen count is low. Most of the pollen is herbaceous and suggests alpine tundra was prevalent. The next zone (LR2; 12,425–10,732 years ago) shows increased forests or woodlands as the climate became warmer and wetter. Pine became more common in the region, likely at low elevations, as did some oak. Birch likely grew at high elevations. The third zone (LR3; 10,732–9,853 years) reveals a decline in forests and reversion to open herbaceous and dwarf shrub vegetation during the colder climate of the Younger Dryas period. The fourth zone (LR4; 9,853–2,726 years ago) is indicative of expanded forests as the climate again warmed and precipitation increased, but with a marked shift from the previous pine forests to oaks. At the beginning of the zone, oak pollen rapidly increased in abundance and remained at high levels, as did birch pollen. Pine pollen did not recover to previously high levels, and pine was likely no longer a major presence in the forests around the lake. Thereafter (LR5; 2,726–1,232 years ago), the presence of pine further declined while oak remained high. Oak forests were extensive in the region, and other deciduous taxa also were present. The overall percentage of tree pollen declined, while heath and tall shrub vegetation likely increased. Forests increased in the uppermost zone (LR6; 1,232 years ago to present), but pine was mostly absent. Oak pollen also decreased while birch pollen increased, suggesting that oak forests decreased in the region and birch woodlands increased at higher elevations.

Forest Succession

The understanding of forest responses to past climates presented in the previous section is one in which climate controls the geographic distribution of vegetation. That is certainly true at long timescales spanning centuries to millennia and over broad continental scales. However, over shorter timescales of a few decades and at the smaller spatial scale of a forest stand, the number of trees, the diameter of their trunks, their height, and species composition vary depending on local site conditions such as soil moisture or soil nutrients. The moist cove forests of the Smoky Mountains are much different from the dry sites (Figure 17.2). In addition, disturbance initiates a process known as forest succession that controls forest composition and temporal changes over periods of decades to centuries. A theory of forest dynamics and community organization holds that the landscape is a successional mosaic of different types of forest created by disturbances that restart forest growth anew at different places and times. These disturbances can be from wildfires, hurricanes, or windthrows that kill trees over a large area or from smaller-scale disturbances such as the death of a large individual tree that creates a gap in the canopy and allows light to reach the forest floor.

Plant succession refers to temporal change in community composition and ecosystem structure following disturbance. It is a progressive change in dominance by particular groups of species, biomass accumulation, and nutrient cycling over periods of several decades to centuries.[27] The 100–200-year-long process by which abandoned fields in the North Carolina Piedmont of the southeastern United States reestablish as pine forests before longer-lived oaks and hickories dominate the site presents a classic study of old-field succession.[28] The northern hardwood forests of New Hampshire are a well-studied example of forest succession.[29] Fast-growing but short-lived pin cherry initially dominates following a disturbance. After about 25–35 years, sugar maple and American beech, which are tolerant of shade and able to survive beneath the dense pin cherry canopy, become dominant (Figure 17.6).[30] Where pin cherry is less dense, it may be codominant with other fast-growing species such as yellow birch and quaking aspen. These longer-lived trees establish within the first few years following disturbance and dominate the canopy for several decades before the slower-growing maple and beech trees reach canopy status. In the boreal forests of interior Alaska, forest communities are a successional mosaic of broadleaf deciduous and spruce trees that reflects recovery from fire.[31] Warm, dry upland sites are recolonized by herbaceous plants, shrubs, and trees. Shrubs, primarily willow, and tree saplings dominate until about 25 years following fire, when the

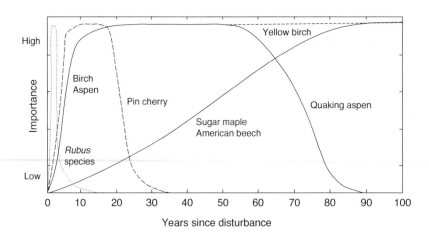

Figure 17.6 Generalized changes in community composition following disturbance in a northern hardwood forest. Shown are the relative importance of various taxa over 100 years. Reproduced from ref. (30).

saplings grow into a dense stand of birch and aspen. These deciduous trees dominate for the next 50 years or so, though white spruce seedlings and saplings grow in the understory. After about 100 years, white spruce grows into the canopy and is the dominant tree.

Threats to Forests

A large body of literature has examined potential changes in the structure and composition of forests over the next several decades in the face of climate change. Predicting the exact type of forest that might grow in a certain region in the future is fraught with difficulty. It depends on the magnitude of changes in temperature, precipitation, and other climate conditions. It depends on myriad ecological processes that determine the pathway of forest succession – the mortality of existing trees, the availability of seeds to replace the existing forest, the conditions under which the seedlings can germinate and under which seedlings and saplings grow into mature trees, and also on unforeseen pests, diseases, and wildfires that may kill the trees. In general, pollen cores and biogeography models show that warmer temperatures will displace cold-adapted species with warmer-climate species. With less rainfall or with conditions that increase atmospheric demand for evaporation, moist forests may transition to dry woodlands or grasslands. Forests in tropical, subtropical, and warm temperate climates are particularly prone to drought mortality. In planning future forests for carbon storage or for other climate protection services, understanding where trees might be capable of growing, the particular species that will thrive in a region, and the threats to forest health is essential.

Diverse forests spanning many different climate zones throughout the world are already stressed and experiencing tree mortality.[32] Drought and warmth across western North America have caused widespread insect outbreaks and wildfires throughout the region over the past decades. The mountain pine beetle has killed billions of trees across millions of hectares of land in the Rocky Mountains extending from British Columbia south to Colorado and other states. Hot, dry conditions make forests more likely to burn, and wildfires scorch vast tracts of forests annually in western lands. European forests, too, are experiencing tree mortality during extreme heat waves and drought.[33] Severe El Niño-related droughts kill trees, reduce forest productivity, and increase wildfires throughout the pantropical region.[34] The smoke from extensive Australian wildfires in 2019–2020 had hemispheric influences on climate, seen in planetary cooling and altered patterns of precipitation.[35] Strong winds associated with severe storms that uproot trees or snap trunks are common in the Amazon, and projected increases in the frequency and intensity of severe windstorms are likely to increase tree mortality.[36] Large hurricanes making landfall along the US Gulf Coast similarly kill hundreds of millions of trees.[37] These and other climate-related disturbances alter the terrestrial carbon sink, the albedo and roughness of the land surface, and the exchanges of heat and moisture with the atmosphere.

Drought, heat stress, insect outbreaks that kill or defoliate trees, diseases, and wildfires are threats to future forests in which climate is different from today.[38] Hotter temperatures, more frequent droughts, longer heat waves, more frequent and severe wildfires, and invasive insects and diseases are expected over the coming decades. The atmosphere is changing in other ways that affect forest productivity. Increasing amounts of ozone damage leaves, reduce photosynthesis, and interfere with stomata and transpiration. Greater amounts of aerosols in the atmosphere increase the proportion of solar radiation that is diffuse rather than direct beam, thereby altering forest productivity. The elevated concentration of CO_2 in the atmosphere has important physiological consequences that alter photosynthesis and water use. Understanding how weather, climate, and atmosphere composition are changing and the associated risks to forests is essential to developing sound policies that utilize forests as natural climate solutions over the coming decades.

■

Forests have existed on the planet far longer than have humans. They have survived ice ages with planetary climates much colder than that of today, and they have persisted in hot, greenhouse-like climates. Forests are durable. There will be forests of the future, just as there have been forests of the past. They will likely differ from the forests that we are accustomed to, and they may not grow in their present locations. A hike through iconic ponderosa pine and Douglas fir forests in the Rocky Mountains of Colorado may become a stroll through a meadow.[39] Old friends will be lost, and new friends found. Whether the forests of the future will provide suitable habitat for wildlife, desired sport and recreation, timber and forest products demanded by societies, clean water needed for towns and cities, and the climate protection we now ask of them is unclear.

18   The Forests before Us

In 1912, after several centuries of intellectual thought culminating in calls to protect, restore, and manage forests for their climate benefits, the forest–climate controversy was nearing its end. "The literature on the subject is somewhat bewildering," wrote a contributor to the scientific journal *Nature*.[1] The practitioners of forest meteorology, battered, bruised, and exhausted, turned away from the influence of forests on large-scale climate to the study of forest microclimates, and the notion of managing climate through forests largely receded from scientific and public inquiry. Imagine, then, the wonder of our letter writer, the perplexed "R.C.", a century later, scanning the pages of the same journal, one of the elite publications of science, to find a commentary with the headline "Nature-based solutions can help cool the planet – if we act now."[2] Chief among the climate solutions advocated by the authors: "Restoration of forest cover is widely considered the most viable near-term opportunity." That forests foster a hospitable climate is a knowledge without end.

There is an enduring connection within the human consciousness between forests and climate whereby forests are understood to influence climate and that clearing the woods or planting trees changes climate. For four centuries spanning the 1500s to the 1900s, the forest–climate question exploded onto public awareness with, first, the belief that the newfound lands of the Americas needed to be cleared of their woods to improve the climate. The optimistic drive for climate betterment gave way to concern for a decline in rainfall as the forests were cleared – an anxiety that grew with European colonial empires – and the nineteenth century saw repeated calls in all reaches of the world to reforest denuded lands to increase rainfall and protect water supplies. Meteorologists, however, rejected these concerns. They dismissed an influence of forests on climate, and the science of forest meteorology was forgotten.

Now, in the twenty-first century, forests are again recognized for their climate benefits. Deforesting large tracts of the tropics, temperate lands, and northern high latitudes changes not only the local microclimate but also the large-scale macroclimate. Like our forebears, we again talk of purposely using forests to improve climate. Protecting existing forests, restoring degraded forests, and planting new forests are seen as critical to solving the climate problem. That the biosphere is fundamental to, not separate from, climate is a core tenet of the newly emerging Earth system science paradigm. This realization is not new. It is borne from the long, controversial chronicle of forests and climate change.

Bridging Two Worlds

The narrative of forest–climate influences, then as now, is one of two distinctly separate scientific cultures. One is the blue marble Earth, with the fluid atmosphere and oceans in motion; the other is the emerald planet, with verdant lands bathed in leafy canopies (Figure 1.1). The fluid world is the domain of climate

224

science; the living world, ecology. Both views are useful in their own right, but finding effective solutions to maintain planetary well-being requires bridging the physical study of the fluid Earth with the biological study of the living Earth. This is the promise of the new Earth system science – that the disciplinary barriers of the past have been surmounted. The challenge is whether the scientific parochialism, and the accompanying overreach of both forest champions and detractors, that doomed the first melding of ecology and climate science has truly been overcome. The evidence is as yet unclear. Today's narrative unfolds in remarkably similar ways to the past.

The history of the forest–climate question is first and foremost a lesson on the need to break down disciplinary boundaries and for effective communication across the physical and biological sciences. Exaggerated and unsupported claims for the influence of forests on climate were justly criticized for lack of scientific rigor, but the inflexible dismissal by meteorologists cast a long shadow with the perception that the study of forest–climate influences was not a real science, lacking the requisite intellectual rigor. With Earth system science, we are moving toward a broad, interdisciplinary understanding of climate. Begun with concern for anthropogenic climate change, Earth system science represents the continuing advancement of climate science, from the atmosphere to the atmosphere and oceans, to the more broadly based climate system including vegetation, and to Earth as a system. The rediscovery that forests influence climate is one part of that scientific revolution. The forest–climate controversy, rather than being an unscientific misstep, was a step in building a comprehensive understanding of climate. The error was the dismissal of forest influences as a component of climate science.

Throughout history it is geophysical scientists who have defined the relevance of ecology for climate – first to the exclusion of forests and now accepting of ecology, though in a limited context. The scientific turnaround is recorded in the evolution from atmospheric general circulation models to climate models and now, Earth system models. Yet there is much ecology still missing from these models. The ecology is mostly framed in terms of fluxes of energy and materials exchanged with the atmosphere. Ecologists, too, can be faulted in not embracing global models of climate as a fundamental tool for their science. Nature-based climate solutions are too often framed contrary to the science of the models. The forest–climate question upends the academic setting in which we learn about our world. Climate change is no longer simply a study of the fluid atmosphere and oceans, but also the biosphere. Climate change is not just about greenhouse gas emissions from fossil fuel combustion; it is also about human uses of the land and its ecosystems. The two broad scientific cultures of the physical geosciences and the biological sciences must meet equally if Earth system science is to be truly transformative.

On Narrow-Mindedness

Ma Yuan's *Scholar Viewing a Waterfall* presents a contemplative study of the natural world and humankind's place in nature (Figure 7.4). It suggests humbleness – that nature, with its complexities, should be approached with humility. The forest–climate controversy, by contrast, is one of hubris. The influences of forests on climate defies absolutes, yet the forest–climate question has invariably been answered with certainty. Since the first claims that deforesting the Americas was improving climate, to later pronouncements that deforestation was desiccating the land, to subsequent insistence that planting trees would ensure plentiful rainfall, and to outright denial of any influence of forests, the science has been contorted to fit the desired narrative.

Rhetorical excesses are still evident today. If the narrative is of European supremacy at the expense of indigenous peoples and local knowledge, then one can confidently state that "the prevailing consensus is that forests generally do not significantly affect rainfall."[3] On the other hand, forest enthusiasts boldly proclaim that "the climate-regulating functions of forests – atmospheric moisture production, rainfall and temperature control at local and regional scale – should be recognized as their principal contribution."[4] The science, however, is less unconditional; it is better told in nuances, and the truth is somewhere in between these extremes. We would be better served by approaching the study of nature with less hubris, less confidence in our academic prowess, and less scorn for other disciplines. Rather than the disciplinary chauvinism and the scientific contempt that comes with single-mindedness, we instead need to foster greater interdisciplinary humility.

The narrowing of the sciences throughout the nineteenth century, and the accompanying intellectual hubris, troubled some scholars. One was the Austrian botanist Anton Kerner von Marilaun, considered the Austrian equivalent of Alexander von Humboldt.[5] Kerner decried the specialization of science: "In our age, which clings to the principle of the division of labor, it has almost become a rule that each researcher progresses only along a single very narrow path. However, narrowness too often has hubris as its consequence. Thus, the paths that others simultaneously tread are frequently arrogantly undervalued."[6] Kerner was not the first to speak of the perils of narrow-mindedness. In his 1802 *Hydrogéologie*, the first of an uncompleted trilogy of works on a grand theory of Earth including its meteorology and biology, Jean-Baptiste Lamarck complained of narrow-minded scientists seeking the "precision and scrupulous exactitude" of the sciences, but which can restrict ideas and produce only "little things."[7] Narrow-minded disciplinary bias was precisely the danger that Antoine-César Becquerel, the French physicist, warned of in his investigations of forests and climate. In explaining the many divergent thoughts on forest–climate influences, Becquerel criticized those who focused on only one part of the science. Without comprehensive study, "we run the risk of expressing an opinion that does not agree with that of another scholar who did not take the same point of view, or who only embraced part of the question."[8]

Exactness, absoluteness, and certainty of knowledge are more common than not when considering the forest–climate question. The central actors throughout the historical controversy were confident in their knowledge and unafraid to say so. Vagueness, doubt, and uncertainty are unwelcome in scientific prediction, then and still so today. Less common was the voice of hesitation. When the forest–climate question was at the forefront of intellectual, political, and social being in nineteenth century France, the respected physicist François Arago was reluctant to take a definitive stand. It was his nature as a scientist to be circumspect in opinion. At best, he would merely admit that further study was needed. To forest friends or foes seeking confirmation for their positions, he could only say, "When it is needed, I do not know."[9] When Austrians sought answers to whether deforestation was harming rainfall, and if forests should be regulated to protect the rains, the meteorologist Julius Hann replied with embarrassment that the science could not provide a precise recommendation to such an important question.[10] Like Arago and Hann more than one and a half centuries ago, the science still cannot be entirely precise in its assertions. That the science may provide multiple answers, that there is a diversity of plausible results rather than a right and a wrong, is less accepted.

The manner in which uncertainty is conveyed shapes public perception of science and its readiness to provide solutions to pressing problems.[11] This is part of the challenge in communicating modern-day climate science to the public. The dilemma faced by Arago and Hann was whether the forests of France and Austria, for the social good, should be appropriated to preserve the rains and protect against drought, or whether private ownership of forests provides the path to socioeconomic betterment. The scientific evidence for forest–rainfall influences was then shaky. Scientists, governments, and the public today face the same quandary when weighing policies to limit planetary warming against the risk of doing nothing. One of those interventions is to protect, restore, and manage forests.

Narrow-mindedness, inability to think beyond a limited perspective, overconfidence in knowledge, the quest for exactitude – these are the challenges to the interdisciplinary study of Earth as a system; the forest–climate controversy is the proof. The past failing was an inability to think beyond a limited perspective and succumbing to disciplinary biases. Despite many advances over the past few decades in scientific methods and countless scientific publications of observations, experiments, and theories, our disciplinary perspective continues to shape our view of forests.

Beyond the Utilitarian Forest

Brocéliande is a legendary forest from medieval times. Featured in Arthurian romances, it is a place of mystery and magic. The historian Wace wrote of the forest in his *Roman de Rou*, begun in the year 1160, which chronicled the history of the Norman people and their rise to rule England. He placed the forest in Brittany and described rumors of magic and fairies found there. One story was of a fountain whose water made rain "in the forest and all around."[12] Wace traveled to the forest of Brocéliande to see the wonders himself but found nothing and returned disillusioned, saddened by his gullibility in believing unfounded tales: "I went there in search of marvels; I saw the forest and the land and looked for marvels, but found none. I came back as a fool and went as a fool. I went as a fool and came back as a fool. I sought foolishness and considered myself a fool." Are we, like Wace, to be disheartened if forests do not meet our hopes to solve the climate problem?

Science says many things of forests, and there is not a single answer to the forest–climate question. Various diametrically opposed views can be found, and the science has been, and continues to be, exploited for parochial gain. Nineteenth century forest conservationists were too willing to link deforestation to a decline in rain, even the demise of ancient civilizations, despite little firm evidence. European governments used fear of drought to impose colonial rule on indigenous populations in their empires. That planting trees increases rainfall provided governmental officials and land speculators an opportunity to develop the arid western lands of the United States, fulfilling the promise of manifest destiny. Foresters found in the forest–climate question a way to lend stature to their profession and to promote forestry as a science. To the contrary, forests had no beneficial influences on rain for the landowners seeking to extract wealth from forests, or to meteorologists touting the rigor of their science. To today's advocates, forests are nature's technology to remove CO_2 from the atmosphere, provide rainwater, and offer relief from heatwaves. To naysayers, however, forests present a problem rather than a solution to environmental ills – at least to Ronald Reagan, the 40th president of the United States. During his 1980 presidential campaign, he famously alluded to trees as a cause of air pollution.[13]

At the heart of the dilemma is a need to define what a forest is – that forests exist to serve humankind, that we are the possessors and masters of forests. The answer to the forest–climate question may, then, require not just a blending of ecology and climate science, but also looking beyond the utilitarian purposing of forests to encompass many perspectives. It is not just a problem for science to solve. The rationalism of science must be balanced with the romanticism of forests. Artistic devices show the natural world in ways that science cannot. The material services of forests for climate and water can be quantified, as well as their medicinal, recreational, and economic benefits,[14] but not so their enrichment of life – human and other – and the destitution without them. A world without trees would be impoverished – culturally, morally, and biologically. Forests, the literature professor Robert Pogue Harrison explained, are part of the cultural memory of humankind; if we lose forests, we lose that memory.[15] Science can tell us the facts and the details and the nuances of forests, but visual art and creative prose provide meaning and symbolism and empowerment. Science allows us to see the trees; humanities allows us to see the forest.

Natural Laws

The famed nineteenth-century naturalist and geographer Alexander von Humboldt has been extolled for "the invention of nature" – a visionary whose study of the natural world as a single interconnected system predates our modern understanding of Earth system science.[16] Moreover, in *Views of Nature* and *Cosmos*, he articulated not just a scientific understanding of nature, but also one in which natural science and the humanities are intimately intertwined.[17] Humboldt's broad, multidisciplinary view of Earth was largely forgotten, and natural science advanced along increasingly disciplinary trajectories throughout the nineteenth and twentieth centuries. Today, we study nature within an expansive spectrum of distinct scientific settings. Two opposing points on this spectrum are climate and ecology. Climate science relies heavily on complex models of atmospheric physics and dynamics. It is a quantitative science framed in mathematical equations that codify theory and observations to test and inform the theory. Terrestrial ecosystems, with their microbes, plants, and diversity of life, do not conform to the mathematics of the physics and fluid dynamics that underpins atmospheric models, yet they are essential to understanding climate and climate change. We need an intellectual framework that balances disciplinary scientific expertise with cross-disciplinary prowess, but one that is also not devoid of moral context and our inalienable responsibilities to the planet. The answer to the forest–climate question requires a "reinvention of nature."

Climate and life are two profound forces on the planet; the forest–climate question brings them together. What, then, are the laws of science needed to understand the workings of the planet? A climate scientist studying the fluid planet will be well grounded in physics and mathematics, in the Navier–Stokes equations and the so-called primitive equations that are the foundation for atmospheric models. An ecologist studying the living planet will be well versed in evolutionary biology, natural selection, niche theory, and myriad other principles governing organisms, their assemblage into populations and communities, and the functioning of ecosystems. It is a science that may question whether it has foundational laws at all, but, if so, hold them in comparison with those of physics.[18] To these essential laws of the individual sciences, we can add collective laws of nature.

The forest–climate question is a reflection of our views of nature and our role as stewards of the planet. In rejecting forest influences on climate at the end of the nineteenth century, the notion that our actions change climate was also rejected.[19] Now, with atmospheric CO_2 concentration relentlessly increasing above values not seen during the past two million years and the planet warming at a rate unseen over at least the last two thousand years, science tells us differently – that our actions can, and do, change climate.[20] It is not just emission of greenhouse gases in fossil fuel combustion; how we manage forests, grow food, and build our cities changes the climate where we live. We are not separate from climate. We mold the climate around us, either deliberately or inadvertently. Clearing woods or planting trees is one way in which that is done. This is one of the natural laws.

Forests shape the land that is inhabited, the freshwater for hydration, the air to breathe, and the climate in which to thrive. Tropical rain forests lessen the sultry equatorial heat, a verdant Sahara bolsters its greenness, pine forests in the arid Israeli landscape dissipate the desert fieriness, and boreal forests warm the frozen north by absorbing sunlight. The predicament for us, now and before, is what to do with that knowledge. We can discount the boreal forest from nature-based climate solutions because it warms the planet, or we can marvel at nature's intricacies by which the forest creates a climate favorable to its own growth. In the pursuit of forests as a public utility, we have forgotten to appreciate the very forests before us. We have forgotten to see the forest for the trees. Human mastery of climate is a long-standing conceit, and it began with forests. There are natural laws that transcend the fundamental laws of physics, chemistry, and biology. That we have superiority over climate and can control climate for our own purposes is not a natural law.

Natural laws do not need scientific prodding and measuring and mathematics to be made meaningful. Indigenous knowledge knows so from experience and traditions. Davi Kopenawa, a Yanomami shaman, certainly does not need modern science to understand the workings of the Amazon rain forest.[21] His people knew the spirits of the forests long ago. Yanomami traditions and knowledge tell that trees "call the rain" – not all trees, and certainly not the crop trees planted by the settlers who clear the forest, just the tall kapok and Brazil nut trees.[22] "It is these big trees that make the rainwater come." This knowledge is their "ancient words," given "at the beginning of time."[23] The Yanomami who inhabit the forest do not need science to know not to wantonly clear the forest. When the trees are cut down, the waters dry out, "the earth starts to bake," the cool winds stop blowing, and "a stifling heat settles everywhere."[24] Those who clear the forest do so because "their ancestors did not give them any words of wisdom about the forest."[25] Modern science can provide the wisdom that is lacking, but without the necessary traditions and ethic, who will heed those words? Before science can provide the answers, we may, as Davi Kopenawa believes, need to "become friends with the forest."[26]

Why We Plant Trees

Tree-planting is a symbolic long-term investment, a hope for the future. John Evelyn aptly captured this sentiment over 350 years ago in *Sylva*, his treatise on trees. In calling for the reforestation of the British Isles, he wrote of forests: "For what more august, more charming and useful, than the culture and preservation of such goodly plantations: that shade to our grand-children give?"[27] That planting trees is an investment in our future and a statement of hope is a belief that arcs across more than two thousand years of history, to the Roman poet Virgil and his four-book poem *Georgics* on farming and land stewardship, from whom Evelyn gained his sentimentality.[28] It is a view that has stood the test of time and is still relevant to the modern world. A newer adage, referencing the slow growth and longevity of trees both as a caution to invest in the future and as a reminder that it is never too late to take action or make a change, tells us: "The best time to plant a tree was twenty years ago. The second best time is now."[29] In today's world, these sayings can be used as a literal call to plant trees now to combat climate change. Tree planting is also an atonement for past transgressions. The British agriculturalist Thomas Hale encouraged tree planting in a 1758 exposition on farming. As well as a national necessity and of economic value, planting trees "is of all methods the best in which a man can make atonement to his successors for his own extravagance."[30] So, too, did François-Antoine Rauch speak of the need to "reconcile with the offended nature" in his fight against deforestation and his call to protect the forests of France.[31] A world without forests, where we know trees only from art and literature, arboretums and tree museums, is not an option.

A moving testament to why we plant trees can be found in the early days of the United States' entry into World War II. After US President Franklin Roosevelt issued Executive Order 9066 on February 19, 1942, more than 100,000 Japanese–American men, women, and children living along the West Coast were relocated and incarcerated in government camps hastily constructed in remote western regions of the country. One family forced to vacate their home, possessions, and livelihood in the San Francisco Bay area of California was that of Juzaburo Furuzawa. Yet in the desolate high desert landscape of the Topaz camp in Utah, surrounded by barbed wire and watchtowers with searchlights and armed guards, living in makeshift barracks exposed to the extremes of winter cold and snow, the scorching heat of summer, and powerful windstorms, Juzaburo Furuzawa planted the seeds of Japanese black pine in tin cans.[32] One of those seeds germinated, grew into a small seedling, and today, more than seventy-five years old and stately in its age, the tree is cared for at the Pacific Bonsai Museum (Figure 18.1). The tree is a poignant expression of one man's hope for a better future, but it is also a symbol of redemption. On February 9, 2020, in advance of an exhibition on Japanese American experiences during World War II, the tree was stolen from the museum, only to be returned anonymously in less than 72 hours, perhaps with regret, after much public outcry.[33]

Figure 18.1 A Japanese black pine bonsai grown by Juzaburo Furuzawa while interned at Topaz (Utah) during World War II. Photo courtesy of Pacific Bonsai Museum (Federal Way, Washington).

For more than 500 years, there has been a lasting knowledge that forests influence the temperature and rainfall of a region, that deforestation changes climate, and that planting trees improves climate. The specifics have changed with time as better understanding has been gained. The last few decades have seen a revolution in observations, theory, and learning. The science of forests and climate has been, and continues to be, contentious. The history of the controversy is open to many interpretations. That the forest–climate question recurs throughout the ages, however, speaks to our relationship, and unease, with nature. The notion of planting trees for climate betterment, like Juzaburo Furuzawa's bonsai, is at its core a story of hope and redemption – hope for a future that is not ravaged by climate change; redemption for our callous disregard of nature. Should we plant trees and safeguard forests for climate protection? That is but one of the many reasons to care for forests.

Notes

1 The Forest–Climate Question

1. Historical climate data for station USW00093725 (Washington, DC) in the Global Historical Climatology Network Daily (GHCND) dataset. The dataset is available from the National Centers for Environmental Information, National Oceanic and Atmospheric Administration; www.ncdc.noaa.gov/cdo-web.
2. Cronon (1983), Williams (1989).
3. Cronon (1983), Abrams (2001), Cogbill et al. (2002), Hall et al. (2002), Thompson et al. (2013).
4. Williams (1989).
5. *Bulletin of the Philosophical Society of Washington*, 11, p. 520 (Washington, DC: Judd & Detweiler, 1892).
6. "The influence of forests on the quantity and frequency of rainfall," *Science*, 12(303), 242–244 (November 23, 1888).
7. Fernow (1894: 30, 32).
8. Gannett (1888a).
9. *Science*, 12(303), p. 241 (November 23, 1888). For Fernow's response, see Fernow (1889: 603–607).
10. *Bulletin of the Philosophical Society of Washington*, 11, p. 521 (Washington, DC: Judd & Detweiler, 1892).
11. Willis and Hooke (2006).
12. Abbe (1893, 1899).
13. Abbe (1889: 687).
14. Hawken (2017).
15. The formal designation for the "blue marble" image in the archives of the National Aeronautics and Space Administration (NASA) is AS17-148-22727. See also "Blue Marble: Next Generation," by Reto Stöckli (October 13, 2005); earthobservatory.nasa.gov.
16. Glacken (1967: 656–663, 681–693), Cronon (1983: 122–126, 198n25), Zilberstein (2016). See also the references that follow.
17. See Ford (2016: 138–163) and Fressoz and Locher (2020: 200–203) for the specific case of reforesting Algeria. See also Davis (2016a) and Fressoz and Locher (2020: 208–210) for drylands.
18. See Coen (2018: 239–273) and Fressoz and Locher (2020) for case studies of Austria-Hungary and France, respectively.
19. Thompson (1981), Fleming (1998), Beattie (2009), Moon (2010, 2013).
20. Brown (1951: 228).
21. Kittredge (1948: 12).
22. Grove (1995).
23. Thompson (1980: 60).
24. Williams (1989: 387; 2003: 431).
25. Whitney (1994: 264).
26. Bennett and Barton (2018).
27. Coen (2018: 241).
28. Andréassian (2004).
29. Malhi (2017), Steffen et al. (2011, 2020).
30. For a similar perspective, see also Locher and Fressoz (2012), Fressoz (2015), Bonneuil and Fressoz (2016), Fressoz and Locher (2020).
31. Brückner (1890: 194): "Durch ein wahres Labyrinth sind wir gewandert, ohne dass uns ein Ariadnefaden geleitet hätte." For the translation, see Stehr and von Storch (2000: 115). See Saint (2021) for a modern telling of the story.
32. Moreau de Jonnès (1825), title page: "La nature attacha, par des liens secrets, le destin des mortels à celui des forêts."
33. Bonneuil and Fressoz (2016), Fressoz and Locher (2020).
34. Harris (1710). See the entry "natural history."
35. oecologie; Haeckel (1866: vol. 2, 286). See also vol. 1, pp. 8n1, 237–238.
36. Tansley (1935), McIntosh (1985), Golley (1993).
37. Suess (1875: 158–159, "Biosphäre"), Vernadsky (1998).

2 Tempering the Climate, c. 1600–1840

1. Theophrastus (1990: 152–159). See Book V, 14.2–14.3, 14.5–14.6.
2. Vitruvius (1914: 229).
3. Pliny (1963: 410–413). See Book 31, chapter 30.
4. Glacken (1967: 129–130).
5. Glacken (1967: 130, 314–316).
6. Glacken (1967: 270), Tilmann (1971: 89–90), Egerton (2003).
7. Kupperman (1982), Fleming (1998: 21–23), Golinski (2008).
8. Glacken (1967), Fleming (1998: 11–24), Fressoz and Locher (2020: 46–52).
9. Chinard (1945), Cronon (1983), Williams (1989), Whitney (1994).
10. John Cracroft Wilson, in *New Zealand Parliamentary Debates. Fourth Session of the Fifth Parliament. Legislative Council and House of Representatives. Comprising the Period from the Third Day of July to the Thirty-First Day of August, 1874*, vol. 16, p. 360 (Wellington: G. Didsbury, 1874).
11. Turner (1849: 562–566).
12. Kalm (1770–71: vol. 2, 194; 1937: vol. 1, 308). The modern translation replaces "lavish" with "hostile."
13. Weld (1799: 23).
14. Chinard (1945: 452).
15. See also Thompson (1980, 1981) and Fleming (1998: 21–32, 45–54) for an introduction to the vast colonial writings on forests and climate. Fressoz and Locher (2020: 31–34) discuss the views of French explorers and their motivations.
16. Lescarbot (1609a: 624): "si ce n'est que nous disions … d'échauffer la terre." See also Lescarbot (1609b: 114).
17. Brown (1951: 220), Coates and Degroot (2015), Fressoz and Locher (2020: 31–33).
18. Biard (1616: 25): "Par ainsi de ces terres ne se peuvent eslever, que des vapeurs froides, mornes & relentes." See also Thwaites (1897a: vol. 3, 60, 61) for the translation.
19. Le Jeune (1634: 106): "L'experience nous fait voir que les bois engendrent les frimas & les gelées." See also Thwaites (1897b: vol. 5, 182, 183) for the translation.
20. Denys (1672: vol. 2, 8–12). See p. 12 for the quote: "la nouvelle France peut tout produire … pour travailler au défrichement."
21. Duhamel du Monceau (1746: 91): "Les habitans du Canada … la grande quantité de terre qu'on a défrichée."
22. Whitbourne (1620: 56, 57).
23. Kalm (1770–71: vol. 3, 249–250).
24. *A Genuine Account of Nova Scotia: Containing, a Description of its Situation, Air, Climate, Soil and its Produce; also Rivers, Bays, Harbours, and Fish, with which they abound in very great Plenty* (London, 1750), p. 4.
25. Vincent (1637).
26. Johnson (1910: 84).
27. Blome (1672: 141–142).
28. Boyle (1671: 13–14). See also Fressoz and Locher (2020: 35–36).
29. Mather (1721: 74).
30. Williamson (1771: 272).
31. Williamson (1771: 277).
32. Williamson (1811: 175–177).
33. *American Husbandry: Containing an Account of the Soil, Climate, Production and Agriculture, of the British Colonies in North-America and the West-Indies*, 2 vols. (London: J. Bew, 1775), vol. 1, p. 46.
34. Franklin (1784: 43–44).
35. Nicholson (1676: 648, 649). See also Vogel (2011).
36. Mitchell (1767: 166).
37. Meusel (1781: 76). For the translation, see also Schöpf (1875: 26).
38. Glacken (1967: 537–543).
39. Herder (1800: 176).
40. Herder (1800: 186–187).
41. Dubos (1719: vol. 2, 268–269). See also Glacken (1967: 554–562).
42. Pelloutier (1740: 124): "On fait aussi, que la Celtique … un plus grand degré de chaleur." See also Fressoz and Locher (2020: 48–49).
43. Hume (1752).
44. Hume (1752: 246).
45. Dunbar (1780: 338).
46. Franklin (1966: 265).
47. Franklin (1755: 14; 1905: 73).
48. Williamson (1771: 275), Jefferson (1954: 525–526).
49. Glacken (1967: 605–610, 684–685), Golinski (2008).

50. Robertson (1777: vol. 1, 450).
51. Williamson (1771: 278).
52. Williamson (1771: 279).
53. Mayr (1982: 101–102, 180–182, 260–263, 329–337, 440–442).
54. Glacken (1967: 587–591, 663–685).
55. Williamson (1773).
56. Buffon (1778: 240–246, 597–599).
57. Buffon (1778: 241–242): "où le soleil peut à peine pénétrer."
58. Buffon (1778: 243): "tant que les arbres . . . plus froide qu'elle descend de plus haut."
59. Buffon (1778: 244): "l'homme peut modifier . . . au point qui lui convient."
60. Jefferson (1787: 72–94), Jefferson (1955).
61. Jefferson (1787: 124).
62. Jefferson (1787: 134).
63. Jefferson (1787: 128).
64. Jefferson (2006).
65. Jefferson (1954). See Le Roy (1954) and Chastellux (1963: vol. 2, 395, 577n14) for the context of Jefferson's letter.
66. Bernardin de Saint-Pierre (1784: vol. 1, 268–269). See also Fressoz and Locher (2020: 72–74).
67. Holyoke (1793).
68. Holyoke (1793: 73).
69. Holyoke (1793: 88).
70. Cleaveland (1809).
71. Dwight (1821: vol. 1, 60).
72. "Change of Climate in North America and Europe," undated; Box 1, Folder 10 in "Papers of Samuel Williams, 1752–1794." HUM 8, Harvard University Archives; hollisarchives.lib.harvard.edu.
73. Williams (1794: 57, 58).
74. Williams (1794: 58–59, 380–385).
75. Williams (1794: 61–65).
76. Williams (1794: 60–61).
77. Williams (1794: 54).
78. Webster (1809).
79. Webster (1799).
80. Webster (1810).
81. Webster (1810: 37).
82. Webster (1810: 40).
83. Webster (1790: 374).
84. Webster (1790: 370).
85. Webster (1810: 44).
86. Webster (1810: 42).
87. Webster (1810: 46).
88. Webster (1810: 68).
89. Dunbar (1809).
90. Imlay (1793: 126).
91. Volney (1804: 266).
92. Volney (1804: 266–268).
93. Volney (1804: 269).
94. Volney (1804: 210).
95. Ramsay (1809: vol. 2, 64–70).
96. Jefferson (1903).
97. Smith (1906: 12). See also Hayes (2008: 467), Meacham (2012: 396).
98. Ure (1821).
99. Lyell (1834: vol. 3, 120).
100. Malte-Brun (1834: vol. 1, 141, 143, 144, 145).
101. Malte-Brun (1834: vol. 1, 144).
102. Malte-Brun (1834: vol. 1, 143).
103. Barton (1807, 62–64).
104. Dwight (1821: vol. 1, 58–71). For the quote, see p. 62.
105. Brewster (1830: vol. 1, 613–614).
106. Leslie (1819). See also Leslie (1804: 181–182).
107. Wulf (2015).
108. Humboldt and Bonpland (2009).
109. Rugendas (1827). For "Chasse dans une forêt vierge" see 3ᵉ division: Moeurs et Usages des Indiens et des Européens, No. 5, plate 23. For "Defrichement d'une forêt" see 4ᵉ division: Moeurs et Usages des Nègres, No. 2, plate 6.
110. Humboldt and Bonpland (1814–29: vol. 4, 143).
111. Humboldt (1817): "l'inclinaison, la nature chimique, la couleur, la force rayonnante et l'évaporation du sol" (pp. 470–471); "la nudité du sol, l'humidité des forêts" (p. 582). See Humboldt (1820–21) for an English translation.
112. Humboldt (2009: 280): "Le premier et le plus noble but de sciences . . . de la force intellectuelle de l'homme." For contemporaneous English translations of Humboldt's speech, see *The Edinburgh Journal of Science*, New Series, vol. 2, no. 2, pp. 286–300 (April 1830); *The Edinburgh New Philosophical Journal*, July–October 1830, pp. 97–111.
113. Humboldt (2009: 276): "C'est aux corps scientifiques . . . ce qui est variable dans l'économie de la nature."
114. Humboldt (2009: 276): "mais qui touchent de près aux besoins matériels de la vie."

115. Humboldt (2009: 278, 280): "l'étude de la configuration du sol . . . avec la destruction des forêts."
116. Humboldt (1831: vol. 2, 397–564).
117. Humboldt (1831: vol. 2): "théorie mathématique des climats" (p. 564); "de l'état de la surface du sol . . . par l'effet du rayonnement" (pp. 401–402); "à l'ombre d'épaisses forêts et dans des plaines couvertes de gazon" (p. 415); "Quelles différences d'effets entre . . . les forêts" (pp. 497–498); "l'examen des changemens que l'homme produit à la surface des continens, en abattant les forêts" (p. 554).
118. Humboldt (1831: vol. 2, 440–442).
119. Humboldt (1831: vol. 2, 497–520). See also Humboldt (1843: vol. 3, 191–208) for the same text.
120. Humboldt (1831: vol. 2, 507–512).
121. Humboldt (1831: vol. 2, 512): "les traînées de vapeurs . . . entre la cime des arbres."
122. Humboldt (1831: vol. 2, 513–516).
123. Humboldt (1831: vol. 2, 507): "Ces triples effets . . . est un des élémens numériques les plus intéressans de la climatologie d'un pays."
124. Humboldt (1843: vol. 3, 199): "les plus intéressants et les plus négligés."
125. Humboldt (1808: 172, 232–234).
126. Humboldt (1849: vol. 1, 158–160; 1808: 92–94; 1826: vol. 1, 114–116). For English translations of the third edition, see Humboldt (1850: 98–99) and Humboldt (2014: 83–84).
127. Gould (1989).
128. Humboldt (1846–58: vol. 1, 279).
129. Humboldt (1846–58: vol. 1, 315–316).
130. Boussingault (1833: 238): "L'abondance des forêts . . . au contraire à en augmenter la chaleur."
131. Thompson (1981), Fleming (1998: 33–45, 47–51).
132. Lovell (1826: 3).
133. Forry (1842).
134. Forry (1842: vi).
135. Forry (1842: 81–82, 101, 108). See page 108 for "extremely subordinate."
136. Forry (1842: 96–99).
137. Forry (1842: 104–105).
138. Forry (1842: 108–109).
139. Forry (1844: 237).
140. Forry (1842: 98).
141. Forry (1842: 96).
142. Napier (1842: vol. 17, 544).
143. Humboldt (1850: 103–104). The discussion borrows from Forry (1844: 99). See also Humboldt (2014: 86).

3 Destroying the Rains, c. 1500–1830

1. Fernández de Oviedo y Valdés (1851: vol. 1, 239–241 [Book 6, Chapter 46]). I am indebted to the Oviedo Project at Vassar College for the translation (translated by Madeline Seibel Dean, edited by Lizabeth Paravisini-Gebert and Michael Aronna); pages.vassar.edu/oviedo/book-vi-chapter-46.
2. Colón (1571: 118): "Il cielo, & la disposition dell'aria . . . non si generano tanti nembi, & pioggie, quate si generavano avanti." For the translation, see Colón (1959: 142–143).
3. Paso y Troncoso (1905: 63): "E al presente ay grandes alcabucos e breñas . . . los ayres la bañen y enjuguen como antiguamente." For the translation, see Glacken (1967: 359) and Grove (1995: 154).
4. Pigafetta (1906: vol. 1, 33). See also Pigafetta (1985: 56): "sino que, al mediodía . . . destilando entonces sus hojas y ramas aqua a placer."
5. See p. 277 in *The Voyages & Travels of J. Albert de Mandelslo*, bound as a supplement to Olearius (1662).
6. Fressoz and Locher (2020: 22–24, 39–42).
7. Stubbe (1667: 498).
8. Evelyn (1679: 273).
9. Clayton (1693: 788).
10. Humboldt (1843: vol. 1, 537–538).
11. Grew (1682: 153). For the illustration, see Table XLVIII.
12. Nash (1952: 21).
13. Woodward (1699: 208).
14. Halley (1686).
15. Halley (1691).
16. Woodward (1699: 208–209).
17. Evelyn (1706: 13).
18. Hales (1727: 88).
19. Hales (1727: 4–6).
20. Humboldt (1831: vol. 2, 511–512): "Des torrens de vapeurs . . . de cette transpiration (exhalation) aqueuse des feuilles."
21. Grove (1995: 160–161, 164–165).

22. *Encyclopédie, ou dictionnaire universel raisonné des connoissances humaines*, vol. 34 (Yverdon, 1744), pp. 173, 174: "On peut encore regarder les forêts ... transpirent une grande quantité de vapeurs"; "Les forêts sont toujours ... par leur élévation dans l'atmosphere."

23. Prévost (1744: vol. 2: 92): "puisqu'à mesure qu'on rase ... les pluies deviennent plus rares & moins épaisses."

24. Long (1774: vol. 1, 357–358; vol. 3, 601, 648, 650, 651).

25. Williamson (1771: 280).

26. Williams (1786: 121–122; 1794: 62, 74–75).

27. Williams (1794: 27).

28. Williams (1794: 50).

29. Volney (1804: 26).

30. Evelyn (1706: 13).

31. Hales (1727: 325).

32. Conant (1950: 13–16), Nash (1952: 42–43).

33. Hales (1727: 206).

34. Priestley (1772: 166).

35. Priestley (1772: 193).

36. Priestley (1772: 198–199).

37. Nash (1952: 11–120), Egerton (2008).

38. Holyoke (1793).

39. Priestley (1772: 199–200).

40. Williams (1794: 77).

41. Rush (1786: 209).

42. Sloan (1981), Grove (1995: 160–161).

43. Buffon (1778: 242): "Il en est de même de la quantité ... moins abondantes & moins continues."

44. Buffon (1778: 245): "mais il est bien plus aisé d'abattre des forêts ... toutes les douceurs d'un climat tempéré."

45. Humboldt (1808: 171–172, 232–234). Humboldt (1849: vol. 2, 15, 101–102) uses the same text. See Humboldt (1850: 217, 266) for a contemporaneous English translation of the third edition. For the modern translation given here, also from the third edition, see Humboldt (2014: 159, 189).

46. Humboldt (1843: vol. 1, 537–538).

47. Humboldt (1846–58: vol. 2, 283n438).

48. Humboldt (1831: vol. 2, 511): "Des torrens de vapeurs s'élèvent audessus d'un pays équinoxial couvert de forêts."

49. Williamson (1811: 24–25).

50. Volney (1804: 25, 26).

51. Vitruvius (1914: 229), Pliny (1963: 410–413).

52. Douglass (1749–51: vol. 2, 54).

53. Williams (1794: 61).

54. Williams (1786: 121–122; 1794: 62, 74–75).

55. White (1789: 205).

56. Webster (1790: 371–372).

57. Volney (1804: 25).

58. Dwight (1821: vol. 2, 92).

59. Fabre (1797: 64–65, 131). For an English translation, see Brown (1876: 55, 57).

60. Surell (1841: 129–133). See also Brown (1876: 45–54).

61. Surell (1841: 180): "Reboiser les parties élevées des montagnes."

62. Brown (1876), Andréassian (2004), Ford (2016: 66–91).

63. Humboldt and Bonpland (1814–29: vol. 4, 142–144, 148).

64. Boussingault (1837). For an English translation, see Boussingault (1838).

65. Boussingault (1837: 113): "C'est une question importante ... peuvent modifier le climat d'un pays."

66. Boussingault (1837: 114): "Ces faits sembleraient indiquer ... de pluie qui tombe sur une contrée"; "Ces dernières observations ... des bois diminue la quantité annuelle de pluie."

67. Boussingault (1837: 140–141). See his list 1°–7°.

68. Boussingault (1851: vol. 2, 730–759).

69. Poivre (1768: 31): "Les pluyes qui dans cette isle ... ne tombent plus sur les terres défrichées." See also Grove (1995: 185–186, 189n57, 196–197, 197n79).

70. Brouard (1963: 16), Grove (1995: 184–185), Fressoz and Locher (2020: 252n9).

71. Grove (1995: 184–222, 255–258), Fressoz and Locher (2020: 66–71).

72. Poivre (1797: 199–232). See also Grove (1995: 202–203).

73. Poivre (1797: 209): "mais de la perte immense ... qui en est la suite?"

74. Poivre (1797: 210): "ils n'ont laissé ... abandonnées par les pluies."

75. Poivre (1797: 210): "La nature a tout ... y ont tout détruit."

76. Poivre (1797: 210): "Les forêts magnifiques ... résoudre en une pluie féconde."

77. Poivre (1797: 210–211): "le ciel, en leur refusant ... à la nature et à la raison."

78. Poivre (1797: 211): "Encore quelques années de destruction ... il faudrait l'abandonner."
79. "Failure of springs in the east," *Chambers's Journal of Popular Literature, Science, and Arts*, ser. 3, vol. 20, no. 496, pp. 1–3 (July 4, 1863). See also Grove (1995: 261).
80. Rogers (1873).
81. Fressoz and Locher (2015; 2020: 127–140), Ford (2016: 16–42).
82. Rauch (1792: 113): "et leur influence visible ... des rivières qu'elles alimentent."
83. Rauch (1802) and revised by Rauch (1818).
84. Rauch (1802: vol. 1, 12): "de leurs cimes attractives ... rafraîchir les vertes prairies."
85. Rauch (1802: vol. 1, 13): "comme la nature ... les brouillards et de nouveaux nuages."
86. Rauch (1802: vol. 1, 15): "Les forêts ... des fleuves commerçants et navigables."
87. Rauch (1802: vol. 1), frontispiece: "La France régénérée vous demande à recréer cette belle nature sur toute sa surface."
88. Fowler (1774: 26).
89. Grove (1995: 271).
90. Beard (1949: 30), Grove (1995: 293–296).
91. Howard and Howard (1983: 37), Grove (1995: 301).
92. Clarke (1835: 480).
93. Murray (1831).
94. Janisch (1908: 236).
95. Grove (1995: 342–346, 356–359).
96. Beatson (1816: 88).
97. Webster (1834: vol. 1, 368–369).
98. Duffey (1964: 227), Grove (1995: 364).

4 Planting Trees for Rain, c. 1840–1900

1. Herschel (1859: 614), Herschel (1861: 244).
2. *Gardeners' Chronicle and Agricultural Gazette for 1859*, April 2, 1859, pp. 287–288.
3. Jamieson (1860: 34–38).
4. Tristram (1868: 32).
5. "The use of forests," *Chambers's Journal of Popular Literature, Science, and Arts*, ser. 4, vol. 13, no. 663, pp. 590–592 (September 9, 1876).
6. Grove (1995: 380–473), Barton (2002), Rajan (2006).
7. Newbold (1839).
8. Grove (1995: 428–462), Rajan (2006: 64–68).
9. Butter (1839: 9).
10. Grove (1995: 431–441), Rajan (2006: 212), Rodrigues (2007).
11. Cleghorn et al. (1852: 102), Grove (1995: 435), Rodrigues (2007: 659).
12. Grove (1995: 441–450), Rajan (2006: 210).
13. Balfour (1849: 403).
14. Balfour (1849: 412).
15. Dalhousie (1868: 2).
16. Dalzell (1863, 1869). See also Rajan (2006: 211–212).
17. Dalzell (1863: 15).
18. Dalzell (1863: 18).
19. Schlich (1889: 25–50).
20. Schlich (1889: 87–90).
21. Nisbet (1894). See also Nisbet (1893).
22. Ribbentrop (1900: 37).
23. Cleghorn et al. (1852: 78).
24. Voeikov (1878).
25. Voeikov (1885a). For an English translation, see Voeikov (1886).
26. Voeikov (1886: 36–38).
27. "The influence of forests on climate," *Nature*, 32, 115–116 (1885).
28. Blanford (1886). See also Blanford (1887, 1888) for a reprint.
29. Blanford (1886: 140).
30. Hann (1888).
31. Brandis (1887: 376): "Wenn es gelänge, durch Waldpflege und Aufforstung ... so würden die Vortheile für das Land und seine Bevölkerung unberechenbar sein." See also Brandis (1888) for an English translation.
32. Fernow (1889: 613–615), and "Forests and rainfall," *The Indian Forester*, 15, 39–41 (1889). See also Harrington (1893: 116–118).
33. Good, Thomas, 1823–1907: Bush clearing near Oeo, 1893. Ref. A-329–005. Alexander Turnbull Library, Wellington, New Zealand; natlib.govt.nz/records/23106356.
34. Legg (2014, 2018).
35. Boussingault (1845: 689–690, note).
36. Clarke (1876).
37. Clarke (1835).
38. Clarke (1876: 187).
39. Mueller (1867: 18).
40. Clarke (1876: 212).

41. *Nature*, vol. 16, p. 217 (July 12, 1877). See also "Is the Climate of Australia Changing?" *The South Australian Register* (Adelaide), May 18, 1877, p. 4.

42. Beattie (2003, 2009), Beattie and Star (2010).

43. "The forests of the colony," in *New Zealand Parliamentary Debates. Third Session of the Fourth Parliament. Legislative Council and House of Representatives. Comprising the Period from the Twenty-Fifth Day of September, to the Twentieth Day of October, 1868*, vol. 4, pp. 188–193 (Wellington: G. Didsbury, 1868). See p. 190 for the quote.

44. Travers (1870: 327).

45. Beattie (2003, 2009).

46. "Conservation of forests," in *New Zealand Parliamentary Debates. Third Session of the Fifth Parliament. Legislative Council and House of Representatives. Comprising the Period from the Twenty-Eighth Day of August, to the Third Day of October, 1873*, vol. 15, pp. 1545–1547 (Wellington: G. Didsbury, 1873). See p. 1545 for the quote.

47. Firth (1874: 181).

48. "New Zealand forests bill," in *New Zealand Parliamentary Debates. Fourth Session of the Fifth Parliament. Legislative Council and House of Representatives. Comprising the Period from the Third Day of July to the Thirty-First Day of August, 1874*, vol. 16, pp. 79–94, 350–381, 399–426 (Wellington: G. Didsbury, 1874). For the quotes, see pp. 419, 94.

49. Campbell Walker (1876a: 199–200; 1876b; 1877: 47–49).

50. Peppercorne (1879: 32), Lecoy (1879: 7).

51. Beattie (2003, 2009).

52. Wilson (1865a). See also Wilson (1865b) for the discussion.

53. Wilson (1865a: 118–119).

54. Grove (1989).

55. Brown (1887: 96, 113).

56. Brown (1887: 90–92).

57. Brown (1875).

58. Brown (1875: 208).

59. Brown (1875: 209).

60. Brown (1877b).

61. Brown (1876).

62. Brown (1883).

63. Brown (1877a).

64. Brown (1877a: 103–104, 117).

65. Brown (1877a: 1–2).

66. Brown (1877a: 42).

67. Brincken (1828: 6–7). See also Clarke (1835).

68. Humboldt (2009).

69. Costlow (2003), Moon (2010; 2013: 109, 119–121, 176), Loskutova (2020).

70. Murchison et al. (1845: vol. 1, 579).

71. Costlow (2003).

72. Chekhov (2011: 124). Used by permission of the publisher, Stanford University Press (sup.org). See also Costlow (2003: 115), Moon (2010: 254).

73. Moon (2010; 2013: 109–110, 119–134, 173–179, 191–194).

74. Thomas (1956), Moon (2010; 2013: 69, 128–130, 133, 178, 192).

75. Voeikov (1901: 198, 199, 203, 204).

76. Voeikov (1885a, 1886, 1888).

77. Voeikov (1885b).

78. Voeikov (1887: vol. 1, 270–295).

79. Schlich (1889: 33–35), Fernow (1889: 611–613), Harrington (1893: 91–94, 115–116), Zon (1912: 209–210, 218).

80. Emerson (1846: 4–6).

81. Gregg (1844: vol. 2, 202–203).

82. Lee (1850: 41).

83. Piper (1855: 47–54).

84. Watson (1866), Kedzie et al. (1866), Kedzie (1867), Wood (1875), Edge (1878). See also "Benefits of forests and screens," in *Transactions of the Wisconsin State Horticultural Society*, vol. 7, pp. 187–195 (Madison: David Atwood, 1877).

85. Lapham et al. (1867: 3).

86. Marsh (1864).

87. Marsh (1864: 214–215).

88. Marsh (1864: 181, 182).

89. Marsh (1864: 216).

90. Emmons (1971), Kutzleb (1971), Thompson (1980).

91. Henry (1886: vol. 2, 19–20).

92. Becquerel (1872).

93. Starr (1866).

94. Starr (1866: 230).

95. Wilson (1868: 173–198).

96. Wilson (1868: 174).

97. Wilson (1868: 197).

98. Hayden (1867: 131, 135–136; 1869: 141–142), Elliot (1871a, 1872).

99. Hayden (1867: 135).
100. Wilson (1867: 106–107).
101. Capron (1869: 448; 1870: 516; 1871: 516), Watts (1872: 233, 354–355, 373–374; 1874: 316–332).
102. *The Statutes at Large and Proclamations of the United States of America, from March 1871 to March 1873, and Treaties and Postal Conventions*, vol. 17, pp. 605–606 (Boston: Little, Brown & Co., 1873).
103. *The Congressional Globe: Containing the Debates and Proceedings of the Second Session Forty-Second Congress; With an Appendix, Embracing the Laws Passed at that Session*, p. 4464 (Washington, DC: F. & J. Rives; George A. Bailey, 1872).
104. *The Statutes at Large the United States, from December, 1873, to March, 1875, and Recent Treaties, Postal Conventions, and Executive Proclamations*, vol. 18, pp. 21–22 (Washington, DC: Government Printing Office, 1875).
105. *The Statutes at Large of the United States of America, from October, 1877, to March, 1879, and Recent Treaties, Postal Conventions, and Executive Proclamations*, vol. 20, pp. 113–115 (Washington, DC: Government Printing Office, 1879).
106. *The Statutes at Large of the United States of America, from December, 1889, to March, 1891, and Recent Treaties, Conventions, and Executive Proclamations*, vol. 26, p. 1095 (Washington, DC: Government Printing Office, 1891).
107. Hough (1874).
108. "Cultivation of timber and the preservation of forests, H.R. Report No. 259, 43rd Congress, 1st Session, March 17, 1874," in *Reports of the Committees of the House of Representatives for the First Session of the Forty-Third Congress, 1873–74* (Washington, DC: Government Printing Office, 1874).
109. See p. 7 of the report.
110. See p. 77 of the report.
111. Hough (1878a: 190).
112. *The Statutes at Large of the United States of America, from December, 1875, to March, 1877, and Recent Treaties, Postal Conventions, and Executive Proclamations*, vol. 19, p. 167 (Washington, DC: Government Printing Office, 1877).
113. Hough (1878b: 221–336).
114. Hough (1885).
115. Becquerel (1878).
116. "Trees and rain," *Bulletin of the Torrey Botanical Club*, 3, 38 (1872).
117. Greeley (1871: 52).
118. Cooper (1876: 15).
119. Schofield (1875), Oswald (1877), "Rainfall and forests," *Scientific American*, 41(20), 312 (1879).
120. Egleston (1896: 9–12, 14), Williams (1989: 382–383).
121. Elliott (1871a: 457).
122. Elliott (1871a,b, 1872).
123. Elliott (1883: 304).
124. Elliott (1883: 299–324).
125. Elliott (1883: 324).
126. *Science*, vol. 12(303), p. 241 (November 23, 1888).
127. R. S. Elliott to George D. Hall, January 9, 1871. Richard Smith Elliott Papers, Vol. 3: 1868 August 13–1872 March 30, Missouri History Museum Archives, St. Louis.
128. Emmons (1971).
129. Aughey (1878: 84).
130. *Harper's Weekly*, vol. 12, no. 593, p. 292 (May 9, 1868). Image from the US Library of Congress; loc.gov/item/90712909.
131. Aughey (1880: 44–45).
132. Wilber (1881: 68).
133. Wilber (1881: 63).
134. Holzman (1937).
135. Holzman (1937: 11–12).
136. Holzman (1937: 14).

5 Making a Science: Forest Meteorology, c. 1850–1880

1. See Löffelholz-Colberg (1872) and Zon (1912) for their literature reviews.
2. Harrington (1893: 25).
3. Glacken (1967: 491–494), Locher and Fressoz (2012), Fressoz and Locher (2015, 2020), Matteson (2015), Ford (2016: 43–65).
4. Mirbel (1815: vol. 1, 448): "Les forêts arrêtent et condensent . . . le soleil ne réchauffe jamais la terre qu'elles ombragent."
5. Mirbel (1815: vol. 1, 452): "Le voyageur qui parcourt . . . ils sont effacés de la terre."

6. Locher and Fressoz (2012), Fressoz and Locher (2015; 2020: 141–158).

7. *Annales européennes*, vol. 6, pp. 389 (1824): "l'étude de ces phénomènes ... ou de nos forêts." See also Fressoz and Locher (2020: 148).

8. Moreau de Jonnès (1825). See also Ford (2016: 52–54), Fressoz and Locher (2020: 155–156, 157).

9. Moreau de Jonnès (1825: 30–34).

10. Moreau de Jonnès (1825: 34–39).

11. Moreau de Jonnès (1825: 47, 49): "on remarque qu'il ne s'opère pas exactement de la même manière, sous la zône torride et sous les zônes tempérées"; "Dans les pays situés entre les tropiques ... par l'existence des bois." See also summary points 6°–8° (pp. 62–63).

12. Moreau de Jonnès (1825: 30): "Trente mille observations météorologiques ... en divers climats."

13. Moreau de Jonnès (1825: 184–185), summary points 14°–18°; "sont incomparablement moins utiles."

14. Moreau de Jonnès (1825), title page: "La nature attacha, par des liens secrets, le destin des mortels à celui des forêts."

15. Moreau de Jonnès (1825: 182–183), summary points 2°, 4°.

16. Moreau de Jonnès (1825: 185), summary point 18°.

17. Becquerel (1853: v–vi; 1865: 78–80), Arago (1859), Fressoz and Locher (2020: 164–166).

18. Bosson (1825).

19. Bosson (1825), title page: "S'il faut une étendue de génie ... l'environne le plus immédiatement." See Lamarck (1794: vol. 2, 1–2).

20. Lamarck (1801/02: 137): "L'influence de la lumière du soleil ... que sur celle qui domine le sol nud."

21. Larmarck (1802: 8). See also Lamarck (1964: 18) for an English translation.

22. Lamarck (1820: 154–155n1).

23. Becquerel and Becquerel (1847: 166–167, 179–180).

24. Becquerel (1853: v–vi; 1865: 78–80), Arago (1859), Fressoz and Locher (2020: 164–166).

25. Becquerel (1853: vi): "en mettant de côté toute idée préconçue sur la question."

26. Becquerel (1872, 1878).

27. Becquerel (1853).

28. Becquerel (1853: 177): "Pour bien étudier un climat ... la végétation d'un pays."

29. Becquerel (1853: 359–360): "Les faits nombreux et variés ... dans la discussion."

30. Becquerel (1853: 136–148).

31. Becquerel (1853: 139): "Il est hors de doute que ... sont plus ou moins éloignées."

32. Becquerel (1860).

33. Becquerel (1860: 262): "Il est bien prouvé maintenant ... une cause de refroidissement."

34. Becquerel (1860: 262–263): "Les bois, les forêts ... encore dont je n'ai pas à m'occuper."

35. Becquerel (1865).

36. Becquerel (1865: 71–111). See Becquerel (1872, 1878) for an English translation.

37. Becquerel (1865: 86): "L'eau aspirée provient ... d'eau à la végétation herbacée."

38. Becquerel (1865: 88): "Nous ajouterons, comme nous l'avons ... à la périphérie des arbres."

39. Becquerel (1865: 110–111): "Nous ferons encore ... aussi grande qu'on l'avait pensé."

40. Becquerel (1865: 73): "Quelle que soit l'action ... comme le fait une grande masse."

41. Becquerel (1865: 72–73): "Ces questions, ne pouvant ... qu'une partie de la question."

42. Becquerel and Becquerel (1866, 1867, 1869a,b), Becquerel (1867).

43. Becquerel (1867: 839): "L'expérience démontre que les arbres ... et par suite sur les phénomènes aqueux." See also Hough (1878: 333).

44. Guyot (1898: 249–250).

45. Mathieu (1855, 1877).

46. Mathieu (1878).

47. Matthieu (1878: 17): "et que les feuilles des arbres ... que celles des plantes hersées."

48. Fressoz and Locher (2020: 177).

49. Mathieu (1878: 4–6).

50. Matthieu (1878: 7): "Dans l'impossibilité ... au boisement ou à la nudité du terrain."

51. Matthieu (1878: 6): "Cette question très-controversée ... et que ce moyen est impraticable."

52. Clavé (1875: 637–638): "L'influence des forêts ... enfin une action mécanique."

53. Renou (1866: 104): "mais cette opinion ... une seule observation incontestable à son appui."

54. Renou (1866: 104–105): "L'opinion de l'influence ... sans en avoir la moindre preuve."

55. Renou (1866: 105): "Autrefois, on a voulu rattacher ... des phénomènes de l'atmosphère."

56. Renou (1866: 106): "Les forêts n'ont aucune influence sur la pluie."
57. Clavé (1875: 648): "M. Marié-Davy, trop préoccupé . . . les forêts exercent une autre action."
58. Cézanne (1870–72).
59. Cézanne (1870–72: vol. 2, 41–42): "le vieux grief du déboisement . . . fonds de la tradition populaire."
60. Cézanne (1870–72: vol. 2, 69): "Comment s'assurer que les forêts . . . un obstacle au vent pluvial?"
61. Cézanne (1870–72: vol. 2, 82): "Il faut reconnaître que . . . c'est l'effet du reboisement!"
62. Cézanne (1870–72: vol. 2, 84): "on peut dès aujourd'hui reléguer . . . petits de la météorologie."
63. Cotta (1832: 6): "Bei zu wenig Wald . . . und Grenada, vorzüglich aber mit Persien."
64. Fraas (1847: 9–11, 60–70). See also Glacken (1956: 79), Warde (2018: 329).
65. Schleiden (1848b: 304, 306). For the original German, see Schleiden (1848a: 281, 283).
66. Liebig (1862: vol. 1, 95): "Wegen ihres Einflusses auf das Klima . . . unter öffentlicher Aufsicht." See also Warde (2018: 330).
67. Dove (1855). For an English translation, see Dove (1855–56).
68. Dove (1855: 54): "zuletzt die immer dichter werdende Bevölkerung . . . welkend dabin sterben wird." For the translation, see Dove (1855–56: vol. 21, 114).
69. Berger (1865). For Berger, see Möller (2020: 61n113).
70. Lorenz von Liburnau (1878: 126, 128).
71. Berger (1865: 541): "Bei Tage wird . . . und den Kreislauf aufs neue zu beginnen."
72. Berger (1865: 548): "Wenn die *Abhänge und Höhen eines Thales* . . . entgegengesetzter Richtung wehen."
73. Berger (1865: 559): "Wenden wir diess mit Berücksichtigung . . . *zwischen Wald und Feld*."
74. Berger (1865: 562–563): "Es werden aber auch die allgemeinen . . . eine Verminderung eintritt."
75. Berger (1865: 563): "Diese Abwechselung herzustellen . . . der Landwirthschaft seyn müssen."
76. Ebermayer (1873).
77. Ebermayer (1873: v): "Schon von den verschiedensten Seiten . . . Ländern ausgeführt werden müssen."
78. Ebermayer (1873: v–vi): "Durch die bereitwillige Genehmigung . . . täglich zweimal vorgenommen wurden."
79. Reifsnyder (1973).
80. Ebermayer (1873: 2–4, 10–22).
81. Ebermayer (1873: 4–5).
82. Ebermayer (1873: 127, 137, 175).
83. Ebermayer (1873: 109, 122): "In einem gut geschlossenen . . . am höchsten in der Baumkrone"; "In den Blättern findet der Temperaturwechsel . . . in die Bäume eindringt."
84. Ebermayer (1873: 183): "Durch ihre meist tiefgehende . . . ebenfalls gefristet wird."
85. Ebermayer (1873: 110–112).
86. Ebermayer (1873: 175): "Wenn im Freien 100 Volumtheile . . . an die Atmosphäre a b."
87. Ebermayer (1873: 87–88, 117, 175–176).
88. Ebermayer (1873: 151): "Es dürfte schon jetzt . . . sich dieser Einfluss bemerkbar."
89. Ebermayer (1873: 200–203).
90. Ebermayer (1873: 203): "Leider ist es uns bis jetzt . . . auf den Regenfall festzustellen."
91. Ebermayer (1873: 225): "Verschiedene bekannte Thatsachen . . . werden nicht ausbleiben."
92. Schlich (1889: 25–45), Harrington (1893: 27–40). For example studies, see Nördlinger (1885), Lorenz von Liburnau (1890).
93. Lorenz (1877: ii, 35–39), Lorenz von Liburnau (1879). See also Harrington (1893: 35–40), Coen (2018: 244).
94. For the network of sites, see Müttrich (1890). For the afforestation study, see Müttrich (1892: 42), Harrington (1893: 113–114), Zon (1912: 217).
95. Hough (1878b: 230–264).
96. Coen (2018: 239–273).
97. Zötl (1831: 60): "Wem daher die Behandlung der Wälder . . . Klima einer Gegend beyzutragen."
98. Hohenstein (1860: 4–41).
99. Hohenstein (1860: 13): "Um den Einfluss genau zu ermitteln . . . fehlen gleichfalls noch."
100. Hohenstein (1860: 14): "Dem Anscheine nach sollte sich die Frage . . . Einflüsse geltend machen könnten."
101. Hohenstein (1860: 40): "Ich habe das Recht . . . so weise, so mütterlich bedacht hat."
102. Hohenstein (1860: 154): "Der Saft, fast reines Wasser . . . der Flüsse und Quellen?"
103. Hohenstein (1860: 156): "So sind wir also . . . ihrem Schmucke zu erhalten."
104. Hann (1867: 129–130).

105. Hann (1867: 131–132): "Dennoch wäre es ungerechtfertigt … sonst anzunehmen geneigt war."

106. Hann (1867: 133): "Der eben besprochene Einfluss … ist dies in der gemässigten Zone."

107. Hann (1867: 136): "Es ist sehr zu wünschen … sicheren Schlüssen gelangen will."

108. Hann (1867: 134–135).

109. Hann (1869: 21): "Noch immer kann der Landwirth … schwer genug empfundene Thatsache."

110. Wex (1873).

111. Brown (1877a: 175–204, 299), Hough (1878b: 294–297), Wex (1880, 1881), Sargent et al. (1897).

112. Coen (2018: 78, 248–253).

113. Lorenz (1877: 4): "Die Rückwirkung der Vegetationsdecke … ihrer Lösung zugeführt warden."

114. Lorenz and Rothe (1874).

115. Lorenz and Rothe (1874: 230–311).

116. Lorenz von Liburnau (1878: 49): "Es kommt dabei eben auf den Standpunkt des Beurtheilers an."

117. Lorenz and Rothe (1874: 300): "Wie die Wasserfläche über ihre … so auch der Wald."

118. Lorenz von Liburnau (1878: 279): "großen Waldfrage."

119. Lorenz von Liburnau (1878: 279): "Da die Wirkung und Bedeutung … zu charakterisiren ist."

120. Lorenz von Liburnau (1878: 97): "Dabei muß gleich vom Anfange … vom 'Walde überhaupt' behaupten könnte."

121. Lorenz von Liburnau (1878: 96–97): "Was nun die Wirkung des Waldes … d. h. des ganzen Waldes, zu erkennen."

122. Lorenz von Liburnau (1878: 283): "daß nicht alle Wälder … und weitere Umgebung üben."

123. Lorenz von Liburnau (1878: 274–278). See also Coen (2018: 251–253).

124. Lorenz von Liburnau (1878: 284): "um, mit Einem Worte … der Natur und der Menschen."

125. Lorenz von Liburnau (1878: 155–156): "Wir müssen also der Frage … auch auf seine Umgebung übertrage."

126. Lorenz von Liburnau (1878: 90–92, 124–129, 157–159).

127. Lorenz von Liburnau (1878: 159–181). For the figures, see pp. 165, 167, 168, 175, 176.

128. Lorenz von Liburnau (1878: 174): "Die Einrichtung solcher Stationen … grundsäßlich richtig erkennen."

129. Coen (2018: 249).

130. Coen (2018: 253–267).

131. Purkyně (1875, 1876, 1877).

132. Purkyně (1875: 519): "kolportirten Erzählungen … sind Fabeln"; Purkyně (1876: 138): "Phantasien über den Einfluß der Wälder auf das Klima"; Purkyně (1876: 176): "Fabeln sind … erfunden wurden"; Purkyně (1876: 242): "ich bin leider so eingerichtet … Studium und Beobachtung unmittelbar ergibt"; Purkyně (1876: 243): "Ich halte aber den Kampf … Anfeindungen zu erwarten habe."

133. Purkyně (1876: 243): "Ueberhaupt ist Meteorologie … in die Oeffentlichkeit zu bringen."

134. Purkyně (1876: 243–244): "Ich habe, um solch schädlichen … selbstständige Meinung Ernst ist."

135. Purkyně (1876: 146): "wurden auch Stimmen laut … ein Steppenklima bekommen."

136. Lorenz von Liburnau (1890: 4): "ein wohlmeinender Praktiker … wie jene des Eiskellers und des Backofens."

137. Purkyně (1876: 179–204, 209–251, 267–291, 327–349, 405–426, 473–498; 1877: 102–143).

138. Purkyně (1875: 481–482): "Die Aenderungen der Witterung … gleich ein nasses und umgekehrt."

139. Purkyně (1875: 482): "und sind die Oststaaten … weit regenärmer sind."

140. Purkyně (1876: 414): "und selbst die allgemeinen … kalte Winter und geringe Regen hat."

141. Purkyně (1876: 147): "daß die, welche über Steppen- … Dünste selbstständig schaffen kann." See also Coen (2018: 257).

142. Purkyně (1875: 520–521): "Es ist unumgänglich nothwendig … Vorgänge einen Einsfluß ausübt."

143. Purkyně (1875: 506–507): "Sie beweisen, wie eigenthümlich das Klima … ist von dem nahen Walde"; Purkyně (1876: 139–140): "welche es gar nicht studirt … auf die Temperatur 2c. eines Landes."

144. Purkyně (1875: 508): "und wie leicht sich der Mensch … Einflusse des Waldes spüren lassen"; Purkyně (1876: 138): "wie die klimatologischen …

Länder mit einem Male zerstörten"; Purkyně (1876: 414): "Endlich haben die Ebermayer'schen ... darauf einwirken kann."

145. Purkyně (1875: 508): "Es verhält sich das Klima ... Bedingungen der Aenderung stattfinden."

146. Purkyně (1875: 508): "einer Unkenntniß der einfachsten physikalischen Geseze."

147. Purkyně (1876: 414): "Die Luftströme, welche bald feucht ... Luftmassen ausüben kann, ist verschwindend."

148. Lorenz von Liburnau (1890: 4): "und dass eine Reihe exacter Beobachtungen ... gründlicher Forschung zu gelangen."

149. Purkyně (1875: 506): "Das Interesse des Waldes ... von Forstmännern."

150. Purkyně (1876: 143): "so hat dieß doch für die Forstwirthschaft ... Kahlhiebe drohenden Gefahren."

151. Purkyně (1876: 146): "Wenn aber Forstmänner ... sonst ist alles verloren."

152. Purkyně (1876: 243): "Ich halte aber den Kampf ... für meine Pflicht."

153. Purkyně (1876: 243): "Maßt sich doch dieser meteorologische ... und des Privateigenthums bedrohen."

154. Purkyně (1875: 492): "Napoleon III. wußte ... Murren vollzogen werden konnten."

155. Purkyně (1876: 243): "Deutschland müsse darauf sehen ... die Elbe versiegen würde."

156. Purkyně (1876: 142): "Denn es liegt wohl auf der Hand ... dicht geschlossener Wald thun könnte."

157. Galvez and Gaillardet (2012).

158. Mirbel (1815: vol. 1, 451): "Tout est lié dans le vaste système de notre monde."

159. Bosson (1825: 7–8): no. 10.

160. Boussingault (1834: 172–173): "Il y a plusieurs causes ... l'acide carbonique atmosphérique."

161. Ébelmen (1845). See also Berner (2012), Galvez and Gaillardet (2012).

162. Ébelmen (1845: 66): "Les variations dans la nature ... sur cette importante question."

163. Hibberd (1855: 90). See also Mabey (2016: 183).

164. Schacht (1853: 347): "Die Waldungen sind mit dem Wohl ... von großer Bedeutung."

165. Hohenstein (1860: 155–156): "Ist es wohl nach diesem Allen ... aller Forstwirthe abhängt?"

166. Clavé (1862: 6): "Plus qu'aucune plante, l'arbre mérite notre reconnaissance."

167. Clavé (1862: 13): "Elles avaient dépouillé l'atmosphère ... en air respirable."

168. Emerson (1870: 128).

169. Ebermayer (1873: 237): "Es ist eine sehr weise Einrichtung ... Sauerstofffabrik bilden, die wir kennen."

170. Purkyně (1876: 176, note): "Es gibt fast nichts, was man dem Walde ... von Kohlensäure und hauche Sauerstoff aus."

171. Becquerel and Becquerel (1847: 300): "On ignore si les variations ... comme éléments météorologiques."

172. Becquerel (1853: 4, 5): "On a trouvé aussi qu'il y a moins d'acide carbonique ... sous l'influence solaire"; "Pour se faire une idée de la quantité ... à 2000 kilogrammes." For the calculation, see Becquerel and Becquerel (1847: 137–138n1).

173. Fleming (1998: 55–74), Hulme (2009), Jackson (2020).

174. Tyndall (1863: 204–205).

175. Ebermayer (1873: 152): "Die feuchtere Luft ... Klima so häufig vorkommen." See also p. 144: "Bei vollkommen trockener Luft ... die der Frost tödtet (Tyndall)."

176. Anders (1878a,b, 1882).

177. Anders (1878b: 805).

178. Anders (1878b: 807).

179. Anders (1887).

6 American Meteorologists Speak Out, c. 1850–1910

1. Lyell (1834: vol. 3, 116–119).

2. Thompson (1981), Fleming (1998: 33–45, 47–51).

3. Blodget (1857: 405–406, 481–492).

4. Blodget (1857: 482).

5. Loomis (1868: 157–158), Schott (1872: 136; 1876: 311). See also Thompson (1981), Fleming (1998: 50–51).

6. Egleston (1896: 52).

7. Blodget (1874).

8. Gannett (1888a).

9. Gannett (1888a: 3).

10. Gannett (1888b).
11. "The influence of forests on the quantity and frequency of rainfall," *Science* 12(303), 242–244 (November 23, 1888). See also Fernow (1889: 603–607).
12. Abbe (1893, 1899).
13. Abbe (1889).
14. Abbe (1889: 687).
15. Abbe (1894a: 57).
16. Abbe (1894a: 45).
17. Abbe (1894a: 45).
18. Abbe (1891a), Todd and Abbe (1891), Harrington (1893: 121).
19. Abbe (1905: 8).
20. Ferrel (1889).
21. Ferrel (1889: 434).
22. Egleston (1883: 453–455; 1884: 157–164; 1885: 184, 186, 192–196).
23. Fernow (1887: 151–154; 1888: 607; 1889: 602–618).
24. Fernow (1889: 603–607).
25. Gannett (1888a: 3).
26. Fernow (1889: 603).
27. Fernow (1893a: 9).
28. Fernow (1893a: 10).
29. Fernow (1889: 603).
30. Fernow (1893a: 22).
31. Hinsdale and Demmon (1906: 256–257).
32. Harrington (1876, 1877).
33. Harrington (1876).
34. Harrington (1893).
35. Harrington (1893: 75).
36. Harrington (1893: 101).
37. Harrington (1893: 105).
38. Harrington (1893: 89–95, 111–118).
39. Harrington (1887; 1893: 116).
40. Harrington (1893: 110).
41. Fernow (1889: 608).
42. Fernow (1893b).
43. Fernow (1889: 606–607).
44. Abbe (1893: 175).
45. *American Meteorological Journal*, 12(1), 1–4 (1895); *Smithsonian Meteorological Tables* (Washington, DC: Smithsonian Institution, 1893).
46. Curtis (1893: 187).
47. Curtis (1893: 191).
48. Curtis (1893: 189–190).
49. Sargent et al. (1897: 5).
50. Sargent et al. (1897: 6, 34).
51. Sargent (1882: 386).
52. Sargent (1882: 395).
53. Sargent (1882: 395).
54. Moore (1910: 3).
55. Moore (1910: 37).
56. Moore (1910: 7).
57. Moore (1910: 5).
58. Moore (1910: 7).
59. Moore (1910: 8).
60. Moore (1910: 14).
61. Moore (1910: 33).
62. Chittenden (1909).
63. Schiff (1962), Dobbs (1969), Saberwal (1998).
64. Fernow (1910).
65. Zon (1913).
66. Zon (1913: 148).
67. Zon (1912).
68. Brückner (1890: 179). For the translation, see Stehr and von Storch (2000: 98).
69. Brückner (1890: 185), Stehr and von Storch (2000: 104).
70. Brückner (1890: 194–195), Stehr and von Storch (2000: 116).
71. Hann (1897, vol. 1: 193–198). For an English translation, see Hann (1903: 192–197).
72. Hann (1903: 194–195).
73. Lowell (1906, 1908), Davis (2016b).
74. R. D. (1908). For an English translation, see "The forests of the planet Mars," *The Indian Forester*, 34, 725–726 (1908).
75. R. C. (1912).
76. Shapiro (2014).
77. Frankenfield (1910), Bates and Henry (1928).
78. Brooks (1928), Nicholson (1929), Hursh (1948).
79. Moon (2013: 124–125), Coen (2018: 270), Fressoz and Locher (2020: 179, 197).
80. Legg (2018).
81. Abbe (1891b: 67).
82. Abbe (1891b: 78).
83. Abbe (1891b: 79).
84. Abbe (1894b, 1895).
85. Abbe (1895: 712).
86. Geiger (1927, 1950). For an English translation of the 1927 edition, see Geiger (1942).
87. Fernow (1893a: 13–14).
88. Baudrillart (1823: 2–4). See p. 3 for the quote: "elles exercent sur l'atmosphère la plus heureuse influence." For Baudrillart, see Guyot (1898: 11–12).

89. Baudrillart (1823: 27): "la France ne seroit plus qu'un vaste désert."

90. Lorentz and Parade (1837: 16; 1883: 19): "de la plus haute importance pour l'état climatérique."

91. Lorey (1888: 22–59), Schlich (1889: 25–50), Nisbet (1905: vol. 1, 68–79).

92. *The Statutes at Large of the United States of America, from December, 1875, to March, 1877, and Recent Treaties, Postal Conventions, and Executive Proclamations*, vol. 19, p. 167 (Washington, DC: Government Printing Office, 1877).

93. Hough (1882: 2).

94. Fernow (1891: 9).

95. Pinchot (1905a: 7, 8).

96. Pinchot (1905a: 56).

97. Pinchot (1905a: 67).

98. Pinchot (1905b: 11; 2017: 42).

99. Munns (1930: 24).

100. Kotok (1940: 384).

101. Kotok (1940: 402).

102. Kittredge (1948: 13).

103. Thornthwaite (1956).

104. Thornthwaite (1956: 568).

105. Curtis (1956).

106. Willis and Hooke (2006), Lynch (2008).

107. Richardson (1922).

108. Richardson (1922: 111).

109. Brown and Escombe (1905), Brown and Wilson (1905).

110. Richardson (1922: 112).

111. Richardson (1922: 113).

112. Richardson (1922: 107–111).

113. Manabe et al. (1965), Kasahara and Washington (1967, 1971), Manabe (1969). See also Edwards (2011).

114. SCEP (1970), SMIC (1971).

115. SCEP (1970: 18–19, 76–77, 96–99), SMIC (1971: 18–19, 60–66, 166–179), Schneider and Dickinson (1974).

116. Lorenz (1970: 329).

117. Joseph Novitski, "Brazil Is Challenging a Last Frontier," *The New York Times*, July 7, 1970, pp. 1, 8.

118. Matthews et al. (1971: 447–448), Newell (1971).

119. Charney et al. (1975), Sagan et al. (1979), Shukla and Mintz (1982).

120. Charney (1975), Charney et al. (1975).

121. Bonan (2016a: 8).

122. Schneider and Dickinson (1974), Dickinson (1983, 1984).

123. Dickinson et al. (1986, 1993), Sellers et al. (1986, 1996b).

124. Dickinson et al. (1981).

125. Hack et al. (1993).

126. Dickinson and Henderson-Sellers (1988).

127. Bolin (1977).

128. Bonan (2016a, 2019), Bonan and Doney (2018), Blyth et al. (2021).

129. National Research Council (1986: 19).

130. Steffen et al. (2020).

7 Views of Forests

1. Mabey (2016).

2. Harrison (1992).

3. Williams (2003: 430).

4. Evelyn (1670: 225).

5. Klein (1995), Henshilwood and Marean (2003), Stringer (2016), Stringer and Galway-Witham (2017), Galway-Witham and Stringer (2018).

6. Stein et al. (2012, 2020), Boyce and Lee (2017), Berry (2019).

7. Crowther et al. (2015).

8. Pan et al. (2013).

9. Bar-On et al. (2018).

10. Crowther et al. (2015), Bar-On et al. (2018).

11. Uno et al. (2016).

12. Roberts et al. (2016).

13. Harrison (1992).

14. Clavé (1862: 12): "Le rôle des forêts . . . à recevoir son maître."

15. Photographs of the aforementioned trees are found in Pakenham (2002).

16. Strutt (1830: 116–118).

17. Rooke (1790: 13–14). See also plate 9.

18. Crane (2013: 22, 289n19, 313n3), Stafford (2016: 187–188), Drori (2018: 38).

19. Sallon et al. (2008, 2020).

20. Crowther et al. (2015).

21. See, for example, *Scientific American*, vol. 50, no. 14, p. 218 (April 5, 1884); Mann and Twiss (1910: 235).

22. Berkeley (1734: 36, 53, 62, 71). See also pp. 232–233, 285, 298–300, 323, 329 for further mentions of trees in the subsequent publication *Three Dialogues between Hylas and Philonous*.

23. FAO (2018: 4, 8).

24. Glacken (1967: 333–334), Young (1979: 1–6), Harrison (1992: 69–75).

25. Manwood (1598: 1), FitzNigel (1983: 60). See also Young (1979: 3, 22), James (1981: 4), Rackham (1986: 129–136).

26. Manwood (1598: 14).

27. Young (1979: 60–73), Robinson (2014), Van Bueren (2015).

28. Young (1979: 74–113), James (1981: 1–32), Rackham (1986: 136, 138).

29. Young (1979: 149–172), James (1981: 118–160).

30. Glacken (1967: 322–345, 491–494), James (1996), Warde (2006).

31. Glacken (1967: 484–491), James (1981: 130–131, 162–165), Hemery and Simblet (2014: 1–17).

32. Evelyn (1664). The quote is from the 2nd ed. See Evelyn (1670: 178).

33. Evelyn (1776).

34. Hemery and Simblet (2014).

35. Le Roy (1757).

36. Harrison (1992: 115–122).

37. Duhamel du Monceau (1755, 1758, 1760, 1764, 1767). See also Duhamel du Monceau (1767: vii): "Je n'aurois probablement … l'immensité d'opérations que je serois obligé d'exécuter."

38. Albion (1926: 231–280), Malone (1964).

39. Cronon (1983), Williams (1989: 9–21), Stegner (1990).

40. See the entry for June 15, 1756, in *John Adams Diary 1, 18 November 1755–29 August 1756, Adams Family Papers: An Electronic Archive*, Massachusetts Historical Society, Boston; masshist.org/digitaladams/archive/index.

41. Stegner (1990).

42. Grove (1995), Barton (2002).

43. Marsh (1864).

44. Totman (1989), Menzies (1994), Williams (2003).

45. Schlich (1910: 645).

46. Costanza et al. (1997, 2017).

47. Gould (1989: 106).

48. See the plate in Evelyn (1776: 500–501).

49. "The Cowthorpe Oak," *The Illustrated London News*, vol. 30, *Jan. to June 1857*, p. 83 (January 31,

1857); "Cowthorpe Oak," *Transactions and Proceedings of the Botanical Society of Edinburgh*, 22 (part III), 396–414 (1904); "The Cowthorpe Oak," *Nature*, 72, 43–44 (May 11, 1905).

50. Photograph of the "Cowthorpe Oak" by Benjamin Stone (1908), accession number E.1135–2001, Victoria and Albert Museum, London.

51. Evelyn (1664: 86–87).

52. See plates in Evelyn (1776: 502–503). See also Rooke (1790), plate 5.

53. "The Greendale Oak and Welbeck Abbey," *Garden and Forest: A Journal of Horticulture, Landscape Art and Forestry*, vol. 3, *January to December, 1890*, pp. 233–234, 239 (May 14, 1890); "Cowthorpe Oak," *Transactions and Proceedings of the Botanical Society of Edinburgh*, 22 (part III), 396–414 (1904).

54. Strutt (1822), p. 12, plate XII. For the photograph, see Pakenham (1996: 19).

55. Strutt (1822, 1830), "Introduction."

56. Payne (2017).

57. Pakenham (1996, 2002).

58. Skea (2015).

59. Skea (2013).

60. Stegner (1990).

61. Gould (1989: 97).

62. Humboldt (1846–58: vol. 1, xviii).

63. Humboldt (1846–58: vol. 2, 74–91).

64. Gould (1989).

65. Lazarus and Pardoe (2003).

66. Michaux (1819).

67. Loudon (1838).

68. Albert et al. (2019).

69. Chianese (2013).

70. Robin Pogrebin, "Maya Lin's 'Ghost Forest' Is a Climate Warning," *The New York Times*, November 14, 2019, C3; Holland Cotter, "In a Ghosted Forest, the Trees Talk Back," *The New York Times*, July 2, 2021, C11; Zachary Small, "Dying Trees Revived as Art, Then Crafts," *The New York Times*, November 25, 2021, C3.

71. See plates in Gilpin (1791: vol. 1, 106–107).

72. Leonardi and Stagi (2019: 154).

73. Darwin (1859: 129).

74. Geis (1981: 12–14, 168–177).

75. Hearn (2000: 210n4), Hogan (2014: 129–131, 170, 274–275).

76. Richard Powers, *The Overstory: A Novel* (W. W. Norton, New York, 2018).

77. Peter Wohlleben, *The Hidden Life of Trees: What They Feel, How They Communicate – Discoveries from a Secret World* (Greystone Books, Vancouver, 2016).

78. Suzanne Simard, *Finding the Mother Tree: Discovering the Wisdom of the Forest* (Alfred A. Knopf, New York, 2021).

79. Fiona Stafford, *The Long, Long Life of Trees* (Yale University Press, New Haven, 2016).

80. David George Haskell, *The Songs of Trees: Stories from Nature's Great Connectors* (Viking, New York, 2017).

81. Jonathan Drori, *Around the World in 80 Trees* (Laurence King, London, 2018).

82. Stegner (1990: 38).

83. Gilpin (1791).

84. Richard Fortey, *The Wood for the Trees: One Man's Long View of Nature* (Alfred A. Knopf, New York, 2016).

85. David George Haskell, *The Forest Unseen: A Year's Watch in Nature* (Viking, New York, 2012).

86. Sara Maitland, *Gossip from the Forest: The Tangled Roots of Our Forests and Fairytales* (Granta, London, 2012).

87. For Evelyn, see Higgins (2017: 76). For Michaux, see Thoreau (2009: 38–39, 250, 279) and Higgins (2017: 76). For Loudon, see Thoreau (2009: 39n187, 39n190, 141n225, 190n97, 250n192) and Thoreau (1863a: 146, 151, 155).

88. Thoreau (2009: 140).

89. Thoreau (1863a). See also Higgins (2017: 5, 77).

90. Higgins (2017).

91. See entry January 4, 1853, in Thoreau (1906: vol. 4, 448). See also Higgins (2017: 20).

92. Thoreau (1863b: 252). See also Higgins (2017: 60).

93. See entry March 19, 1859, in Thoreau (1906: vol. 12, 63, 64). See also Higgins (2017: 24–25).

94. Thoreau (2009: 213, 214).

95. Thoreau (2009: 111, 112). See also Higgins (2017: 110–111, 124).

96. Melville (1987: 227–229), Fantham (2004: 57, 76, 162). See also Carey (2012: 49), Barton (2017: 23).

97. Hunt (2016: 196–199).

98. Melville (1987: 228–229), Fantham (2004: 77, 162).

99. Baum (1900: 222).

100. Saunders (1993: 58–94).

101. Zipes (1988: 43–61), Harrison (1992: 167–176).

102. Zipes (1988: 43).

103. Zipes (1988: 45).

104. Irving (1848: 423–462).

105. Irving (1848: 452, 453).

106. Barton (2017).

107. A. A. Milne, *Winnie-the-Pooh* (E. P. Dutton, New York, 1926).

108. Harrison (1992: x).

109. Strassberg (1994: 281), from "The Wind-in-the-Pines Pavilion I." Reproduced with permission of University of California Press.

110. Hey (1837: vii).

111. Hey (1837: vi).

112. Hey (1837: 44).

113. Longfellow (1863: 220).

114. See entry September 16, 1857, in Thoreau (1906: vol. 10, 33). See also Higgins (2017: 120–121).

115. See entry January 2, 1859, in Thoreau (1906: vol. 11, 384). See also Higgins (2017: 200).

116. See entry May 17, 1860, in Thoreau (1906: vol. 13, 299). See also Higgins (2017: 192).

117. Muir (1894: 250).

118. Abu-Izzeddin (2013). See also Carey (2012: 78–81), Drori (2018: 75–76).

119. Schmidt (2019).

120. Harrison (1992: 13–18), Abu-Izzeddin (2013: 13, 71–73).

121. See, for example, Psalms 92:12, Song of Solomon 5:15, and Ezekiel 31:3 (New King James Version).

122. Douglas (1951: 185, 186). See also Abu-Izzeddin (2013: 56, 139).

123. Abu-Izzeddin (2013).

124. Oberthaler et al. (2018: 214–241).

125. Weppelmann (2019: 8).

126. Kemp (2008).

127. Sellink (2019: 342).

128. Oberthaler et al. (2018: 213n4).

129. Lamb (1977: 275–276; 1995: 233–235).

130. Forman and Godron (1986: 5–6).

131. Schuster (1973).

132. Lusardi (1973). See p. 1042 for the specific writing that More critiqued: "the church is a spiritual thing and no exterior thing but invisible from carnal eyes (I say not that they be invisible that be of the church / but that holy church in herself is invisible) as faith is." For analysis, see Parker (2008).

133. More (1973: 845).

134. Habenicht (1963).
135. Habenicht (1963: 29–32, 63–78).
136. Heywood (1546). See the second part, fourth chapter. See also Habenicht (1963: 145) for the original text.
137. An explanation of English proverbs is found in Bailey (1736). "Plenty is no dainty" means that we undervalue what we have in excess. The entry for "you cannot see wood for trees" says it is "a proverb spoken to them who over look things that are just before them." For other uses of the phrase, see Tilley (1950: 749–750).
138. Habenicht (1963: 214).
139. Ridgway (1872: 660, 661).

8 Global Physical Climatology

1. Bonan (2016a: 47).
2. Bonan (2016a: 79).
3. Bonan (2016a: 89–102).
4. Adapted from FAO (2020), Figure 5 (p. 14).
5. Bonan (2016a: 119); from original data in Jouzel et al. (2007) and Lüthi et al. (2008).
6. Data provided by Dr. Pieter Tans (Global Monitoring Laboratory, National Oceanic and Atmospheric Administration, Boulder, Colorado) and Dr. Ralph Keeling (Scripps Institution of Oceanography, La Jolla, California). The dataset is available from the NOAA Global Monitoring Laboratory (gml.noaa.gov/ccgg/trends/).
7. Arias et al. (2021).
8. Morice et al. (2021); updated through 2021 and available from the Met Office Hadley Centre (metoffice.gov.uk/hadobs/hadcrut5/).
9. Arias et al. (2021).
10. Otto-Bliesner et al. (2016).
11. O'Neill et al. (2016, 2017), Riahi et al. (2017).
12. Gidden et al. (2019), Meinshausen et al. (2020).

9 Forest Biometeorology

1. Park et al. (2013).
2. Evelyn (1776).
3. Bonan (2019: 153); from original data in Gates (1963).
4. Bonan (2016a: 197).
5. Miller et al. (2011).
6. Urbanski et al. (2007).
7. Thomas et al. (2009).
8. Goulden et al. (2006).
9. Bonan (2016a: 199); from original data in Bonan et al. (1997).
10. Bonan (2016a: 201); from original data in Jackson et al. (2008).
11. Shaw and Pereira (1982), Raupach (1994), Massman (1997), Nakai et al. (2008), Hu et al. (2020).
12. Jackson et al. (2008).
13. Leuzinger and Körner (2007), Aubrecht et al. (2016), Still et al. (2019, 2021).

10 Scientific Tools

1. Ryan (2013).
2. Medhurst et al. (2006), Barton et al. (2010), Ryan (2013).
3. Aubinet et al. (2012), Baldocchi (2014).
4. Pastorello et al. (2020).
5. Bonan (2019: 38); from original data in Whittaker et al. (1974).
6. Phillips et al. (2009).
7. Poyatos et al. (2016).
8. McLeod and Long (1999), Norby and Zak (2011), Jiang et al. (2020).
9. Meir et al. (2015).
10. Bonan (2016a: 175); from original data in Likens et al. (1977).
11. Schimel et al. (2019).
12. Bonan and Doney (2018). See also Bonan (2016a: 8).
13. Bonan (2016a: 467).

11 Forest Microclimates

1. Geiger (1927, 1950). For an English translation of the 1927 edition, see Geiger (1942).
2. Wright et al. (2004).
3. Niinemets and Anten (2009), Niinemets et al. (2015).
4. Hutchison et al. (1986).
5. Smith et al. (2019a).
6. Gu et al. (2003), Mercado et al. (2009).
7. Ellsworth and Reich (1993).

8. Raynor (1971).
9. Raupach (1988).
10. Geiger (1950: 343).
11. von Arx et al. (2012).
12. von Arx et al. (2013).
13. Raynor (1971), Matsuda et al. (1987), Chen et al. (1993), Carlson and Groot (1997), Morecroft et al. (1998), Davis et al. (2019).
14. De Freene et al. (2019).
15. Haesen et al. (2021).
16. De Frenne et al. (2013, 2019, 2021), Scheffers et al. (2013).
17. Geiger (1927: 153).
18. Bonan (2019: 269); from original data in Staudt et al. (2011).
19. Roberts et al. (1990). Tropical forest drawing adapted from Beard (1949), figure 42.
20. Thompson and Pinker (1975).
21. Lee et al. (2011), Zhang et al. (2014).

12 Water Yield

1. "The influence of forests on the quantity and frequency of rainfall," *Science*, 12(303), 242–244 (November 23, 1888).
2. Wilson (1898: 811).
3. Sheil and Murdiyarso (2009), Ellison et al. (2012, 2017), Sheil (2014).
4. Calder (2002, 2007), Calder et al. (2004).
5. Bennett and Barton (2018).
6. Bonan (2016a: 50).
7. Halley (1691).
8. Bonan (2016a: 156).
9. Dawson (1998).
10. Bruijnzeel et al. (2011).
11. Miralles et al. (2010), Carlyle-Moses and Gash (2011).
12. Bonan (2016a: 159).
13. Frankenfield (1910), Bates and Henry (1928), Andréassian (2004), McGuire and Likens (2011).
14. Bonan (2016a: 176); adapted from Borman and Likens (1979: 85). See also Hornbeck et al. (1970).
15. Postel and Thompson (2005), Foley et al. (2005).
16. Hibbert (1967), Bosch and Hewlett (1982), Andréassian (2004), Brown et al. (2005), Jackson et al. (2005).

17. Swank and Miner (1968), Swank and Douglass (1974), Swank et al. (1988), Ford et al. (2011).
18. Malmer et al. (2010).
19. Zhang et al. (2001), Oudin et al. (2008), Li et al. (2017), Zhang et al. (2017).
20. Stoy et al. (2006), Juang et al. (2007).
21. Chen et al. (2018).
22. Teuling et al. (2010).
23. Williams et al. (2012).
24. Teuling (2018).
25. Beven (2006).

13 Carbon Sequestration

1. Anderson-Teixeira et al. (2021).
2. Baldocchi and Penuelas (2019).
3. Graven et al. (2013), Forkel et al. (2016), Canadell et al. (2021).
4. Bonan (2016a: 416); adapted from Houghton (2005).
5. Bonan (2016a: 54); updated for Friedlingstein et al. (2020), Canadell et al. (2021).
6. Atmospheric CO_2 can be reported as a concentration (parts per million; ppm), the mass of CO_2 (petagram; Pg), or the mass of C. These are related as follows: 1 Pg C = 3.664 Pg CO_2; 1 ppm CO_2 = 2.124 Pg C.
7. Houghton (2005, 2013), Pongratz et al. (2014).
8. Ciais et al. (2019), Gaubert et al. (2019), Tagesson et al. (2020), Canadell et al. (2021).
9. Xu et al. (2021).
10. Arora et al. (2020), Canadell et al. (2021).
11. Walker et al. (2021).
12. Walker et al. (2021), Canadell et al. (2021), Keenan et al. (2021).
13. Medlyn et al. (2015), Canadell et al. (2021).
14. Canadell et al. (2021).
15. Humphrey et al. (2018, 2021), Gentine et al. (2019a).
16. Phillips et al. (2009), Lewis et al. (2011), Yang et al. (2018).
17. Gatti et al. (2021).
18. Friedlingstein et al. (2020).
19. Canadell et al. (2021).
20. Arora et al. (2020), Canadell et al. (2021).
21. Sitch et al. (2007), Lombardozzi et al. (2015).
22. van der Werf et al. (2017).

23. Kurz et al. (2008), Hicke et al. (2012), Berner et al. (2017).
24. Thornton et al. (2009), Zaehle et al. (2010, 2014), Wieder et al. (2015).
25. Galloway et al. (2003, 2008).
26. Gu et al. (2003), Mercado et al. (2009).
27. Cowan (1977), Cowan and Farquhar (1977).
28. Medrano et al. (2012).
29. McDowell et al. (2008, 2013), Adams et al. (2017).
30. Baldocchi and Penuelas (2019).
31. Jackson et al. (2005).
32. Keenan and Williams (2018), Walker et al. (2021), Canadell et al. (2021).
33. Priestley (1772: 199–200).
34. SCEP (1970: 74–75), Broecker (1970).
35. Hales (1727: 325).
36. Manning and Keeling (2006).
37. Keeling et al. (1996), Battle et al. (2000), Manning and Keeling (2006).

14 Forest Macroclimates

1. Mathieu (1878: 6): "Cette question très-controversée ... et que ce moyen est impraticable."
2. Geiger (1927:150): "Einzelne wissenschaftliche Versuche lassen sich, wie diese Fragestellung zeigt, dazu überhaupt nicht machen." For the translation, see Geiger (1942: 105).
3. Geiger (1927:150): "Man kann versuchen, das Problem theoretisch zu lösen."
4. Ferrel (1889: 433).
5. Shukla and Mintz (1982).
6. Shukla and Mintz (1982: 1498).
7. Kleidon et al. (2000).
8. Brovkin et al. (2009).
9. Bonan (2019).
10. Bonan (2016a: 11; 2016b).
11. Davin and de Noblet-Ducoudré (2010).
12. Bonan (2016a: 527); from original data in Davin and de Noblet-Ducoudré (2010).
13. Boiser et al. (2012), de Noblet-Ducoudré et al. (2012), Lejeune et al. (2017), Devaraju et al. (2018), Laguë et al. (2019), Boysen et al. (2020).
14. Bonan (2008), Jia et al. (2019).
15. Snyder et al. (2004), Li et al. (2016), Perugini et al. (2017), Devaraju et al. (2018), Jia et al. (2019).
16. Winckler et al. (2019b).
17. Boysen et al. (2020).
18. de Noblet-Ducoudré et al. (2012), Lejeune et al. (2017), Boysen et al. (2020), Davin et al. (2020).
19. Laguë et al. (2019), Boysen et al. (2020).
20. Teuling et al. (2017), Duveiller et al. (2021), Cerasoli et al. (2021).
21. Swann et al. (2010), Boysen et al. (2020), Laguë et al. (2021b).
22. Laguë et al. (2021a).
23. Davin and de Noblet-Ducoudré (2010), Swann et al. (2010).
24. Hansen et al. (2013), Song et al. (2018).
25. Lee et al. (2011).
26. Zhang et al. (2014).
27. Li et al. (2015), Alkama and Cescatti (2016), Bright et al. (2017), Schultz et al. (2017), Duveiller et al. (2018).
28. Juang et al. (2007), Burakowski et al. (2018), Zhang et al. (2020a).
29. Chen and Dirmeyer (2019), Winckler et al. (2019c), Boysen et al. (2020), Breil et al. (2020), Novick and Katul (2020).
30. Lee et al. (2011), Schultz et al. (2017), Burakowski et al. (2018).
31. Juang et al. (2007), Bright et al. (2017), Duveiller et al. (2018), Zhang et al. (2020a).
32. Laguë et al. (2019), Chen and Dirmeyer (2020).
33. Winckler et al. (2019a), Chen and Dirmeyer (2020), Pongratz et al. (2021).
34. Winckler et al. (2019a).
35. Williamson (1771), Jefferson (1954).
36. Anthes (1984), Segal et al. (1988), Taylor et al. (2007), Pielke et al. (2011), Mahmood et al. (2014).
37. Seth and Giorgi (1996).
38. Pielke et al. (1993).
39. Mahrt and Ek (1993).
40. Pielke et al. (1997).
41. Swann et al. (2012).
42. Bonan (2016a: 549); original figure from Swann et al. (2014).
43. Laguë and Swann (2016).
44. Devaraju et al. (2015).
45. Werth and Avissar (2002), Snyder (2010), Medvigy et al. (2013), Badger and Dirmeyer (2016), Jia et al. (2019).
46. Li et al. (2021).
47. Devaraju et al. (2018).

48. Chen and Dirmeyer (2017), Swann et al. (2018).
49. Li et al. (2020).
50. Lorenz et al. (2016).
51. Lejeune et al. (2018).
52. Findell et al. (2017).
53. Parsons et al. (2021).
54. Teuling et al. (2010).
55. Zaitchik et al. (2006).
56. Seneviratne et al. (2010), Berg et al. (2016), Miralles et al. (2019).
57. Fischer et al. (2007), Hirschi et al. (2011), Miralles et al. (2014), Dirmeyer et al. (2021).
58. Seager and Hoerling (2014), Schubert et al. (2016).
59. Koster et al. (2004).
60. Schubert et al. (2004).
61. Cook et al. (2009, 2011, 2016), Cowan et al. (2020).
62. Cook et al. (2008, 2009, 2011).
63. Sellers et al. (1996a).
64. Skinner et al. (2018), Schwingshackl et al. (2019), Zarakas et al. (2020).
65. Skinner et al. (2018), Zarakas et al. (2020).
66. Gedney et al. (2006), Betts et al. (2007), Lemordant et al. (2018), Fowler et al. (2019).
67. Kooperman et al. (2018).
68. Schwartz and Karl (1990), Hogg et al. (2000), Levis and Bonan (2004).
69. Zhu et al. (2016), Piao et al. (2020).
70. Zeng et al. (2017).
71. Chen et al. (2020).
72. Forzieri et al. (2017).
73. Zeng et al. (2017), Piao et al. (2020).
74. Xu et al. (2020).
75. Claussen et al. (2001), Brovkin et al. (2004), Matthews et al. (2004), Bala et al. (2007), Pongratz et al. (2010), Arora and Montenegro (2011), Boysen et al. (2014, 2020), Davies-Barnard et al. (2014), Simmons and Matthews (2016).
76. Betts (2000).
77. Anderson-Teixeira et al. (2012), Windisch et al. (2021).
78. Lathière et al. (2010), Unger (2013).
79. Unger (2014), Scott et al. (2018).
80. Scott et al. (2018).
81. Rap et al. (2018).
82. Shukla et al. (2019: 49).
83. Shukla et al. (2019: 47).

15 Case Studies

1. Matthews et al. (1971: 447–448), Newell (1971).
2. Dickinson and Henderson-Sellers (1988).
3. Lawrence and Vandecar (2015), Spracklen et al. (2018). See Swann et al. (2015), Pitman and Lorenz (2016), Jiang et al. (2021), and Li et al. (2022a) for specific modeling studies of Amazonian deforestation.
4. Eltahir and Bras (1994), Marengo et al. (2018), Staal et al. (2018), Gentine et al. (2019b).
5. Staal et al. (2018).
6. Wright et al. (2017).
7. Antonio Donato Nobre, "The magic of the Amazon: a river that flows invisibly all around us," TEDxAmazonia, November 2010; ted.com.
8. Worden et al. (2021).
9. Spracklen et al. (2012).
10. Baker and Spracklen (2019).
11. Davidson et al. (2012).
12. Spracklen and Garcia-Carreras (2015).
13. Lawrence and Vandecar (2015), Spracklen et al. (2018), Gentine et al. (2019b). See Wang et al. (2009) and Khanna et al. (2017) for specific case studies.
14. Khanna et al. (2017). Tropical forest drawing adapted from Beard (1949), figure 42.
15. Gentine et al. (2019b).
16. Gash et al. (1996), Gash and Nobre (1997), Nobre et al. (2004), von Randow et al. (2004), Restrepo-Coupe et al. (2021).
17. Alkama and Cescatti (2016), Duveiller et al. (2018), Baker and Spracklen (2019), Vargas Zeppetello et al. (2020).
18. Chapter 5: Berger (note 69), Hann (note 108), Lorenz (notes 125–127).
19. Spracklen et al. (2018: 206).
20. Martin et al. (2010), Pöschl et al. (2010), Artaxo et al. (2013, 2022).
21. Andreae et al. (2004), Koren et al. (2004, 2008), Spracklen et al. (2018), Thornhill et al. (2018), Liu et al. (2020), Herbert et al. (2021).
22. Williams et al. (2002), Artaxo et al. (2022).
23. Buffon (Chapter 2, note 58), Williamson (Chapter 3, note 49), Humboldt (Chapter 2, note 110).
24. Chapter 5, note 68.
25. Lenton et al. (2008, 2019), Steffen et al. (2018).

26. Oyama and Nobre (2003), Nepstad et al. (2008), Nobre and Borma (2009), Davidson et al. (2012), Lovejoy and Nobre (2018).
27. Lovejoy and Nobre (2018).
28. Zemp et al. (2017), Staal et al. (2020).
29. Hubau et al. (2020), Gatti et al. (2021).
30. Li et al. (2022a).
31. Boulton et al. (2022).
32. Lovelock and Margulis (1974), Lovelock (1979).
33. Beerling (2007).
34. Bonan (2016a: 393); from original data in Amiro et al. (2006).
35. Kuusinen et al. (2014), Halim et al. (2019), Hovi et al. (2019).
36. Bonan et al. (1992), Foley et al. (1994), Betts (2000), Snyder et al. (2004), Davin and de Noblet-Ducoudré (2010), Boysen et al. (2020).
37. Snyder et al. (2004), Davin and de Noblet-Ducoudré (2010), Boysen et al. (2020).
38. de Noblet et al. (1996), Gallimore and Kutzbach (1996), Levis et al. (1999), Meissner et al. (2003), Claussen et al. (2006), Davies-Barnard et al. (2017).
39. Foley et al. (1994), Claussen (2009).
40. Betts (2000), Windisch et al. (2021).
41. Baldocchi et al. (2000).
42. Swann et al. (2010).
43. Tunved et al. (2006).
44. Petäjä et al. (2022).
45. Kurtén et al. (2003), Lihavainen et al. (2009, 2015).
46. Spracklen et al. (2008), Scott et al. (2014).
47. Kulmala et al. (2004, 2014, 2020).
48. Paasonen et al. (2013), Yli-Juuti et al. (2021).
49. Kulmala et al. (2014, 2020).
50. Bonan (2016a: 620).
51. Kurtén et al. (2003), Kulmala et al. (2020).
52. Kalliokoski et al. (2020).
53. Liu et al. (2005), Amiro et al. (2006), Liu and Randerson (2008).
54. Randerson et al. (2006).
55. Rogers et al. (2013).
56. Potter et al. (2020).
57. Rogers et al. (2015).
58. Stuenzi and Schaepman-Strub (2020).
59. Lenton et al. (2008, 2019), Steffen et al. (2018).
60. Walker et al. (2019), Wang et al. (2021).
61. Larsen (1980), Bonan and Shugart (1989), Shugart et al. (1992), Hall et al. (2004).
62. Bonan et al. (1992).
63. Lejeune et al. (2017), Winckler et al. (2019c).
64. Lejeune et al. (2017), Boysen et al. (2020).
65. Boysen et al. (2020).
66. Lee et al. (2011), Zhang et al. (2014), Li et al. (2015), Alkama and Cescatti (2016), Bright et al. (2017), Duveiller et al. (2018).
67. Davin et al. (2020), Breil et al. (2020).
68. Teuling et al. (2017), Duveiller et al. (2021).
69. Meier et al. (2021).
70. Bonan (1997, 1999), Oleson et al. (2004).
71. Mueller et al. (2016), Alter et al. (2018).
72. Burakowski et al. (2016).
73. Juang et al. (2007), Zhang et al. (2020a).
74. Burakowski et al. (2018).
75. Chen et al. (2018).
76. Williams et al. (2021).
77. Edburg et al. (2012), Dhar et al. (2016), Negrón and Cain (2019).
78. Kurz et al. (2008), Mikkelson et al. (2013).
79. O'Halloran et al. (2012), Bright et al. (2013), Vanderhoof et al. (2013, 2014), Li et al. (2022b).
80. Bright et al. (2013), Maness et al. (2013), Vanderhoof and Williams (2015), Li et al. (2022b).
81. Biederman et al. (2015), Slinski et al. (2016), Goeking and Tarboton (2020).
82. Ren et al. (2021). See also Goeking and Tarboton (2020).
83. Dickerson-Lange et al. (2021).
84. Abatzoglou and Williams (2016), Willams et al. (2019), Goss et al. (2020), Parks and Abatzoglou (2020), Higuera and Abatzoglou (2021), Higuera et al. (2021), Zhuang et al. (2021), Iglesias et al. (2022).
85. Gleason et al. (2013, 2019).
86. Twohy et al. (2021).
87. Makar et al. (2021).
88. Liu et al. (2008), Feng et al. (2021).
89. Zhang et al. (2016b), Chen et al. (2019).
90. Peng et al. (2014), Ge et al. (2019).
91. Li et al. (2020), Yu et al. (2020).
92. Li et al. (2018b).
93. Feng et al. (2016), Ge et al. (2020).
94. Zhang et al. (2016a, 2020b).
95. Bonan (2016a: 530).
96. Chapter 3, note 44.
97. Chapter 3, note 45.
98. Marsh (1864: 182).

99. Chapter 5, note 31.
100. Lenton et al. (2008, 2019), Steffen et al. (2018).
101. Charney (1975), Charney et al. (1975).
102. Bonan (2016a: 531).
103. Charney et al. (1977).
104. Xue and Shukla (1993), Dirmeyer and Shukla (1996), Clark et al. (2001), Xue et al. (2004), Xue (2006), Wang et al. (2016), Chilukoti and Xue (2021).
105. Xue and Shukla (1996), Bamba et al. (2019).
106. Zeng et al. (1999), Wang and Eltahir (2000b,c), Wang et al. (2004), Zeng and Yoon (2009), Kucharski et al. (2013).
107. Foley et al. (2003), Giannini et al. (2008), Nicholson (2013), Pausata et al. (2020).
108. Folland et al. (1986), Giannini et al. (2003), Rodríguez-Fonseca et al. (2015).
109. Zeng et al. (1999).
110. Joussaume et al. (1999), Braconnot et al. (2000), Tierney et al. (2017).
111. Hoelzmann et al. (1998), Prentice et al. (2000), Bartlein et al. (2011).
112. Kutzbach et al. (1996), Levis et al. (2004), Claussen (2009), Swann et al. (2014), Chandan and Peltier (2020).
113. Brovkin et al. (1998), Claussen et al. (1999, 2003), Claussen (2009), Dallmeyer et al. (2020), Hopcroft and Valdes (2021).
114. Wang and Eltahir (2000a).
115. Black (1963), Black and Tarmy (1963).
116. Anthes (1984).
117. Ornstein et al. (2009), Kemena et al. (2018), Yosef et al. (2018), Branch and Wulfmeyer (2019), Ellison and Ifejika Speranza (2020).
118. Li et al. (2018a), Lu et al. (2021).
119. Li et al. (2018a).
120. Rotenberg and Yakir (2010, 2011).
121. Banerjee et al. (2017), Brugger et al. (2019).
122. Eder et al. (2015), Banerjee et al. (2018), Kröniger et al. (2018).
123. Rohatyn et al. (2018).

16 Climate-Smart Forests

1. Rogelj et al. (2018a,b), Jia et al. (2019), Roe et al. (2019).
2. Grassi et al. (2017), Griscom et al. (2017, 2020), Fargione et al. (2018), Roe et al. (2019, 2021), Drever et al. (2021).
3. Griscom et al. (2017).
4. Griscom et al. (2020).
5. Fargione et al. (2018).
6. Griscom et al. (2017: 11649).
7. Pan et al. (2013), Watson et al. (2018).
8. *USA Today*, December 8, 1992, 16A. Original graphic by Julie Stacey (*USA Today*), based on Bonan et al. (1992).
9. Kreidenweis et al. (2016), Griscom et al. (2017).
10. Drever et al. (2021).
11. Fargione et al. (2018).
12. Stephens et al. (2019), Ellis (2021).
13. Hurtt et al. (2020).
14. Crowther et al. (2015).
15. Bastin et al. (2019).
16. Riahi et al. (2017).
17. Popp et al. (2017), Riahi et al. (2017).
18. Hurtt et al. (2020).
19. Rogelj et al. (2018a,b).
20. See, for example, Griscom et al. (2017, 2020), Roe et al. (2021).
21. Feddema et al. (2005), Brovkin et al. (2013), Boysen et al. (2014), Davies-Barnard et al. (2014), Sonntag et al. (2016), Quesada et al. (2017), Hirsch et al. (2018).
22. Arora and Montenegro (2011).
23. Betts (2000), Randerson et al. (2006), Rotenberg and Yakir (2010), O'Halloran et al. (2012), Zhao and Jackson (2014), Bright et al. (2015), Bright and Lund (2021), Kalliokoski et al. (2020).
24. Anderson-Teixeira et al. (2012).
25. Davin et al. (2007).
26. Li et al. (2015), Alkama and Cescatti (2016), Bright et al. (2017), Duveiller et al. (2018, 2020).
27. Windisch et al. (2021).
28. Windisch et al. (2022).
29. Luyssaert et al. (2014), Erb et al. (2017).
30. Nabuurs et al. (2017), Yousefpour et al. (2018), Bowditch et al. (2020), Verkerk et al. (2020), Santopuoli et al. (2021).
31. Bowditch et al. (2020).
32. Bormann and Likens (1979), Shugart (1984, 1987, 1998).
33. Lewis et al. (2019), Moomaw et al. (2019), Forster et al. (2021).

34. Ter-Mikaelian et al. (2015), Hudiburg et al. (2019), Churkina et al. (2020), Schulze et al. (2020), Forster et al. (2021), Hurmekoski et al. (2021), Keith et al. (2021).
35. Fargione et al. (2008), Searchinger et al. (2009).
36. Smith et al. (2016, 2019b, 2022), Canadell et al. (2021).
37. Smith et al. (2016).
38. Canadell et al. (2021), Roe et al. (2021).
39. Watson et al. (2018), Strassburg et al. (2020), Smith et al. (2022).
40. De Frenne et al. (2013, 2019, 2021).
41. Vargas Zeppetello et al. (2022).
42. Ridgwell et al. (2009), Irvine et al. (2011), Caldeira et al. (2013), Seneviratne et al. (2018), Genesio et al. (2021).
43. Thompson et al. (2009), Lutz and Howarth (2014), Lutz et al. (2016), Favero et al. (2018), Lintunen et al. (2022).
44. Thompson et al. (2009), Lutz and Howarth (2014).
45. Matthies and Valsta (2016).
46. Astrup et al. (2018).
47. Shaw and Pereira (1982), Raupach (1994), Massman (1997), Nakai et al. (2008), Hu et al. (2020).
48. Zhang et al. (2022).
49. Zhang et al. (2021).
50. Jackson et al. (2005).
51. Swank and Miner (1968), Swank and Douglass (1974), Swank et al. (1988), Stoy et al. (2006), Ford et al. (2011).
52. Anderegg et al. (2019).
53. Teuling et al. (2010).
54. Cheng et al. (2017), Canadell et al. (2021).
55. Naudts et al. (2016).
56. Schwaab et al. (2020).
57. Luyssaert et al. (2018).
58. Lintunen et al. (2022).
59. Kellomäki et al. (2021).
60. Matthies and Valsta (2016).
61. Astrup et al. (2018), Pukkala (2018), Huuskonen et al. (2021).
62. Kalliokoski et al. (2020).
63. Bright et al. (2014).
64. Bright et al. (2017).
65. Bright et al. (2020).
66. Venäläinen et al. (2020).

17 Forests of the Future

1. Bonan (2016a: 424). Redrawn from Whittaker (1975: 167).
2. Giraud-Soulavie (1783: 265), Fressoz and Locher (2020: 62).
3. Humboldt and Bonpland (2009).
4. Morueta-Holme et al. (2015), Moret et al. (2019).
5. Bonan (2016a: 323). Redrawn from Whittaker (1956).
6. Holdridge (1967).
7. Boyce and Lee (2017), Berry (2019).
8. Stein et al. (2012, 2020).
9. Boyce and Lee (2017), Wilson et al. (2017).
10. Rothwell et al. (1997), Spencer et al. (2015), Leslie et al. (2018).
11. Rothwell et al. (2012), Leslie et al. (2018).
12. Crane (2013).
13. Boyce and Lee (2017), Soltis et al. (2018).
14. Franks et al. (2013, 2014), Rae et al. (2021).
15. McCurry et al. (2022).
16. Ferguson and Knobloch (1998). See also Crane (2013: 156–157).
17. Williams et al. (2003), Jahren (2007), Eberle and Greenwood (2012).
18. Beerling and Osborne (2002), Royer et al. (2003).
19. Pross et al. (2012).
20. Klages et al. (2020).
21. Bonan (2016a: 436). Redrawn from Solomon et al. (1981).
22. Overpeck et al. (1992), Webb et al. (1993), Williams et al. (2004).
23. Davis (1981), McLachlan (2005).
24. Bonan (2016a: 440). Redrawn from Davis (1981).
25. Huntley and Prentice (1993), Brewer et al. (2017).
26. Allen et al. (1996).
27. Bormann and Likens (1979), West et al. (1981), Shugart (1984, 1998).
28. Billings (1938), Oosting (1942), Keever (1950, 1983), Christensen and Peet (1981).
29. Likens et al. (1977), Bormann and Likens (1979).
30. Bonan (2016a: 384). Redrawn from Marks (1974).
31. Van Cleve and Viereck (1981), Van Cleve et al. (1983, 1986), Chapin et al. (2006).
32. Allen et al. (2010), Anderegg et al. (2020).
33. Senf et al. (2018, 2020).
34. McDowell et al. (2018), Brando et al. (2019).

35. Fasullo et al. (2021).
36. Negrón-Juárez et al. (2018).
37. Negrón-Juárez et al. (2010).
38. Anderegg et al. (2020), McDowell et al. (2020).
39. Rodman et al. (2020).

18 The Forests before Us

1. R. C. (1912).
2. Girardin et al. (2021).
3. Davis (2016a: 18).
4. Ellison et al. (2017: 57).
5. Coen (2018: 276).
6. Translation from Coen (2018: 299). For the original German text, see Kerner von Marilaun (1888: 18–19). For a contemporary translation, see Kerner von Marilaun (1894: 19).
7. Lamarck (1802: 6): "cette précision et cette scrupuleuse exactitude . . . et par l'habitude qu'il leur donne de ne voir et de ne s'occuper que de petites choses." See also Lamarck (1964: 17) for an English translation.
8. Becquerel (1865: 72–73): "Ces questions, ne pouvant être résolues . . . qu'une partie de la question."
9. Arago (1859: 443): "qu'elle est la conséquence inévitable de la nature de mes études . . . quand il le faut, je ne sais pais."
10. Hann (1869: 21): "Noch immer kann der Landwirth . . . schwer genug empfundene Thatsache."
11. Howe et al. (2019), van der Bles et al. (2020).
12. Burgess (2004: xxxii, 162). See also Saunders (1993: 204).
13. Rutkow (2012: 320, 377n320). See also Douglas E. Kneeland, "Teamsters Back Republican; Reagan Defends His Record on the Environment Issues," *The New York Times*, October 10, 1980, A1, D14.
14. Costanza et al. (1997, 2017).
15. Harrison (1992).
16. Wulf (2015).
17. Gould (1989).
18. Lawton (1999), Colyvan and Ginzburg (2003), Lange (2005), Linquist et al. (2016), Travassos-Britto et al. (2021).
19. Bonneuil and Fressoz (2016), Fressoz and Locher (2020).
20. Allan et al. (2021).
21. Kopenawa and Albert (2013: 381–400).
22. Kopenawa and Albert (2013: 385).
23. Kopenawa and Albert (2013: 393).
24. Kopenawa and Albert (2013: 385).
25. Kopenawa and Albert (2013: 385–386).
26. Kopenawa and Albert (2013: 399).
27. Evelyn (1670: 244; 1776: 645).
28. Lembke (2005: 23). Virgil wrote that a tree growing from seed in the wild "develops slowly, at last giving shade to our grandchildren."
29. Crane (2013: 8).
30. Hale (1758, vol. 1: 287).
31. *Annales européennes*, vol. 8, p. 341 (1825): "réconcilier avec la nature offensée." See also Fressoz and Locher (2015: 58).
32. For a description of the camp, see Arrington (1997).
33. James Doubek, "'Priceless' Bonsai Trees Stolen from Museum in Washington State," National Public Radio, February 11, 2020 (npr.org); Nicole Brodeur, "'They Are My Family': Stolen Bonsai Trees Mysteriously Returned to Federal Way Museum," *Seattle Times*, February 12, 2020 (seattletimes.com).

References

Abatzoglou, J. T., and Williams, A. P. (2016). Impact of anthropogenic climate change on wildfire across western US forests. *Proceedings of the National Academy of Sciences USA*, 113, 11770–11775.

Abbe, C. (1889). Is our climate changing? *The Forum*, 6(February), 678–688.

Abbe, C. (1891a). Cloud observations at sea. *American Meteorological Journal*, 8(6), 250–264.

Abbe, C. (1891b). A plea for terrestrial physics. *Proceedings of the American Association for the Advancement of Science*, 39, 65–79.

Abbe, C. (1893). Determination of the true amount of precipitation and its bearing on theories of forest influences. In *Forest Influences* (US Department of Agriculture, Forestry Division Bulletin Number 7), edited by B. E. Fernow. Washington, DC: Government Printing Office, pp. 175–186.

Abbe, C. (1894a). The relation of forests to climate and health. *Proceedings of the American Forestry Association*, 10, 45–57.

Abbe, C. (1894b). Schools of meteorology. *Nature*, 50, 576–577.

Abbe, C. (1895). Meteorology in the university. *Science*, 2(48), 709–714.

Abbe, C. (1899). The rain gage and the wind. *Monthly Weather Review*, 27, 464–468.

Abbe, C. (1905). *A First Report on the Relations between Climates and Crops*, US Department of Agriculture, Weather Bureau Bulletin Number 36. Washington, DC: Government Printing Office.

Abrams, M. D. (2001). Eastern white pine versatility in the presettlement forest. *BioScience*, 51, 967–979.

Abu-Izzeddin, F. (2013). *Memoirs of a Cedar: A History of Deforestation; A Future of Conservation*. Lebanon: Shouf Biosphere Reserve.

Adams, H. D., Zeppel, M. J. B., Anderegg, W. R. L., et al. (2017). A multi-species synthesis of physiological mechanisms in drought-induced tree mortality. *Nature Ecology and Evolution*, 1, 1285–1291.

Albert, B., Chandès, H., and Gaudefroy, I. (2019). *Trees*. New York: Thames & Hudson.

Albion, R. G. (1926). *Forests and Sea Power: The Timber Problem of the Royal Navy, 1652–1862*. Cambridge, MA: Harvard University Press.

Alkama, R., and Cescatti, A. (2016). Biophysical climate impacts of recent changes in global forest cover. *Science*, 351, 600–604.

Allan, R. P., Arias, P. A., Berger, S., et al. (2021). Summary for Policymakers. In *Climate Change 2021: The Physical Science Basis. Contribution of Working Group I to the Sixth Assessment Report of the Intergovernmental Panel on Climate Change*, edited by V. Masson-Delmotte, P. Zhai, A. Pirani, et al. Cambridge, UK: Cambridge University Press, pp. 3–32.

Allen, C. D., Macalady, A. K., Chenchouni, H., et al. (2010). A global overview of drought and heat-induced tree mortality reveals emerging climate change risks for forests. *Forest Ecology and Management*, 259, 660–684.

Allen, J. R. M., Huntley, B., and Watts, W. A. (1996). The vegetation and climate of northwest Iberia over the last 14 000 yr. *Journal of Quaternary Science*, 11, 125–147.

Alter, R. E., Douglas, H. C., Winter, J. M., and Eltahir, E. A. B. (2018). Twentieth century regional climate change during the summer in the central United States attributed to agricultural intensification. *Geophysical Research Letters*, 45, 1586–1594.

Amiro, B. D., Orchansky, A. L., Barr, A. G., et al. (2006). The effect of post-fire stand age on the boreal forest energy balance. *Agricultural and Forest Meteorology*, 140, 41–50.

Anderegg, W. R. L., Trugman, A. T., Badgley, G., et al. (2020). Climate-driven risks to the climate mitigation potential of forests. *Science*, 368, eaaz7005, DOI: https://doi.org/10.1126/science.aaz7005.

Anderegg, W. R. L., Trugman, A. T., Bowling, D. R., Salvucci, G., and Tuttle, S. E. (2019). Plant functional traits and climate influence drought intensification and land–atmosphere feedbacks. *Proceedings of the National Academy of Sciences USA*, 116, 14071–14076.

Anders, J. M. (1878a). On the transpiration of plants. *American Naturalist*, 12, 160–171.

Anders, J. M. (1878b). The beneficial influence of plants. *American Naturalist*, 12, 793–807.

Anders, J. M. (1882). Forests: Their influence upon climate and rainfall. *American Naturalist*, 16, 19–30.

Anders, J. M. (1887). *House-Plants as Sanitary Agents; or, the Relation of Growing Vegetation to Health and Disease*. Philadelphia: J. B. Lippincott.

Anderson-Teixeira, K. J., Herrmann, V., Morgan, R. B., et al. (2021). Carbon cycling in mature and regrowth forests globally. *Environmental Research Letters*, 16, 053009, DOI: https://doi.org/10.1088/1748-9326/abed01.

Anderson-Teixeira, K. J., Snyder, P. K., Twine, T. E., et al. (2012). Climate-regulation services of natural and agricultural ecoregions of the Americas. *Nature Climate Change*, 2, 177–181.

Andreae, M. O., Rosenfeld, D., Artaxo, P., et al. (2004). Smoking rain clouds over the Amazon. *Science*, 303, 1337–1342.

Andréassian, V. (2004). Waters and forests: From historical controversy to scientific debate. *Journal of Hydrology*, 291, 1–27.

Anthes, R. A. (1984). Enhancement of convective precipitation by mesoscale variations in vegetative covering in semiarid regions. *Journal of Climate and Applied Meteorology*, 23, 541–554.

Arago, F. (1859). De l'influence du déboisement sur les climats. In *Oeuvres complètes de François Arago*, vol. 12, edited by J.-A. Barral. Paris: Gide, pp. 432–443.

Arias, P. A., Bellouin, N., Coppola, E., et al. (2021). Technical Summary. In *Climate Change 2021: The Physical Science Basis. Contribution of Working Group I to the Sixth Assessment Report of the Intergovernmental Panel on Climate Change*, edited by V. Masson-Delmotte, P. Zhai, A. Pirani, et al. Cambridge, UK: Cambridge University Press, pp. 33–144.

Arora, V. K., and Montenegro, A. (2011). Small temperature benefits provided by realistic afforestation efforts. *Nature Geoscience*, 4, 514–518.

Arora, V. K., Katavouta, A., Williams, R. G., et al. (2020). Carbon–concentration and carbon–climate feedbacks in CMIP6 models and their comparison to CMIP5 models. *Biogeosciences*, 17, 4173–4222.

Arrington, L. J. (1997). *The Price of Prejudice: The Japanese-American Relocation Center in Utah during World War II*, 2nd ed. Logan: Utah State University.

Artaxo, P., Hansson, H.-C., Andreae, M. O., et al. (2022). Tropical and boreal forest–atmosphere interactions: A review. *Tellus B*, 74, 24–163.

Artaxo, P., Rizzo, L. V., Brito, J. F., et al. (2013). Atmospheric aerosols in Amazonia and land use change: From natural biogenic to biomass burning conditions. *Faraday Discussions*, 165, 203–235.

Astrup, R., Bernier, P. Y., Genet, H., Lutz, D. A., and Bright, R. M. (2018). A sensible climate solution for the boreal forest. *Nature Climate Change*, 8, 11–12.

Aubinet, M., Vesala, T., and Papale, D. (2012). *Eddy Covariance: A Practical Guide to Measurement and Data Analysis*. Dordrecht: Springer.

Aubrecht, D. M., Helliker, B. R., Goulden, M. L., et al. (2016). Continuous, long-term, high-frequency thermal imaging of vegetation: Uncertainties and recommended best practices. *Agricultural and Forest Meteorology*, 228–229, 315–326.

Aughey, S. (1878). Geology of Nebraska. In *Fourth Annual Report of the President and Secretary of the Nebraska State Board of Agriculture*. Lincoln: Journal Company, pp. 67–85.

Aughey, S. (1880). *Sketches of the Physical Geography and Geology of Nebraska*. Omaha: Daily Republican Book and Job Office.

Badger, A. M., and Dirmeyer, P. A. (2016). Remote tropical and sub-tropical responses to Amazon deforestation. *Climate Dynamics*, 46, 3057–3066.

Bailey, N. (1736). *Dictionarium Britannicum: Or a More Compleat Universal Etymological English Dictionary Than Any Extant*, 2nd ed. London: T. Cox.

Baker, J. C. A., and Spracklen, D. V. (2019). Climate benefits of intact Amazon forests and the biophysical consequences of disturbance. *Frontiers in Forests and Global Change*, 2, 47, DOI: https://doi.org/10.3389/ffgc.2019.00047.

Bala, G., Caldeira, K., Wickett, M., et al. (2007). Combined climate and carbon-cycle effects of large-scale deforestation. *Proceedings of the National Academy of Sciences USA*, 104, 6550–6555.

Baldocchi, D. (2014). Measuring fluxes of trace gases and energy between ecosystems and the atmosphere – the state and future of the eddy covariance method. *Global Change Biology*, 20, 3600–3609.

Baldocchi, D., and Penuelas, J. (2019). The physics and ecology of mining carbon dioxide from the atmosphere by ecosystems. *Global Change Biology*, 25, 1191–1197.

Baldocchi, D., Kelliher, F. M., Black, T. A., and Jarvis, P. (2000). Climate and vegetation controls on boreal zone energy exchange. *Global Change Biology*, 6(s1), 69–83.

Balfour, E. (1849). Notes on the influence exercised by trees in inducing rain and preserving moisture. *Madras Journal of Literature and Science*, 15(36), 402–448.

Bamba, A., Diallo, I., Touré, N. E., et al. (2019). Effect of the African greenbelt position on West African summer climate: A regional climate modeling study. *Theoretical and Applied Climatology*, 137, 309–322.

Banerjee, T., De Roo, F., and Mauder, M. (2017). Explaining the convector effect in canopy turbulence by means of large-eddy simulation. *Hydrology and Earth System Sciences*, 21, 2987–3000.

Banerjee, T., Brugger, P., De Roo, F., et al. (2018). Turbulent transport of energy across a forest and a semiarid shrubland. *Atmospheric Chemistry and Physics*, 18, 10025–10038.

Bar-On, Y. M., Phillips, R., and Milo, R. (2018). The biomass distribution on Earth. *Proceedings of the National Academy of Sciences USA*, 115, 6506–6511.

Bartlein, P. J., Harrison, S. P., Brewer, S., et al. (2011). Pollen-based continental climate reconstructions at 6 and 21 ka: A global synthesis. *Climate Dynamics*, 37, 775–802.

Barton, A. (2017). *The Shakespearean Forest*. Cambridge, UK: Cambridge University Press.

Barton, B. S. (1807). *A Discourse on Some of the Principal Desiderata in Natural History, and on the Best Means of Promoting the Study of this Science, in the United-States*. Philadelphia: Denham & Town.

Barton, C. V. M., Ellsworth, D. S., Medlyn, B. E., et al. (2010). Whole-tree chambers for elevated atmospheric CO_2 experimentation and tree scale flux measurements in south-eastern Australia: The Hawkesbury Forest Experiment. *Agricultural and Forest Meteorology*, 150, 941–951.

Barton, G. A. (2002). *Empire Forestry and the Origins of Environmentalism*. Cambridge, UK: Cambridge University Press.

Bastin, J.-F., Finegold, Y., Garcia, C., et al. (2019). The global tree restoration potential. *Science*, 365, 76–79.

Bates, C. G., and Henry, A. J. (1928). Forest and stream-flow experiment at Wagon Wheel Gap, Colo.: Final report, on completion of the second phase of the experiment. *Monthly Weather Review*, Supplement Number 30, 1–79.

Battle, M., Bender, M. L., Tans, P. P., et al. (2000). Global carbon sinks and their variability inferred from atmospheric O_2 and $\delta^{13}C$. *Science*, 287, 2467–2470.

Baudrillart, J.-J. (1823). *Traité général des eaux et forêts, chasses et pêches*. 2nd partie. *Dictionnaire général, raisonné et historique des eaux et forêts*, vol. 1. Paris: Huzard; Artus Bertrand; Warée oncle.

Baum, L. F. (1900). *The Wonderful Wizard of Oz*. Chicago: George M. Hill.

Beard, J. S. (1949). *The Natural Vegetation of the Windward & Leeward Islands*. Oxford: Clarendon Press.

Beatson, A. (1816). *Tracts Relative to the Island of St. Helena; Written During a Residence of Five Years*. London: W. Bulmer.

Beattie, J. (2003). Environmental anxiety in New Zealand, 1840–1941: Climate change, soil erosion, sand drift, flooding and forest conservation. *Environment and History*, 9, 379–392.

Beattie, J. (2009). Climate change, forest conservation and science: A case study of New Zealand, 1860s–1920. *History of Meteorology*, 5, 1–18.

Beattie, J., and Star, P. (2010). Global influences and local environments: Forestry and forest conservation in New Zealand, 1850s–1925. *British Scholar*, 3, 191–218.

Becquerel, A.-C. (1853). *Des climats et de l'influence qu'exercent les sols boisés et non boisés*. Paris: Firmin Didot Frères.

Becquerel, A.-C. (1860). *Recherches sur la température des végétaux et de l'air et sur celle du sol a diverses profondeurs*. Paris: Firmin Didot Frères, Fils et Cie.

Becquerel, A.-C. (1865). *Mémoire sur les forêts et leur influence climatérique*. Paris: Firmin Didot Frères, Fils, et Cie.

Becquerel, A.-C. (1867). Mémoire sur les principales causes qui influent sur les pluies. *Comptes rendus hebdomadaires des séances de l'Académie des sciences*, 64, 837–843.

Becquerel, A.-C. (1872). Forests and their climatic influence. In *Annual Report of the Board of Regents of the Smithsonian Institution, Showing the Operations, Expenditures, and Condition of the Institution for the Year 1869*. Washington, DC: Government Printing Office, pp. 394–416.

Becquerel, A.-C. (1878). Memoir upon forests, and their climatic influence. In *Report upon Forestry*, edited by F. B. Hough. Washington, DC: Government Printing Office, pp. 310–333.

Becquerel, A.-C., and Becquerel, E. (1847). *Éléments de physique terrestre et de météorologie*. Paris: Firmin Didot Frères.

Becquerel, A.-C., and Becquerel, E. (1866). Des pluies dans les lieux boisés et non boisés. *Comptes rendus hebdomadaires des séances de l'Académie des sciences*, 62, 855–858.

Becquerel, A.-C., and Becquerel, E. (1867). Extrait d'un mémoire sur les températures de l'air et les quantités d'eau tombées hors du bois et sous bois. *Comptes rendus hebdomadaires des séances de l'Académie des sciences*, 64, 16–19.

Becquerel, A.-C., and Becquerel, E. (1869a). Mémoire sur la température de l'air sous bois et hors des bois. *Comptes rendus hebdomadaires des séances de l'Académie des sciences*, 68, 677–682.

Becquerel, A.-C., and Becquerel, E. (1869b). Des quantités d'eau tombées près et loin des bois. *Comptes rendus hebdomadaires des séances de l'Académie des sciences*, 68, 789–793.

Beerling, D. (2007). *The Emerald Planet: How Plants Changed Earth's History*. Oxford: Oxford University Press.

Beerling, D. J., and Osborne, C. P. (2002). Physiological ecology of Mesozoic polar forests in a high CO_2 environment. *Annals of Botany*, 89, 329–339.

Bennett, B. M., and Barton, G. A. (2018). The enduring link between forest cover and rainfall: A historical perspective on science and policy discussions. *Forest Ecosystems*, 5, 5, DOI: https://doi.org/10.1186/s40663-017-0124-9.

Berg, A., Findell, K., Lintner, B., et al. (2016). Land-atmosphere feedbacks amplify aridity increase over land under global warming. *Nature Climate Change*, 6, 869–874.

Berger, Dr. (1865). Wald und Witterung. *Annalen der Physik und Chemie*, 124, 528–568.

Berkeley, G. (1734). *A Treatise Concerning the Principles of Human Knowledge… to Which Are Added Three Dialogues between Hylas and Philonous*, in *Opposition to Scepticks and Atheists*. London: Jacob Tonson.

Bernardin de Saint-Pierre, J.-H. (1784). *Etudes de la nature*, 3 vols. Paris: Pierre-François Didot.

Berner, L. T., Law, B. E., Meddens, A. J. H., and Hicke, J. A. (2017). Tree mortality from fires, bark beetles, and timber harvest during a hot and dry decade in the western United States (2003–2012). *Environmental Research Letters*, 12, 065005, DOI: https://doi.org/10.1088/1748-9326/aa6f94.

Berner, R. A. (2012). Jacques-Joseph Ébelmen, the founder of earth system science. *Comptes Rendus Geoscience*, 344, 544–548.

Berry, C. M. (2019). Palaeobotany: The rise of the Earth's early forests. *Current Biology*, 29, R792–R794.

Betts, R. A. (2000). Offset of the potential carbon sink from boreal forestation by decreases in surface albedo. *Nature*, 408, 187–190.

Betts, R. A., Boucher, O., Collins, M., et al. (2007). Projected increase in continental runoff due to plant responses to increasing carbon dioxide. *Nature*, 448, 1037–1041.

Beven, K. (2006). A manifesto for the equifinality thesis. *Journal of Hydrology*, 320, 18–36.

Biard, P. (1616). *Relation de la Nouvelle France, de ses terres, naturel du païs, & de ses habitans*. Lyon: Louys Muguet.

Biederman, J. A., Somor, A. J., Harpold, A. A., et al. (2015). Recent tree die-off has little effect on streamflow in contrast to expected increases from historical studies. *Water Resources Research*, 51, 9775–9789.

Billings, W. D. (1938). The structure and development of old field shortleaf pine stands and certain associated physical properties of the soil. *Ecological Monographs*, 8, 437–499.

Black, J. F. (1963). Weather control: Use of asphalt coatings to tap solar energy. *Science*, 139, 226–227.

Black, J. F., and Tarmy, B. L. (1963). The use of asphalt coatings to increase rainfall. *Journal of Applied Meteorology*, 2, 557–564.

Blanford, H. F. (1886). Influence of forests on rainfall. *Indian Meteorological Memoirs*, 3, 135–145.

Blanford, H. F. (1887). On the influence of Indian forests on the rainfall. *Journal of the Asiatic Society of Bengal*, Part II, 56, 1–15.

Blanford, H. F. (1888). Influence of forests on rainfall. *The Indian Forester*, 14, 34–47.

Blodget, L. (1857). *Climatology of the United States, and of the Temperate Latitudes of the North American Continent*. Philadelphia: J. B. Lippincott.

Blodget, L. (1874). Forest cultivation on the plains: The climate and cultivable capacity of the plains considered in regard to the ameliorations possible through greater protection by forests. In *Report of the Commissioner of Agriculture for the Year 1872*. Washington, DC: Government Printing Office, pp. 316–332.

Blome, R. (1672). *A Description of the Island of Jamaica; with the Other Isles and Territories in America, to which the English are Related*. London: T. Milbourn.

Blyth, E. M., Arora, V. K., Clark, D. B., et al. (2021). Advances in land surface modelling. *Current Climate Change Reports*, 7, 45–71.

Boisier, J. P., de Noblet-Ducoudré, N., Pitman, A. J., et al. (2012). Attributing the impacts of land-cover changes in temperate regions on surface temperature and heat fluxes to specific causes: Results from the first LUCID set of simulations. *Journal of Geophysical Research*, 117, D12116, DOI: https://doi.org/10.1029/2011JD017106.

Bolin, B. (1977). Changes of land biota and their importance for the carbon cycle. *Science*, 196, 613–615.

Bonan, G. B. (1997). Effects of land use on the climate of the United States. *Climatic Change*, 37, 449–486.

Bonan, G. B. (1999). Frost followed the plow: Impacts of deforestation on the climate of the United States. *Ecological Applications*, 9, 1305–1315.

Bonan, G. B. (2008). Forests and climate change: Forcings, feedbacks, and the climate benefits of forests. *Science*, 320, 1444–1449.

Bonan, G. B. (2016a). *Ecological Climatology: Concepts and Applications*, 3rd ed. Cambridge, UK: Cambridge University Press.

Bonan, G. B. (2016b). Forests, climate, and public policy: A 500-year interdisciplinary odyssey. *Annual Review of Ecology, Evolution, and Systematics*, 47, 97–121.

Bonan, G. B. (2019). *Climate Change and Terrestrial Ecosystem Modeling*. Cambridge, UK: Cambridge University Press.

Bonan, G. B., and Doney, S. C. (2018). Climate, ecosystems, and planetary futures: The challenge to predict life in Earth system models. *Science*, 359, eaam8328, DOI: https://doi.org/10.1126/science.aam8328.

Bonan, G. B., and Shugart, H. H. (1989). Environmental factors and ecological processes in boreal forests. *Annual Review of Ecology and Systematics*, 20, 1–28.

Bonan, G. B., Davis, K. J., Baldocchi, D., Fitzjarrald, D., and Neumann, H. (1997). Comparison of the NCAR LSM1 land surface model with BOREAS aspen and jack pine tower fluxes. *Journal of Geophysical Research*, 102D, 29065–29075.

Bonan, G. B., Pollard, D., and Thompson, S. L. (1992). Effects of boreal forest vegetation on global climate. *Nature*, 359, 716–718.

Bonneuil, C., and Fressoz, J.-B. (2016). *The Shock of the Anthropocene: The Earth, History and Us*, translated by D. Fernbach. London: Verso.

Bormann, F. H., and Likens, G. E. (1979). *Pattern and Process in a Forested Ecosystem*. New York: Springer-Verlag.

Bosch, J. M., and Hewlett, J. D. (1982). A review of catchment experiments to determine the effect of vegetation changes on water yield and evapotranspiration. *Journal of Hydrology*, 55, 3–23.

Bosson, M[onsieur] (1825). *Second mémoire en réponse a cette question: Quels sont les changemens que peut occasioner le déboisement de forêts considérables sur les contrées et communes adjacentes.…*. Brussels: P. J. de Mat.

Boulton, C. A., Lenton, T. M., and Boers, N. (2022). Pronounced loss of Amazon rainforest resilience since the early 2000s. *Nature Climate Change*, 12, 271–278.

Boussingault, J.-B. (1833). Mémoire sur la profondeur à laquelle se trouve la couche de température invariable entre les tropiques. Détermination de la température moyenne de la zône torride au niveau de la mer. Observations sur le décroissement de la chaleur dans les Cordilières. *Annales de chimie et de physique*, 53, 225–247.

Boussingault, J.-B. (1834). Recherches sur la composition de l'atmosphère. Premier Mémoire. Sur la possibilité de constater l'existence des miasmes. – Sur la présence d'un principe hydrogéné dans l'air. *Annales de chimie et de physique*, 57, 148–182.

Boussingault, J.-B. (1837). Mémoire sur l'influence des défrichemens dans la diminution des cours d'eau. *Annales de chimie et de physique*, 64, 113–141.

Boussingault, J.-B. (1838). Memoir concerning the effect which the clearing of land has in diminishing the quantity of water in the streams of the district. *Edinburgh New Philosophical Journal*, 24, 85–106.

Boussingault, J.-B. (1845). *Rural Economy, in its Relations with Chemistry, Physics, and Meteorology*, translated by G. Law. London: H. Bailliere.

Boussingault, J.-B. (1851). *Économie rurale considérée das ses rapports avec la chimie, la physique et la météorologie*, 2nd ed., 2 vols. Paris: Béchet Jeune.

Bowditch, E., Santopuoli, G., Binder, F., et al. (2020). What is Climate-Smart Forestry? A definition from a multinational collaborative process focused on mountain regions of Europe. *Ecosystem Services*, 43, 101113, DOI: https://doi.org/10.1016/j.ecoser.2020.101113.

Boyce, C. K., and Lee, J.-E. (2017). Plant evolution and climate over geological timescales. *Annual Review of Earth and Planetary Sciences*, 45, 61–87.

Boyle, R. (1671). Cosmicall suspitions (subjoyned as an appendix to the discourse of the cosmicall qualities of things). In *Tracts*. Oxford: W. H. for R. Davis, pp. 1–28.

Boysen, L. R., Brovkin, V., Arora, V. K., et al. (2014). Global and regional effects of land-use change on climate in 21st century simulations with interactive carbon cycle. *Earth System Dynamics*, 5, 309–319.

Boysen, L. R., Brovkin, V., Pongratz, J., et al. (2020). Global climate response to idealized deforestation in CMIP6 models. *Biogeosciences*, 17, 5615–5638.

Braconnot, P., Joussaume, S., de Noblet, N., and Ramstein, G. (2000). Mid-Holocene and Last Glacial Maximum African monsoon changes as simulated within the Paleoclimate Modelling Intercomparison Project. *Global and Planetary Change*, 26, 51–66.

Branch, O., and Wulfmeyer, V. (2019). Deliberate enhancement of rainfall using desert plantations. *Proceedings of the National Academy of Sciences USA*, 116, 18841–18847.

Brandis, D. (1887). Regen und Wald in Indien. *Meteorologische Zeitschrift*, 4, 369–376.

Brandis, D. (1888). The influence of forests on rainfall. *The Indian Forester*, 14, 10–20.

Brando, P. M., Paolucci, L., Ummenhofer, C. C., et al. (2019). Droughts, wildfires, and forest carbon cycling: A pantropical synthesis. *Annual Review of Earth and Planetary Sciences*, 47, 555–581.

Breil, M., Rechid, D., Davin, E. L., et al. (2020). The opposing effects of reforestation and afforestation on the diurnal temperature cycle at the surface and in the lowest atmospheric model level in the European summer. *Journal of Climate*, 33, 9159–9179.

Brewer, S., Giesecke, T., Davis, B. A. S., et al. (2017). Late-glacial and Holocene European pollen data. *Journal of Maps*, 13, 921–928.

Brewster, D. (1830). *The Edinburgh Encyclopædia*, 4th ed., 18 vols. Edinburgh: William Blackwood; John Waugh; and others.

Bright, B. C., Hicke, J. A., and Meddens, A. J. H. (2013). Effects of bark beetle-caused tree mortality on biogeochemical and biogeophysical MODIS products. *Journal of Geophysical Research: Biogeosciences*, 118, 974–982.

Bright, R. M., Allen, M., Antón-Fernández, C., et al. (2020). Evaluating the terrestrial carbon dioxide removal potential of improved forest management and accelerated forest conversion in Norway. *Global Change Biology*, 26, 5087–5105.

Bright, R. M., Antón-Fernández, C., Astrup, R., et al. (2014). Climate change implications of shifting forest management strategy in a boreal forest ecosystem of Norway. *Global Change Biology*, 20, 607–621.

Bright, R. M., Davin, E., O'Halloran, T., et al. (2017). Local temperature response to land cover and management change driven by non-radiative processes. *Nature Climate Change*, 7, 296–302.

Bright, R. M., and Lund, M. T. (2021). CO_2-equivalence metrics for surface albedo change based on the radiative forcing concept: A critical review. *Atmospheric Chemistry and Physics*, 21, 9887–9907.

Bright, R. M., Zhao, K., Jackson, R. B., and Cherubini, F. (2015). Quantifying surface albedo and other direct biogeophysical climate forcings of forestry activities. *Global Change Biology*, 21, 3246–3266.

Brincken, J. (1828). *Mémoire descriptif sur la forêt impériale de Białowieża, en Lithuanie*. Varsovie: N. Glücksberg.

Broecker, W. S. (1970). Man's oxygen reserves. *Science*, 168, 1537–1538.

Brooks, C. E. P. (1928). The influence of forests on rainfall and run-off. *Quarterly Journal of the Royal Meteorological Society*, 54, 1–17.

Brouard, N. R. (1963). *A History of Woods and Forests in Mauritius*. Port Louis, Mauritius: J. E. Félix.

Brovkin, V., Boysen, L., Arora, V. K., et al. (2013). Effect of anthropogenic land-use and land-cover changes on climate and land carbon storage in CMIP5 projections for the twenty-first century. *Journal of Climate*, 26, 6859–6881.

Brovkin, V., Claussen, M., Petoukhov, V., and Ganopolski, A. (1998). On the stability of the atmosphere-vegetation system in the Sahara/Sahel region. *Journal of Geophysical Research*, 103D, 31613–31624.

Brovkin, V., Raddatz, T., Reick, C. H., Claussen, M., and Gayler, V. (2009). Global biogeophysical interactions between forest and climate. *Geophysical Research Letters*, 36, L07405, DOI: https://doi.org/10.1029/2009GL037543.

Brovkin, V., Sitch, S., von Bloh, W., et al. (2004). Role of land cover changes for atmospheric CO_2 increase and climate change during the last 150 years. *Global Change Biology*, 10, 1253–1266.

Brown, A. E., Zhang, L., McMahon, T. A., Western, A. W., and Vertessy, R. A. (2005). A review of paired catchment studies for determining changes in water yield resulting from alterations in vegetation. *Journal of Hydrology*, 310, 28–61.

Brown, H. T., and Escombe, F. (1905). Researches on some of the physiological processes of green leaves, with special reference to the interchange of energy between the leaf and its surroundings. *Proceedings of the Royal Society of London B*, 76, 29–111.

Brown, H. T., and Wilson, W. E. (1905). On the thermal emissivity of a green leaf in still and moving air. *Proceedings of the Royal Society London B*, 76, 122–137.

Brown, J. C. (1875). *Hydrology of South Africa; or Details of the Former Hydrographic Condition of the Cape of Good Hope, and of Causes of its Present Aridity, with Suggestions of Appropriate Remedies for this Aridity*. London: Henry S. King.

Brown, J. C. (1876). *Reboisement in France: Or, Records of the Replanting of the Alps, the Cevennes, and the Pyrenees with Trees, Herbage, and Bush*. London: Henry S. King.

Brown, J. C. (1877a). *Forests and Moisture; or Effects of Forests on Humidity of Climate*. Edinburgh: Oliver & Boyd.

Brown, J. C. (1877b). *Water Supply of South Africa and Facilities for the Storage of It*. Edinburgh: Oliver & Boyd.

Brown, J. C. (1883). *French Forest Ordinance of 1669; with Historical Sketch of Previous Treatment of Forests in France*. Edinburgh: Oliver & Boyd.

Brown, J. C. (1887). *Management of Crown Forests at the Cape of Good Hope under the Old Regime and under the New*. Edinburgh: Oliver & Boyd.

Brown, R. H. (1951). The seaboard climate in the view of 1800. *Annals of the Association of American Geographers*, 41, 217–232.

Brückner, E. (1890). Klimaschwankungen seit 1700 nebst Bemerkungen über die Klimaschwankungen der Diluvialzeit. *Penck's Geographische Abhandlungen*, 4, 153–484.

Brugger, P., De Roo, F., Kröniger, K., et al. (2019). Contrasting turbulent transport regimes explain cooling effect in a semi-arid forest compared to surrounding shrubland. *Agricultural and Forest Meteorology*, 269–270, 19–27.

Bruijnzeel, L. A., Mulligan, M., and Scatena, F. N. (2011). Hydrometeorology of tropical montane cloud forests: Emerging patterns. *Hydrological Processes*, 25, 465–498.

Buffon, G.-L. Leclerc, comte de (1778). *Histoire naturelle, générale et particulière*. Supplément, tome cinquième. *Des époques de la nature*. Paris: Imprimerie royale.

Burakowski, E. A., Ollinger, S. V., Bonan, G. B., et al. (2016). Evaluating the climate effects of reforestation in New England using a Weather Research and Forecasting (WRF) Model multiphysics ensemble. *Journal of Climate*, 29, 5141–5156.

Burakowski, E., Tawfik, A., Ouimette, A., et al. (2018). The role of surface roughness, albedo, and Bowen ratio on ecosystem energy balance in the Eastern United States. *Agricultural and Forest Meteorology*, 249, 367–376.

Burgess, G. S. (2004). *The History of the Norman People: Wace's "Roman de Rou,"* translated by G. S. Burgess, notes by G. S. Burgess and E. van Houts. Woodbridge, UK: Boydell Press.

Butter, D. (1839). *Outlines of the Topography and Statistics of the Southern Districts of Oud'h, and of the Cantonment of Sultanpur-Oud'h*. Calcutta: G. H. Huttmann.

Caldeira, K., Bala, G., and Cao, L. (2013). The science of geoengineering. *Annual Review of Earth and Planetary Sciences*, 41, 231–256.

Calder, I. R. (2002). Forests and hydrological services: Reconciling public and science perceptions. *Land Use and Water Resources Research*, 2, 1–12.

Calder, I. R. (2007). Forests and water: Ensuring forest benefits outweigh water costs. *Forest Ecology and Management*, 251, 110–120.

Calder, I., Amezaga, J., Aylward, B., et al. (2004). Forest and water policies. The need to reconcile public and science perceptions. *Geologica Acta*, 2, 157–166.

Campbell Walker, I. (1876a). State forestry: Its aim and object. *Transactions and Proceedings of the New Zealand Institute*, 9, 187–203.

Campbell Walker, I. (1876b). The climatic and financial aspect of forest conservancy as applicable to New Zealand. *Transactions and Proceedings of the New Zealand Institute*, 9, Appendix, pp. xxvii–xlix.

Campbell Walker, I. (1877). Report of the conservator of state forests. In *Appendix to the Journals of the House of Representatives of New Zealand*, vol. 1. Wellington: George Didsbury, pp. C3:1–59.

Canadell, J. G., Monteiro, P. M. S., Costa, M. H., et al. (2021). Global carbon and other biogeochemical cycles and feedbacks. In *Climate Change 2021: The Physical Science Basis. Contribution of Working Group I to the Sixth Assessment Report of the Intergovernmental Panel on Climate Change*, edited by V. Masson-Delmotte, P. Zhai, A. Pirani, et al. Cambridge, UK: Cambridge University Press, pp. 673–816.

Capron, H. (1869). *Report of the Commissioner of Agriculture for the Year 1868*. Washington, DC: Government Printing Office.

Capron, H. (1870). *Report of the Commissioner of Agriculture for the Year 1869*. Washington, DC: Government Printing Office.

Capron, H. (1871). *Report of the Commissioner of Agriculture for the Year 1870*. Washington, DC: Government Printing Office.

Carey, F. (2012). *The Tree: Meaning and Myth*. Burlington, VT: Lund Humphries.

Carlson, D. W., and Groot, A. (1997). Microclimate of clear-cut, forest interior, and small openings in trembling aspen forest. *Agricultural and Forest Meteorology*, 87, 313–329.

Carlyle-Moses, D. E., and Gash, J. H. C. (2011). Rainfall interception loss by forest canopies. In *Forest Hydrology and Biogeochemistry: Synthesis of Past Research and Future Directions*, edited by D. F. Levia, D. Carlyle-Moses, and T. Tanaka. Dordrecht: Springer, pp. 407–423.

Cerasoli, S., Yin, J., and Porporato, A. (2021). Cloud cooling effects of afforestation and reforestation at midlatitudes. *Proceedings of the National Academy of Sciences USA*, 118, e2026241118, DOI: https://doi.org/10.1073/pnas.2026241118.

Cézanne, E. (1870–72). *Étude sur les torrents des hautes-alpes par Alexandre Surell, 2e édition, avec une suite par Ernest Cézanne*, 2 vols. Paris: Dunod.

Chandan, D., and Peltier, W. R. (2020). African Humid Period precipitation sustained by robust vegetation, soil, and lake feedbacks. *Geophysical Research Letters*, 47, e2020GL088728, DOI: https://doi.org/10.1029/2020GL088728.

Chapin, F. S., III, Oswood, M. W., Van Cleve, K., Viereck, L. A., and Verbyla, D. L. (2006). *Alaska's Changing Boreal Forest*. Oxford: Oxford University Press.

Charney, J. G. (1975). Dynamics of deserts and drought in the Sahel. *Quarterly Journal of the Royal Meteorological Society*, 101, 193–202.

Charney, J., Stone, P. H., and Quirk, W. J. (1975). Drought in the Sahara: A biogeophysical feedback mechanism. *Science*, 187, 434–435.

Charney, J., Quirk, W. J., Chow, S.-H., and Kornfield, J. (1977). A comparative study of the effects of albedo change on drought in semi-arid regions. *Journal of the Atmospheric Sciences*, 34, 1366–1385.

Chastellux, F.-J., marquis de (1963). *Travels in North America in the Years 1780, 1781 and 1782*, translated with introduction and notes by H. C. Rice, Jr., 2 vols. Chapel Hill: University of North Carolina Press.

Chekhov, A. (2011). *Five Plays. Antov Chekhov*, translated by M. Brodskaya, introduction by T. Wolff. Stanford: Stanford University Press.

Chen, C., Li, D., Li, Y., et al. (2020). Biophysical impacts of Earth greening largely controlled by aerodynamic resistance. *Science Advances*, 6, eabb1981, DOI: https://doi.org/10.1126/sciadv.abb1981.

Chen, C., Park, T., Wang, X., et al. (2019). China and India lead in greening of the world through land-use management. *Nature Sustainability*, 2, 122–129.

Chen, J., Franklin, J. F., and Spies, T. A. (1993). Contrasting microclimates among clearcut, edge, and interior of old-growth Douglas-fir forest. *Agricultural and Forest Meteorology*, 63, 219–237.

Chen, L., and Dirmeyer, P. A. (2017). Impacts of land-use/land-cover change on afternoon precipitation over North America. *Journal of Climate*, 30, 2121–2140.

Chen, L., and Dirmeyer, P. A. (2019). Differing responses of the diurnal cycle of land surface and air temperatures to deforestation. *Journal of Climate*, 32, 7067–7079.

Chen, L., and Dirmeyer, P. A. (2020). Reconciling the disagreement between observed and simulated temperature responses to deforestation. *Nature Communications*, 11, 202, DOI: https://doi.org/10.1038/s41467-019-14017-0.

Chen, L., Dirmeyer, P. A., Guo, Z., and Schultz, N. M. (2018). Pairing FLUXNET sites to validate model representations of land-use/land-cover change. *Hydrology and Earth System Sciences*, 22, 111–125.

Cheng, L., Zhang, L., Wang, Y.-P., et al. (2017). Recent increases in terrestrial carbon uptake at little cost to the water cycle. *Nature Communications*, 8, 110, DOI: https://doi.org/10.1038/s41467-017-00114-5.

Chianese, R. L. (2013). Regeneration on Tree Mountain. *American Scientist*, 101(5), 350–351.

Chilukoti, N., and Xue, Y. (2021). An assessment of potential climate impact during 1948–2010 using historical land use land cover change maps. *International Journal of Climatology*, 41, 295–315.

Chinard, G. (1945). The American Philosophical Society and the early history of forestry in America. *Proceedings of the American Philosophical Society*, 89, 444–488.

Chittenden, H. M. (1909). Forests and reservoirs in their relation to stream flow, with particular reference to navigable rivers. *Transactions of the American Society of Civil Engineers*, 62, 245–546.

Christensen, N. L., and Peet, R. K. (1981). Secondary forest succession on the North Carolina Piedmont. In *Forest Succession: Concepts and Application*, edited by D. C. West, H. H. Shugart, and D. B. Botkin. New York: Springer-Verlag, pp. 230–245.

Churkina, G., Organschi, A., Reyer, C. P. O., et al. (2020). Buildings as a global carbon sink. *Nature Sustainability*, 3, 269–276.

Ciais, P., Tan, J., Wang, X., et al. (2019). Five decades of northern land carbon uptake revealed by the interhemispheric CO_2 gradient. *Nature*, 568, 221–225.

Clark, D. B., Xue, Y., Harding, R. J., and Valdes, P. J. (2001). Modeling the impact of land surface degradation on the climate of tropical North Africa. *Journal of Climate*, 14, 1809–1822.

Clarke, W. B. (1835). Instances of the effects of forest vegetation on climate. *Magazine of Natural History, and Journal of Zoology, Botany, Mineralogy, Geology, and Meteorology*, 8, 473–482.

Clarke, W. B. (1876). Effects of forest vegetation on climate. *Journal and Proceedings of the Royal Society of New South Wales*, 10, 179–235.

Claussen, M. (2009). Late Quaternary vegetation–climate feedbacks. *Climate of the Past*, 5, 203–216.

Claussen, M., Brovkin, V., and Ganopolski, A. (2001). Biogeophysical versus biogeochemical feedbacks of large-scale land cover change. *Geophysical Research Letters*, 28, 1011–1014.

Claussen, M., Brovkin, V., Ganopolski, A., Kubatzki, C., and Petoukhov, V. (2003). Climate change in northern Africa: The past is not the future. *Climatic Change*, 57, 99–118.

Claussen, M., Fohlmeister, J., Ganopolski, A., and Brovkin, V. (2006). Vegetation dynamics amplifies precessional forcing. *Geophysical Research Letters*, 33, L09709, DOI: https://doi.org/10.1029/2006GL026111.

Claussen, M., Kubatzki, C., Brovkin, V., et al. (1999). Simulation of an abrupt change in Saharan vegetation in the mid-Holocene. *Geophysical Research Letters*, 26, 2037–2040.

Clavé, J. (1862). *Études sur l'économie forestière*. Paris: Guillaumin.

Clavé, J. (1875). Étude de météorologie forestière. *Revue des deux mondes*, troisième période, 9(3), 632–649.

Clayton, J. (1693). A letter from Mr. John Clayton Rector of Crofton at Wakefield in Yorkshire to the Royal Society, May 12, 1688, giving an account of several observables in Virginia, and in his voyage thither, more particularly concerning the air. *Philosophical Transactions*, 17(201), 781–795.

Cleaveland, P. (1809). Meteorological observations, made at Bowdoin College. *Memoirs of the American Academy of Arts and Sciences*, 3(1), 119–121.

Cleghorn, H., Royle, J. F., Baird Smith, R., and Strachey, R. (1852). Report of the committee appointed by the British Association to consider the probable effects in an economical and physical point of view of the destruction of tropical forests. In *Report of the Twenty-First Meeting of the British Association for the Advancement of Science; Held at Ipswich in July 1851*. London: John Murray, pp. 78–102.

Coates, C., and Degroot, D. (2015). "Les bois engendrent les frimas et les gelées": compendre le climat en Nouvelle-France. *Revue d'histoire de l'Amérique française*, 68, 197–219.

Coen, D. R. (2018). *Climate in Motion: Science, Empire, and the Problem of Scale*. Chicago: University of Chicago Press.

Cogbill, C. V., Burk, J., and Motzkin, G. (2002). The forests of presettlement New England, USA: Spatial and compositional patterns based on town proprietor surveys. *Journal of Biogeography*, 29, 1279–1304.

Colón, F. (1571). *Historie del S. D. Fernando Colombo; nelle quali s'ha particolare, & vera relatione della vita, & de' fatti dell'Ammiraglio D. Christoforo Colombo, suo padre*. Venice: Francesco de' Franceschi Sanese.

Colón, F. (1959). *The Life of the Admiral Christopher Columbus by His Son Ferdinand*, translated and annotated by B. Keen. New Brunswick, NJ: Rutgers University Press.

Colyvan, M., and Ginzburg, L. R. (2003). Laws of nature and laws of ecology. *Oikos*, 101, 649–653.

Conant, J. B. (1950). *The Overthrow of the Phlogiston Theory: The Chemical Revolution of 1775–1789* (Harvard Case Histories in Experimental Science, Case 2). Cambridge, MA: Harvard University Press.

Cook, B. I., Cook, E. R., Smerdon, J. E., et al. (2016). North American megadroughts in the Common Era: Reconstructions and simulations. *WIREs Climate Change*, 7, 411–432.

Cook, B. I., Miller, R. L., and Seager, R. (2008). Dust and sea surface temperature forcing of the 1930s "Dust Bowl" drought. *Geophysical Research Letters*, 35, L08710, DOI: https://doi.org/10.1029/2008GL033486.

Cook, B. I., Miller, R. L., and Seager, R. (2009). Amplification of the North American "Dust Bowl" drought through human-induced land degradation. *Proceedings of the National Academy of Sciences USA*, 106, 4997–5001.

Cook, B. I., Seager, R., and Miller, R. L. (2011). Atmospheric circulation anomalies during two persistent North American droughts: 1932–1939 and 1948–1957. *Climate Dynamics*, 36, 2339–2355

Cooper, E. (1876). *Forest Culture and Eucalyptus Trees*. San Francisco: Cubery.

Costanza, R., d'Arge, R., de Groot, R., et al. (1997). The value of the world's ecosystem services and natural capital. *Nature*, 387, 253–260.

Costanza, R., de Groot, R., Braat, L., et al. (2017). Twenty years of ecosystem services: How far have we come and how far do we still need to go? *Ecosystem Services*, 28 (part A), 1–16.

Costlow, J. (2003). Imaginations of destruction: The "forest question" in nineteenth-century Russian culture. *The Russian Review*, 62, 91–118.

Cotta, H. (1832). *Grundriß der Forstwissenschaft*. Dresden & Leipzig: Arnold.

Cowan, I. R. (1977). Stomatal behaviour and environment. *Advances in Botanical Research*, 4, 117–228.

Cowan, I. R., and Farquhar, G. D. (1977). Stomatal function in relation to leaf metabolism and environment. In *Integration of Activity in the Higher Plant*, edited by D. H. Jennings.

Cambridge, UK: Cambridge University Press, pp. 471–505.

Cowan, T., Hegerl, G. C., Schurer, A., et al. (2020). Ocean and land forcing of the record-breaking Dust Bowl heatwaves across central United States. *Nature Communications*, 11, 2870, DOI: https://doi.org/10.1038/s41467-020-16676-w.

Crane, P. (2013). *Ginkgo: The Tree That Time Forgot*. New Haven, CT: Yale University Press.

Cronon, W. (1983). *Changes in the Land: Indians, Colonists, and the Ecology of New England*. New York: Hill & Wang.

Crowther, T. W., Glick, H. B., Covey, K. R., et al. (2015). Mapping tree density at a global scale. *Nature*, 525, 201–205.

Curtis, G. E. (1893). Analysis of the causes of rainfall with special relation to surface conditions. In *Forest Influences* (US Department of Agriculture, Forestry Division Bulletin Number 7), edited by B. E. Fernow. Washington, DC: Government Printing Office, pp. 187–191.

Curtis, J. T. (1956). The modification of mid-latitude grasslands and forests by man. In *Man's Role in Changing the Face of the Earth*, edited by W. L. Thomas, Jr. Chicago: University of Chicago Press, pp. 721–736.

Dalhousie, J. A. Broun Ramsay, marquess (1868). Minute by the most noble the Marquis of Dalhousie, K. G., Governor General of India, dated 20th February, 1851. In *Select Papers of the Agri-Horticultural Society of the Punjab, from its Commencement to 1862*. Lahore: Lahore Chronicle Press, pp. 1–5.

Dallmeyer, A., Claussen, M., Lorenz, S. J., and Shanahan, T. (2020). The end of the African humid period as seen by a transient comprehensive Earth system model simulation of the last 8000 years. *Climate of the Past*, 16, 117–140.

Dalzell, N. A. (1863). *Observations on the Influence of Forests, and on the General Principles of Management, as Applicable to Bombay*. Bombay: Education Society's Press.

Dalzell, N. A. (1869). *Extracts on Forests and Forestry*. Bombay: Education Society's Press.

Darwin, C. (1859). *On the Origin of Species by Means of Natural Selection*. London: John Murray.

Davidson, E. A., de Araújo, A. C., Artaxo, P., et al. (2012). The Amazon basin in transition. *Nature*, 481, 321–328.

Davies-Barnard, T., Ridgwell, A., Singarayer, J., and Valdes, P. (2017). Quantifying the influence of the terrestrial biosphere on glacial–interglacial climate dynamics. *Climate of the Past*, 13, 1381–1401.

Davies-Barnard, T., Valdes, P. J., Singarayer, J. S., Pacifico, F. M., and Jones, C. D. (2014). Full effects of land use change in the representative concentration pathways. *Environmental Research Letters*, 9, 114014, DOI: https://doi.org/10.1088/1748-9326/9/11/114014.

Davin, E. L., and de Noblet-Ducoudré, N. (2010). Climatic impact of global-scale deforestation: Radiative versus nonradiative processes. *Journal of Climate*, 23, 97–112.

Davin, E. L., de Noblet-Ducoudré, N., and Friedlingstein, P. (2007). Impact of land cover change on surface climate: Relevance of the radiative forcing concept. *Geophysical Research Letters*, 34, L13702, DOI: https://doi.org/10.1029/2007GL029678.

Davin, E. L., Rechid, D., Breil, M., et al. (2020). Biogeophysical impacts of forestation in Europe: First results from the LUCAS (Land Use and Climate Across Scales) regional climate model intercomparison. *Earth System Dynamics*, 11, 183–200.

Davis, D. K. (2016a). *The Arid Lands: History, Power, Knowledge*. Cambridge, MA: Massachusetts Institute of Technology Press.

Davis, K. T., Dobrowski, S. Z., Holden, Z. A., Higuera, P. E., and Abatzoglou, J. T. (2019). Microclimatic buffering in forests of the future: The role of local water balance. *Ecography*, 42, 1–11.

Davis, M. (2016b). The coming desert: Kropotkin, Mars and the pulse of Asia. *New Left Review*, 97, 23–43.

Davis, M. B. (1981). Quaternary history and the stability of forest communities. In *Forest Succession: Concepts and Application*, edited by D. C. West, H. H. Shugart, and D. B. Botkin. New York: Springer-Verlag, pp. 132–153.

Dawson, T. E. (1998). Fog in the California redwood forest: Ecosystem inputs and use by plants. *Oecologia*, 117, 476–485.

De Frenne, P., Lenoir, J., Luoto, M., et al. (2021). Forest microclimates and climate change: Importance, drivers and future research agenda. *Global Change Biology*, 27, 2279–2297.

De Frenne, P., Rodríguez-Sánchez, F., Coomes, D. A., et al. (2013). Microclimate moderates plant responses to macroclimate warming. *Proceedings of the National Academy of Sciences USA*, 110, 18561–18565.

De Frenne, P., Zellweger, F., Rodríguez-Sánchez, F., et al. (2019). Global buffering of temperatures under forest canopies. *Nature Ecology and Evolution*, 3, 744–749.

de Noblet, N. I., Prentice, I. C., Joussaume, S., et al. (1996). Possible role of atmosphere–biosphere interactions in triggering the last glaciation. *Geophysical Research Letters*, 23, 3191–3194.

de Noblet-Ducoudré, N., Boisier, J.-P., Pitman, A., et al. (2012). Determining robust impacts of land-use-induced land cover changes on surface climate over North America and Eurasia: Results from the first set of LUCID experiments. *Journal of Climate*, 25, 3261–3281.

Denys, N. (1672). *Description geographique et historique des costes de l'Amerique septentrionale*, 2 vols. Paris: Claude Barbin.

Devaraju, N., Bala, G., and Modak, A. (2015). Effects of large-scale deforestation on precipitation in the monsoon regions: Remote versus local effects. *Proceedings of the National Academy of Sciences USA*, 112, 3257–3262.

Devaraju, N., de Noblet-Ducoudré, N., Quesada, B., and Bala, G. (2018). Quantifying the relative importance of direct and indirect biophysical effects of deforestation on surface temperature and teleconnections. *Journal of Climate*, 31, 3811–3829.

Dhar, A., Parrott, L., and Heckbert, S. (2016). Consequences of mountain pine beetle outbreak on forest ecosystem services in western Canada. *Canadian Journal of Forest Research*, 46, 987–999.

Dickerson-Lange, S. E., Vano, J. A., Gersonde, R., and Lundquist, J. D. (2021). Ranking forest effects on snow storage: A decision tool for forest management. *Water Resources Research*, 57, e2020WR027926, DOI: https://doi.org/10.1029/2020WR027926.

Dickinson, R. E. (1983). Land surface processes and climate–surface albedos and energy balance. *Advances in Geophysics*, 25, 305–353.

Dickinson, R. E. (1984). Modeling evapotranspiration for three-dimensional global climate models. In *Climate Processes and Climate Sensitivity*, edited by J. E. Hansen and T. Takahashi. Washington, DC: American Geophysical Union, pp. 58–72.

Dickinson, R. E., and Henderson-Sellers, A. (1988). Modelling tropical deforestation: A study of GCM land-surface parameterizations. *Quarterly Journal of the Royal Meteorological Society*, 114, 439–462.

Dickinson, R. E., Henderson-Sellers, A., Kennedy, P. J., and Wilson, M. F. (1986). *Biosphere–Atmosphere Transfer Scheme (BATS) for the NCAR Community Climate Model*, Technical Note NCAR/TN-275+STR. Boulder, CO: National Center for Atmospheric Research.

Dickinson, R. E., Henderson-Sellers, A., and Kennedy, P. J. (1993). *Biosphere–Atmosphere Transfer Scheme (BATS) Version 1e as Coupled to the NCAR Community Climate Model*, Technical Note NCAR/TN-387+STR. Boulder, CO: National Center for Atmospheric Research.

Dickinson, R. E., Jäger, J., Washington, W. M., and Wolski, R. (1981). *Boundary Subroutine for the NCAR Global Climate Model*, Technical Note NCAR/TN-173+IA. Boulder, CO: National Center for Atmospheric Research.

Dirmeyer, P. A., and Shukla, J. (1996). The effect on regional and global climate of expansion of the world's deserts. *Quarterly Journal of the Royal Meteorological Society*, 122, 451–482.

Dirmeyer, P. A., Balsamo, G., Blyth, E. M., Morrison, R., and Cooper, H. M. (2021). Land-atmosphere interactions exacerbated the drought and heatwave over northern Europe during summer 2018. *AGU Advances*, 2, e2020AV000283, DOI: https://doi.org/10.1029/2020AV000283.

Dodds, G. B. (1969). The stream-flow controversy: A conservation turning point. *Journal of American History*, 56, 59–69.

Douglas, W. O. (1951). *Strange Lands and Friendly People*. New York: Harper.

Douglass, W. (1749–51). *A Summary, Historical and Political, of the First Planting, Progressive Improvements, and Present State of the British Settlements in North-America*, 2 vols. Boston: Rogers & Fowle.

Dove, H. W. (1855). Ueber die Vertheilung der Regen in der gemäfsigten Zone. *Annalen der Physik und Chemie*, 94, 42–59.

Dove, H. W. (1855–56). On the distribution of rain in the temperate zone. *American Journal of Science and Arts*, 20, 397–402; 21, 112–117.

Drever, C. R., Cook-Patton, S. C., Akhter, F., et al. (2021). Natural climate solutions for Canada. *Science Advances*, 7, eabd6034, DOI: https://doi.org/10.1126/sciadv.abd6034.

Drori, J. (2018). *Around the World in 80 Trees*. London: Laurence King.

Dubos, J.-B (1719). *Reflexions critiques sur la poesie et sur la peinture*, 2 vols. Paris: Jean Mariette.

Duffey, E. (1964). The terrestrial ecology of Ascension Island. *Journal of Applied Ecology*, 1, 219–251.

Duhamel du Monceau, H.-L. (1746). Observations botanico-météorologiques faites à Québec pendant les mois d'Octobre, Novembre & Décembre 1744, & les mois de Janvier, Février, Mars, Avril & Mai 1745. *Mémoires de mathématique et de physique, tirés des registres de l'Académie Royale des Sciences, de l'année 1746*, pp.88–97.

Duhamel du Monceau, H.-L. (1755). *Traité des arbres et arbustes qui se cultivent en France en pleine terre*, 2 vols. Paris: H. L. Guerin & L. F. Delatour.

Duhamel du Monceau, H.-L. (1758). *La physique des arbres*, 2 vols. Paris: H. L. Guerin & L. F. Delatour.

Duhamel du Monceau, H.-L. (1760). *Des semis et plantations des arbres, et de leur culture*. Paris: H. L. Guerin & L. F. Delatour.

Duhamel du Monceau, H.-L. (1764). *De l'exploitation des bois*, 2 vols. Paris: H. L. Guerin & L. F. Delatour.

Duhamel du Monceau, H.-L. (1767). *Du transport, de la conservation et de la force des bois*. Paris: L. F. Delatour.

Dunbar, J. (1780). *Essays on the History of Mankind in Rude and Cultivated Ages*. London: W. Strahan.

Dunbar, W. (1809). Meteorological observations. *Transactions of the American Philosophical Society*, 6, 43–55.

Duveiller, G., Caporaso, L., Abad-Viñas, R., et al. (2020). Local biophysical effects of land use and land cover change: Towards an assessment tool for policy makers. *Land Use Policy*, 91, 104382, DOI: https://doi.org/10.1016/j.landusepol.2019 .104382.

Duveiller, G., Filipponi, F., Ceglar, A., et al. (2021). Revealing the widespread potential of forests to increase low level cloud cover. *Nature Communications*, 12, 4337, DOI: https://doi.org/ 10.1038/s41467-021-24551-5.

Duveiller, G., Hooker, J., and Cescatti, A. (2018). The mark of vegetation change on Earth's surface energy balance. *Nature Communications*, 9, 679, DOI: https://doi.org/10.1038/s41467-017-02810-8.

Dwight, T. (1821–22). *Travels; in New-England and New-York*, 4 vols. New Haven, CT: S. Converse.

Ébelmen, J.-J. (1845). Sur les produits de la décomposition des espèces minérales de la famille des silicates. *Annales des mines* (série 4), 7, 3–66.

Eberle, J. J., and Greenwood, D. R. (2012). Life at the top of the greenhouse Eocene world: A review of the Eocene flora and vertebrate fauna from Canada's High Arctic. *Geological Society of America Bulletin*, 124, 3–23.

Ebermayer, E. (1873). *Die physikalischen Einwirkungen des Waldes auf Luft und Boden und seine klimatologische und hygienische Bedeutung*. Aschaffenburg: C. Krebs.

Edburg, S. L., Hicke, J. A., Brooks, P. D., et al. (2012). Cascading impacts of bark beetle-caused tree mortality on coupled biogeophysical and biogeochemical processes. *Frontiers in Ecology and the Environment*, 10, 416–424.

Eder, F., De Roo, F., Rotenberg, E., et al. (2015). Secondary circulations at a solitary forest surrounded by semi-arid shrubland and their impact on eddy-covariance measurements. *Agricultural and Forest Meteorology*, 211–212, 115–127.

Edge, T. J. (1878). The forests of our state: Their value and their influence upon streams, temperature, climate, and rain-fall. In *First Annual Report of the Pennsylvania Board of Agriculture for the Year 1877, with an Appendix*. Harrisburg: Lane S. Hart, pp. 61–77.

Edwards, P. N. (2011). History of climate modeling. *WIREs Climate Change*, 2, 128–139.

Egerton, F. N. (2003). A history of the ecological sciences, part 9. Albertus Magnus: A scholastic naturalist. *Bulletin of the Ecological Society of America*, 84, 87–91.

Egerton, F. N. (2008). A history of the ecological sciences, part 28: Plant growth studies during the 1700s. *Bulletin of the Ecological Society of America*, 89, 159–175.

Egleston, N. H. (1883). Forestry division. In *Report of the Commissioner of Agriculture for the Year 1883*. Washington, DC: Government Printing Office, pp. 444–462.

Egleston, N. H. (1884). Report of chief of the forestry bureau. In *Report of the Commissioner of Agriculture for the Year 1884*. Washington, DC: Government Printing Office, pp. 137–180.

Egleston, N. H. (1885). Report of chief of division of forestry. In *Report of the Commissioner of Agriculture, 1885*. Washington, DC: Government Printing Office, pp. 183–206.

Egleston, N. H. (1896). *Arbor Day: Its History and Observance*. Washington, DC: Government Printing Office.

Elliott, R. S. (1871a). Report on the industrial resources of western Kansas and eastern Colorado. In *Preliminary Report of the United States Geological Survey of Wyoming, and Portions of Contiguous Territories*. Washington, DC: Government Printing Office, pp. 442–458.

Elliott, R. S. (1871b). Climate of Kansas. In *Annual Report of the Board of Regents of the Smithsonian Institution, Showing the Operations, Expenditures, and Condition of the Institution for the Year 1870*. Washington, DC: Government Printing Office, pp. 472–474.

Elliott, R. S. (1872). Experiments in cultivation on the plains along the line of the Kansas Pacific Railway. In *Preliminary Report of the United States Geological Survey of Montana and Portions of Adjacent Territories*. Washington, DC: Government Printing Office, pp. 274–279.

Elliott, R. S. (1883). *Notes Taken in Sixty Years*. St. Louis: R. P. Studley.

Ellis, E. C. (2021). Land use and ecological change: A 12,000-year history. *Annual Review of Environment and Resources*, 46, 1–33.

Ellison, D., and Ifejika Speranza, C. (2020). From blue to green water and back again: Promoting tree, shrub and forest-based landscape resilience in the Sahel. *Science of the Total Environment*, 739, 140002, DOI: https://doi.org/10.1016/j .scitotenv.2020.140002.

Ellison, D., Futter, M. N., and Bishop, K. (2012). On the forest cover–water yield debate: From demand- to supply-side thinking. *Global Change Biology*, 18, 806–820.

Ellison, D., Morris, C. E., Locatelli, B., et al. (2017). Trees, forests and water: Cool insights for a hot world. *Global Environmental Change*, 43, 51–61.

Ellsworth, D. S., and Reich, P. B. (1993). Canopy structure and vertical patterns of photosynthesis and related leaf traits in a deciduous forest. *Oecologia*, 96, 169–178.

Eltahir, E. A. B., and Bras, R. L. (1994). Precipitation recycling in the Amazon basin. *Quarterly Journal of the Royal Meteorological Society*, 120, 861–880.

Emerson, G. B. (1846). *A Report on the Trees and Shrubs Growing Naturally in the Forests of Massachusetts*. Boston: Dutton & Wentworth.

Emerson, R. W. (1870). *Society and Solitude: 12 Chapters*. Boston: Fields, Osgood & Co.

Emmons, D. M. (1971). Theories of increased rainfall and the Timber Culture Act of 1873. *Forest History*, 15(3), 6–14.

Erb, K.-H., Luyssaert, S., Meyfroidt, P., et al. (2017). Land management: Data availability and process understanding for global change studies. *Global Change Biology*, 23, 512–533.

Evelyn, J. (1664). *Sylva, or a Discourse of Forest-Trees, and the Propagation of Timber in His Majesties Dominions*. London: John Martyn & James Allestry.

Evelyn, J. (1670). *Sylva, or a Discourse of Forest-Trees, and the Propagation of Timber in His Majesties Dominions*, 2nd ed. London: John Martyn & James Allestry.

Evelyn, J. (1679). *Sylva, or a Discourse of Forest-Trees, and the Propagation of Timber in His Majesties Dominions*, 3rd ed. London: John Martyn.

Evelyn, J. (1706). *Silva, or a Discourse of Forest-Trees, and the Propagation of Timber in His Majesty's Dominions*, 4th ed. London: Robert Scott; Richard Chiswell; and others.

Evelyn, J. (1776). *Silva: Or, a Discourse of Forest-Trees, and the Propagation of Timber in His Majesty's Dominions*, with notes by A. Hunter. York: A. Ward.

Fabre, J.-A. (1797). *Essai sur la théorie des torrens et des rivières*. Paris: Bidault.

Fantham, E. (2004). *Ovid's Metamorphoses*. New York: Oxford University Press.

FAO (2018). *Global Forest Resources Assessment 2020: Terms and Definitions*, Forest Resources Assessment Working Paper 188. Rome: Food and Agriculture Organization of the United Nations.

FAO (2020). *Global Forest Resources Assessment 2020: Main report*. Rome: Food and Agriculture Organization of the United Nations; https://doi .org/10.4060/ca9825en.

Fargione, J., Hill, J., Tilman, D., Polasky, S., and Hawthorne, P. (2008). Land clearing and the biofuel carbon debt. *Science*, 319, 1235–1238.

Fargione, J. E., Bassett, S., Boucher, T., et al. (2018). Natural climate solutions for the United States. *Science Advances*, 4, eaat1869, DOI: https://doi .org/10.1126/sciadv.aat1869.

Fasullo, J. T., Rosenbloom, N., Buchholz, R. R., et al. (2021). Coupled climate responses to recent Australian wildfire and COVID-19 emissions anomalies estimated in CESM2. *Geophysical Research Letters*, 48, e2021GL093841, DOI: https://doi.org/10 .1029/2021GL093841.

Favero, A., Sohngen, B., Huang, Y., and Jin, Y. (2018). Global cost estimates of forest climate mitigation with albedo: A new integrative policy approach. *Environmental Research Letters*, 13, 125002, DOI: https://doi.org/10.1088/1748-9326/aaeaa2.

Feddema, J. J., Oleson, K. W., Bonan, G. B., et al. (2005). The importance of land-cover change in simulating future climates. *Science*, 310, 1674–1678.

Feng, D., Bao, W., Yang, Y., and Fu, M. (2021). How do government policies promote greening? Evidence from China. *Land Use Policy*, 104, 105389, DOI: https://doi.org/10.1016/j .landusepol.2021.105389.

Feng, X., Fu, B., Piao, S., et al. (2016). Revegetation in China's Loess Plateau is approaching sustainable water resource limits. *Nature Climate Change*, 6, 1019–1022.

Ferguson, D. K., and Knobloch, E. (1998). A fresh look at the rich assemblage from the Pliocene sink-hole of Willershausen, Germany. *Review of Palaeobotany and Palynology*, 101, 271–286.

Fernández de Oviedo y Valdés, G. (1851–55). *Historia general y natural de las Indias, islas y tierra-firme del Mar Océano*, edited by José Amador de los Rios, 4 vols. Madrid: Imprenta de la Real Academia de la Historia.

Fernow, B. E. (1887). Report of chief of forestry division. In *Report of the Commissioner of Agriculture, 1886*. Washington, DC: Government Printing Office, pp. 149–226.

Fernow, B. E. (1888). Report of the chief of forestry division. In *Report of the Commissioner of Agriculture, 1887*. Washington, DC: Government Printing Office, pp. 605–616.

Fernow, B. E. (1889). Report of the chief of forestry division. In *Report of the Commissioner of Agriculture, 1888*. Washington, DC: Government Printing Office, pp. 597–641.

Fernow, B. E. (1891). *What Is Forestry?* U.S. Department of Agriculture, Forestry Division Bulletin Number 5. Washington, DC: Government Printing Office.

Fernow, B. E. (1893a). Forest influences: Introduction and summary of conclusions. In *Forest Influences* (U.S. Department of Agriculture, Forestry Division Bulletin Number 7), edited by B. E. Fernow. Washington, DC: Government Printing Office, pp. 9–22.

Fernow, B. E. (1893b). Relation of forests to water supplies. In *Forest Influences* (U.S. Department of Agriculture, Forestry Division Bulletin Number 7), edited by B. E. Fernow. Washington, DC: Government Printing Office, pp. 123–170.

Fernow, B. E. (1894). Forest conditions and forestry problems in the United States. *Proceedings of the American Forestry Association*, 10, 29–36.

Fernow, B. E. (1910). Current Literature: *The Influence of Forests on Climate and on Floods* by Willis L. Moore. *Forestry Quarterly*, 8, 74–75.

Ferrel, W. (1889). Note on the influence of forests upon rainfall. *American Meteorological Journal*, 5 (10), 433–435.

Findell, K. L., Berg, A., Gentine, P., et al. (2017). The impact of anthropogenic land use and land cover change on regional climate extremes. *Nature Communications*, 8, 989, DOI: https://doi.org/10.1038/s41467-017-01038-w.

Firth, J. C. (1874). On forest culture. *Transactions and Proceedings of the New Zealand Institute*, 7, 181–195.

Fischer, E. M., Seneviratne, S. I., Lüthi, D., and Schär, C. (2007). Contribution of land–atmosphere coupling to recent European summer heat waves. *Geophysical Research Letters*, 34, L06707, DOI: https://doi.org/10.1029/2006GL029068.

FitzNigel, R. (1983). *Dialogus de Scaccario: The Course of the Exchequer*, edited and translated by C. Johnson with corrections by F. E. L. Carter and D. E. Greenway. Oxford: Oxford University Press.

Fleming, J. R. (1998). *Historical Perspectives on Climate Change*. New York: Oxford University Press.

Foley, J. A., Coe, M. T., Scheffer, M., and Wang, G. (2003). Regime shifts in the Sahara and Sahel: Interactions between ecological and climatic systems in northern Africa. *Ecosystems*, 6, 524–539.

Foley, J. A., DeFries, R., Asner, G. P., et al. (2005). Global consequences of land use. *Science*, 309, 570–574.

Foley, J. A., Kutzbach, J. E., Coe, M. T., and Levis, S. (1994). Feedbacks between climate and boreal forests during the Holocene epoch. *Nature*, 371, 52–54.

Folland, C. K., Palmer, T. N., and Parker, D. E. (1986). Sahel rainfall and worldwide sea temperatures, 1901–85. *Nature*, 320, 602–607.

Ford, C. (2016). *Natural Interests: The Contest over Environment in Modern France*. Cambridge, MA: Harvard University Press.

Ford, C. R., Laseter, S. H., Swank, W. T., and Vose, J. M. (2011). Can forest management be used to sustain water-based ecosystem services in the face of climate change? *Ecological Applications*, 21, 2049–2067.

Forkel, M., Carvalhais, N., Rödenbeck, C., et al. (2016). Enhanced seasonal CO_2 exchange caused by amplified plant productivity in northern ecosystems. *Science*, 351, 696–699.

Forman, R. T. T., and Godron, M. (1986). *Landscape Ecology*. New York: Wiley.

Forry, S. (1842). *The Climate of the United States and Its Endemic Influences*. New York: J. & H. G. Langley.

Forry, S. (1844). Researches in elucidation of the distribution of heat over the globe, and especially of the climatic features peculiar to the region of the United States. *American Journal of Science and Arts*, 47, 18–50, 221–241.

Forster, E. J., Healey, J. R., Dymond, C., and Styles, D. (2021). Commercial afforestation can deliver effective climate change mitigation under multiple decarbonisation pathways. *Nature Communications*, 12, 3831, DOI: https://doi.org/10.1038/s41467-021-24084-x.

Forzieri, G., Alkama, R., Miralles, D. G., and Cescatti, A. (2017). Satellites reveal contrasting responses of regional climate to the widespread greening of Earth. *Science*, 356, 1180–1184.

Fowler, J. (1774). *A Summary Account of the Present Flourishing State of the Respectable Colony of Tobago, in the British West Indies*. London: A. Grant.

Fowler, M. D., Kooperman, G. J., Randerson, J. T., and Pritchard, M. S. (2019). The effect of plant physiological responses to rising CO_2 on global streamflow. *Nature Climate Change*, 9, 873–879.

Fraas, C. (1847). *Klima und Pflanzenwelt in der Zeit, ein Beitrag zur Geschichte beider*. Landshut: J. G. Wölfle.

Frankenfield, H. C. (1910). The experiment station at Wagon Wheel Gap, Colo. *Monthly Weather Review*, 38, 1453–1455.

Franklin, B. (1755). Observations concerning the increase of mankind, peopling of countries, etc. In *Observations on the Late and Present Conduct of the French, with Regard to Their Encroachments upon the British Colonies in North America. Together with Remarks on the Importance of These Colonies to Great-Britain*, edited by W. Clarke. Boston: S. Kneeland, pp. Appendix, 1–15.

Franklin, B. (1905). Observations concerning the increase of mankind, peopling of countries, etc. In *The Writings of Benjamin Franklin*, vol. 3: *1750–1759*, edited by A. H. Smyth. New York: Macmillan, pp. 63–73.

Franklin, B. (1966). To Ezra Stiles, May 29, 1763. In *The Papers of Benjamin Franklin*, vol. 10. *January 1, 1762, through December 31, 1763*, edited by L. W. Labaree. New Haven, CT: Yale University Press, pp. 264–267.

Franklin, J. (1784). *The Philosophical and Political History of the Thirteen United States of America*. London: J. Hinton & W. Adams.

Franks, P. J., Adams, M. A., Amthor, J. S., et al. (2013). Sensitivity of plants to changing atmospheric CO_2 concentration: From the geological past to the next century. *New Phytologist*, 197, 1077–1094.

Franks, P. J., Royer, D. L., Beerling, D. J., et al. (2014). New constraints on atmospheric CO_2 concentration for the Phanerozoic. *Geophysical Research Letters*, 41, 4685–4694.

Fressoz, J.-B. (2015). Losing the Earth knowingly: Six environmental grammars around 1800. In *The Anthropocene and the Global Environmental Crisis: Rethinking Modernity in a New Epoch*, edited by C. Hamilton, C. Bonneuil, and F. Gemenne. New York: Routledge, pp. 70–83.

Fressoz, J.-B., and Locher, F. (2015). Régénérer la nature, restaurer les climats: François-Antoine Rauch et les *Annales Européennes de physique végétale et d'economie publique*, 1815-1830. *Le Temps des medias*, 25(2), 52–69.

Fressoz, J.-B., and Locher, F. (2020). *Les révoltes du ciel: une histoire du changement climatique (XVe–XXe siècle)*. Paris: Éditions du Seuil.

Friedlingstein, P., O'Sullivan, M., Jones, M. W., et al. (2020). Global carbon budget 2020. *Earth System Science Data*, 12, 3269–3340.

Gallimore, R. G., and Kutzbach, J. E. (1996). Role of orbitally induced changes in tundra area in the onset of glaciation. *Nature*, 381, 503–505.

Galloway, J. N., Aber, J. D., Erisman, J. W., et al. (2003). The nitrogen cascade. *BioScience*, 53, 341–356.

Galloway, J. N., Townsend, A. R., Erisman, J. W., et al. (2008). Transformation of the nitrogen cycle: Recent trends, questions, and potential solutions. *Science*, 320, 889–892.

Galvez, M. E., and Gaillardet, J. (2012). Historical constraints on the origins of the carbon cycle concept. *Comptes Rendus Geoscience*, 344, 549–567.

Galway-Witham, J., and Stringer, C. (2018). How did *Homo sapiens* evolve? *Science*, 360, 1296–1298.

Gannett, H. (1888a). Do forests influence rainfall? *Science*, 11 (257), 3–5 (January 6, 1888).

Gannett, H. (1888b). Is the rainfall increasing upon the plains? *Science*, 11 (265), 99–100 (March 2, 1888).

Gash, J. H. C., and Nobre, C. A. (1997). Climatic effects of Amazonian deforestation: Some results from ABRACOS. *Bulletin of the American Meteorological Society*, 78, 823–830.

Gash, J. H. C., Nobre, C. A., Roberts, J. M., and Victoria, R. L. (1996). *Amazonian Deforestation and Climate*. New York: Wiley.

Gates, D. M. (1963). Leaf temperature and energy exchange. *Archiv für Meteorologie, Geophysik und Bioklimatologie*, 12B, 321–336.

Gatti, L. V., Basso, L. S., Miller, J. B., et al. (2021). Amazonia as a carbon source linked to deforestation and climate change. *Nature*, 595, 388–393.

Gaubert, B., Stephens, B. B., Basu, S., et al. (2019). Global atmospheric CO_2 inverse models converging on neutral tropical land exchange, but disagreeing on fossil fuel and atmospheric growth rate. *Biogeosciences*, 16, 117–134.

Ge, J., Guo, W., Pitman, A. J., et al. (2019). The nonradiative effect dominates local surface temperature change caused by afforestation in China. *Journal of Climate*, 32, 4445–4471.

Ge, J., Pitman, A. J., Guo, W., Zan, B., and Fu, C. (2020). Impact of revegetation of the Loess Plateau of China on the regional growing season water balance. *Hydrology and Earth System Sciences*, 24, 515–533.

Gedney, N., Cox, P. M., Betts, R. A., et al. (2006). Detection of a direct carbon dioxide effect in continental river runoff records. *Nature*, 439, 835–838.

Geiger, R. (1927). *Das Klima der bodennahen Luftschicht*. Braunschweig: Friedr. Vieweg.

Geiger, R. (1942). *The Climate of the Layer of Air near the Ground*, translated by J. Leighly for restricted official use of the Soil Conservation Service, United States Department of Agriculture. New Philadelphia, OH: Muskingum Climatic Research Center.

Geiger, R. (1950). *The Climate near the Ground*, translation by M. N. Stewart and others of the second German edition of Das Klima der bodennahen Luftschicht with revisions and enlargements by the author. Cambridge, MA: Harvard University Press.

Geis, D. (1981). *Walt Disney's Treasury of Silly Symphonies*. New York: Harry N. Abrams.

Genesio, L., Bassi, R., and Miglietta, F. (2021). Plants with less chlorophyll: A global change perspective. *Global Change Biology*, 27, 959–967.

Gentine, P., Green, J. K., Guérin, M., et al. (2019a). Coupling between the terrestrial carbon and water cycles: A review. *Environmental Research Letters*, 14, 083003, DOI: https://doi.org/10.1088/1748-9326/ab22d6.

Gentine, P., Massmann, A., Lintner, B. R., et al. (2019b). Land–atmosphere interactions in the tropics: A review. *Hydrology and Earth System Science*, 23, 4171–4197.

Giannini, A., Biasutti, M., and Verstraete, M. M. (2008). A climate model-based review of drought in the Sahel: Desertification, the re-greening and climate change. *Global and Planetary Change*, 64, 119–128.

Giannini, A., Saravanan, R., and Chang, P. (2003). Oceanic forcing of Sahel rainfall on interannual to interdecadal time scales. *Science*, 302, 1027–1030.

Gidden, M. J., Riahi, K., Smith, S. J., et al. (2019). Global emissions pathways under different socioeconomic scenarios for use in CMIP6: A dataset of harmonized emissions trajectories through the end of the century. *Geoscientific Model Development*, 12, 1443–1475.

Gilpin, W. (1791). *Remarks on Forest Scenery, and Other Woodland Views*, 2 vols. London: R. Blamire.

Girardin, C. A. J., Jenkins, S., Seddon, N., et al. (2021). Nature-based solutions can help cool the planet: If we act now. *Nature*, 593, 191–194.

Giraud-Soulavie, J.-L. (1783). *Histoire naturelle de la France méridionale. Second partie. Les végétaux*, vol. 1. Paris: Quillau; Merigot l'aîné; Merigot jeune; Belin.

Glacken, C. J. (1956). Changing ideas of the habitable world. In *Man's Role in Changing the Face of the Earth*, edited by W. L. Thomas, Jr. Chicago: University of Chicago Press, pp. 70–92.

Glacken, C. J. (1967). *Traces on the Rhodian Shore*. Berkeley: University of California Press.

Gleason, K. E., McConnell, J. R., Arienzo, M. M., Chellman, N., and Calvin, W. M. (2019). Fourfold increase in solar forcing on snow in western U.S. burned forests since 1999. *Nature Communications*, 10, 2026, DOI: https://doi.org/10.1038/s41467-019-09935-y.

Gleason, K. E., Nolin, A. W., and Roth, T. R. (2013). Charred forests increase snowmelt: Effects of burned woody debris and incoming solar radiation on snow ablation. *Geophysical Research Letters*, 40, 4654–4661.

Goeking, S. A., and Tarboton, D. G. (2020). Forests and water yield: A synthesis of disturbance effects on streamflow and snowpack in western coniferous forests. *Journal of Forestry*, 118, 172–192.

Golinski, J. (2008). American climate and the civilization of nature. In *Science and Empire in the Atlantic World*, edited by J. Delbourgo and N. Dew. New York: Routledge, pp. 153–174.

Golley, F. B. (1993). *A History of the Ecosystem Concept in Ecology: More than the Sum of the Parts*. New Haven, CT: Yale University Press.

Goss, M., Swain, D. L., Abatzoglou, J. T., et al. (2020). Climate change is increasing the likelihood of extreme autumn wildfire conditions across California. *Environmental Research Letters*, 15, 094016, DOI: https://doi.org/10.1088/1748-9326/ab83a7.

Gould, S. J. (1989). Church, Humboldt, and Darwin: The tension and harmony of art and science. In *Frederic Edwin Church*, edited by F. Kelly. Washington, DC: National Gallery of Art, pp. 94–107.

Goulden, M. L., Winston, G. C., McMillan, A. M. S., et al. (2006). An eddy covariance mesonet to measure the effect of forest age on land–atmosphere exchange. *Global Change Biology*, 12, 2146–2162.

Grassi, G., House, J., Dentener, F., et al. (2017). The key role of forests in meeting climate targets requires science for credible mitigation. *Nature Climate Change*, 7, 220–226.

Graven, H. D., Keeling, R. F., Piper, S. C., et al. (2013). Enhanced seasonal exchange of CO_2 by northern ecosystems since 1960. *Science*, 341, 1085–1089.

Greeley, H. (1871). *What I Know of Farming: A Series of Brief and Plain Expositions of Practical Agriculture as an Art Based upon Science*. New York: G. W. Carleton.

Gregg, J. (1844). *Commerce of the Prairies: Or the Journal of a Sante Fé Trader, during Eight Expeditions across the Great Western Prairies, and a Residence of Nearly Nine Years in Northern Mexico*, 2 vols. New York: Henry G. Langley.

Grew, N. (1682). *The Anatomy of Plants, with an Idea of a Philosophical History of Plants*. [London]: W. Rawlins.

Griscom, B. W., Adams, J., Ellis, P. W., et al. (2017). Natural climate solutions. *Proceedings of the National Academy of Sciences USA*, 114, 11645–11650.

Griscom, B. W., Busch, J., Cook-Patton, S. C., et al. (2020). National mitigation potential from natural climate solutions in the tropics. *Philosophical Transactions of the Royal Society London B*, 375, 20190126, DOI: https://doi.org/10.1098/rstb.2019.0126.

Grove, R. (1989). Scottish missionaries, evangelical discourses and the origins of conservation thinking in southern Africa 1820–1900. *Journal of Southern African Studies*, 15, 163–187.

Grove, R. H. (1995). *Green Imperialism: Colonial Expansion, Tropical Island Edens and the Origins of Environmentalism, 1600–1860*. Cambridge, UK: Cambridge University Press.

Gu, L., Baldocchi, D. D., Wofsy, S. C., et al. (2003). Response of a deciduous forest to the Mount Pinatubo eruption: Enhanced photosynthesis. *Science*, 299, 2035–2038.

Guyot, C. (1898). *L'Enseignement forestier en France: l'école de Nancy*. Nancy: Crépin-Leblond.

Habenicht, R. E. (1963). *John Heywood's A Dialogue of Proverbs*, edited, with introduction, commentary, and indexes. Berkeley: University of California Press.

Hack, J. J., Boville, B. A., Briegleb, B. P., et al. (1993). *Description of the NCAR Community Climate Model (CCM2)*, Technical Note NCAR/TN-382+STR. Boulder, CO: National Center for Atmospheric Research.

Haeckel, E. (1866). *Generelle Morphologie der Organismen: allgemeine Grundzüüge der organischen Formen-Wissenschaft, mechanisch begründet durch die von Charles Darwin reformirte Descendenz-Theorie*, 2 vols. Berlin: Georg Reimer.

Haesen, S., Lembrechts, J. J., De Frenne, P., et al. (2021). ForestTemp: Sub-canopy microclimate temperatures of European forests. *Global Change Biology*, 27, 6307–6319.

Hale, T. (1758–59). *A Compleat Body of Husbandry*, 2nd ed., 4 vols. London: T. Osborne; T. Trye; S. Crowder.

Hales, S. (1727). *Vegetable Staticks: Or, an Account of Some Statical Experiments on the Sap in Vegetables: Being an Essay towards a Natural History of Vegetation*. London: W. & J. Innys; T. Woodward.

Halim, M. A., Chen, H. Y. H., and Thomas, S. C. (2019). Stand age and species composition effects on surface albedo in a mixedwood boreal forest. *Biogeosciences*, 16, 4357–4375.

Hall, B., Motzkin, G., Foster, D. R., Syfert, M., and Burk, J. (2002). Three hundred years of forest and land-use change in Massachusetts, USA. *Journal of Biogeography*, 29, 1319–1335.

Hall, F. G., Betts, A. K., Frolking, S., et al. (2004). The boreal climate. In *Vegetation, Water, Humans and the Climate: A New Perspective on an Interactive System*, edited by P. Kabat, M. Claussen, P. A. Dirmeyer, et al. Berlin: Springer-Verlag, pp. 93–114.

Halley, E. (1686). An historical account of the trade winds, and monsoons, observable in the seas between and near the tropicks, with an attempt to assign the phisical cause of the said winds. *Philosophical Transactions*, 16(183), 153–168.

Halley, E. (1691). An account of the circulation of the watry vapours of the sea, and of the cause of springs, presented to the Royal Society. *Philosophical Transactions* 17(192), 468–473.

Hann, J. (1867). Wald und Regen. *Zeitschrift der österreichischen Gesellschaft für Meteorologie*, 2, 129–136.

Hann, J. (1869). Thatsachen und Bemerkungen über einige schädliche Folgen dr Zerstörung des natürlichen Pflanzenkleides der Erdoberfläche. *Zeitschrift der österreichischen Gesellschaft für Meteorologie*, 4, 18–22.

Hann, J. (1888). Wald und Regen in Indien. *Meteorologische Zeitschrift*, 5, 235–237.

Hann, J. (1897). *Handbuch der Klimatologie*, 3 vols. Stuttgart: J. Engelhorn.

Hann, J. (1903). *Handbook of Climatology*, part 1. *General Climatology*, translated by R. De Courcy Ward. New York: Macmillan.

Hansen, M. C., Potapov, P. V., Moore, R., et al. (2013). High-resolution global maps of 21st-century forest cover change. *Science*, 342, 850–853.

Harrington, M. W. (1876). *The Analysis of Plants: Intended for Schools and Colleges and for the Independent Botanical Student*. Ann Arbor: Sheehan.

Harrington, M. W. (1877). The tropical ferns collected by Professor Steere in the years 1870–75. *Journal of the Linnean Society: Botany*, 16 (89), 25–37.

Harrington, M. W. (1887) Is the rain-fall increasing on the plains? *American Meteorological Journal*, 4(8), 369–373.

Harrington, M. W. (1893). Review of forest meteorological observations: A study preliminary to the discussion of the relation of forests to climate. In *Forest Influences* (U.S. Department of Agriculture, Forestry Division Bulletin Number 7), edited by B. E. Fernow. Washington, DC: Government Printing Office, pp. 23–122.

Harris, J. (1710). *Lexicon Technicum: Or, an Universal English Dictionary of Arts and Sciences*, vol. 2. London: Dan. Brown; Tim. Goodwin; and others.

Harrison, R. P. (1992). *Forests: The Shadow of Civilization*. Chicago: University of Chicago Press.

Hawken, P. (2017). *Drawdown: The Most Comprehensive Plan Ever Proposed to Reverse Global Warming*. New York: Penguin Books.

Hayden, F. V. (1867). Ferdinand V. Hayden to Joseph S. Wilson, July 1, 1867. In *Report of the Commissioner of General Land Office, for the Year 1867*. Washington, DC: Government Printing Office, pp. 128–181.

Hayden, F. V. (1869). *Preliminary Field Report of the United States Geological Survey of Colorado and New Mexico*. Washington, DC: Government Printing Office.

Hayes, K. J. (2008). *The Road to Monticello: The Life and Mind of Thomas Jefferson*. New York: Oxford University Press.

Hearn, M. P. (2000). *The Annotated Wizard of Oz*. New York: W. W. Norton.

Hemery, G., and Simblet, S. (2014). *The New Sylva: A Discourse of Forest and Orchard Trees for the Twenty-first Century*. London: Bloomsbury.

Henry, J. (1886). *Scientific Writings of Joseph Henry*, 2 vols. Washington, DC: Smithsonian Institution.

Henshilwood, C. S., and Marean, C. W. (2003). The origin of modern human behavior: Critique of the models and their test implications. *Current Anthropology*, 44, 627–651.

Herbert, R., Stier, P., and Dagan, G. (2021). Isolating large-scale smoke impacts on cloud and precipitation processes over the Amazon with convection permitting resolution. *Journal of Geophysical Research: Atmospheres*, 126, e2021JD034615, DOI: https://doi.org/10.1029/2021JD034615.

Herder, J. G. (1800). *Outlines of a Philosophy of the History of Man*, translated by T. Churchill. London: J. Johnson.

Herschel, J. F. W. (1859). Physical geography. In *The Encyclopædia Britànnica, or Dictionary of Arts, Sciences, and General Literature*, 8th ed., vol. 17. Edinburgh: Adam & Charles Black, pp. 569–647.

Herschel, J. F. W. (1861). *Physical Geography: From the Encyclopædia Britannica*. Edinburgh: Adam & Charles Black.

Hey, R. (1837). *The Spirit of the Woods*. London: Longman, Rees, Orme, Brown, Green, & Longman.

Heywood, J. (1546). *A dialogue conteinyng the nomber in effect of all the prouerbes in the englishe tongue compacte in a matter concernyng two maner of mariages*. London: Thomas Berthelet (Text Creation Partnership, Ann Arbor, Michigan; http://name.umdl.umich.edu/A03168.0001.001).

Hibberd, S. (1855). *Brambles and Bay Leaves: Essays on the Homely and the Beautiful*. London: Longman, Brown, Green, & Longmans.

Hibbert, A. R. (1967). Forest treatment effects on water yield. In *Forest Hydrology: Proceedings of a National Science Foundation Advanced Science Seminar Held at The Pennsylvania State University, University Park, Pennsylvania, Aug 29–Sept 10, 1965*, edited by W. E. Sopper and H. W. Lull. Oxford: Pergamon Press, pp. 527–543.

Hicke, J. A., Allen, C. D., Desai, A. R., et al. (2012). Effects of biotic disturbances on forest carbon cycling in the United States and Canada. *Global Change Biology*, 18, 7–34.

Higgins, R. (2017). *Thoreau and the Language of Trees*. Oakland: University of California Press.

Higuera, P. E., and Abatzoglou, J. T. (2021). Record-setting climate enabled the extraordinary 2020 fire season in the western United States. *Global Change Biology*, 27, 1–2.

Higuera, P. E., Shuman, B. N., and Wolf, K. D. (2021). Rocky Mountain subalpine forests now burning more than any time in recent millennia. *Proceedings of the National Academy of Sciences USA*, 118, e2103135118, DOI: https://doi.org/10.1073/pnas.2103135118.

Hinsdale, B. A., and Demmon, I. N. (1906). *History of the University of Michigan*. Ann Arbor: University of Michigan Press.

Hirsch, A. L., Guillod, B. P., Seneviratne, S. I., et al. (2018). Biogeophysical impacts of land-use change on climate extremes in low-emission scenarios: Results from HAPPI-Land. *Earth's Future*, 6, 396–409.

Hirschi, M., Seneviratne, S. I., Alexandrov, V., et al. (2011). Observational evidence for soil-moisture impact on hot extremes in southeastern Europe. *Nature Geoscience*, 4, 17–21.

Hoelzmann, P., Jolly, D., Harrison, S. P., et al. (1998). Mid-Holocene land-surface conditions in northern Africa and the Arabian peninsula: A data set for the analysis of biogeophysical feedbacks in the climate system. *Global Biogeochemical Cycles*, 12, 35–51.

Hogan, D. J. (2014). *The Wizard of Oz FAQ: All That's Left to Know about Life According to Oz*. Milwaukee, WI: Applause Theatre and Cinema Books.

Hogg, E. H., Price, D. T., and Black, T. A. (2000). Postulated feedbacks of deciduous forest phenology on seasonal climate patterns in the western Canadian interior. *Journal of Climate*, 13, 4229–4243.

Hohenstein, A. (1860). *Der Wald sammt dessen wichtigem Einfluss auf das Klima der Länder, Wohl der Staaten und Völker, sowie der Gesundheit der Menschen*. Vienna: Carl Gerold's Sohn.

Holdridge, L. R. (1967). *Life Zone Ecology*. San Jose, Costa Rica: Tropical Science Center.

Holyoke, E. A. (1793). An estimate of the excess of the heat and cold of the American atmosphere beyond the European, in the same parallel of latitude: To which are added, some thoughts on the causes of this excess. *Memoirs of the American Academy of Arts and Sciences*, 2(1), 65–92.

Holzman, B. (1937). *Sources of Moisture for Precipitation in the United States*, Technical Bulletin No. 589. Washington, DC: U.S. Department of Agriculture.

Hopcroft, P. O., and Valdes, P. J. (2021). Paleoclimate-conditioning reveals a North Africa land–atmosphere tipping point. *Proceedings of the National Academy of Sciences USA*, 118, e2108783118, DOI: https://doi.org/10.1073/pnas.2108783118.

Hornbeck, J. W., Pierce, R. S., and Federer, C. A. (1970). Streamflow changes after forest clearing in New England. *Water Resources Research*, 6, 1124–1132.

Hough, F. B. (1874). On the duty of governments in the preservation of forests. *Proceedings of the American Association for the Advancement of Science*, 22B, 1–10.

Hough, F. B. (1878a). On the preservation of forests and the planting of timber. In *Transactions of the New York State Agricultural Society, 1872–1876*, vol. 32. [Troy, N.Y.]: Jerome B. Parmenter, pp. 177–194, 293.

Hough, F. B. (1878b). *Report upon Forestry*. Washington, DC: Government Printing Office.

Hough, F. B. (1882). *The Elements of Forestry*. Cincinnati: Robert Clarke.

Hough, F. B. (1885). Letter from Dr. Franklin B. Hough, in regard to the effect of forests in increasing the amount of rainfall. In *Report in Regard to the Range and Ranch Cattle Business of the United States*, edited by J. Nimmo, Jr. Washington, DC: Government Printing Office, pp. 130–131.

Houghton, R. A. (2005). Aboveground forest biomass and the global carbon balance. *Global Change Biology*, 11, 945–958.

Houghton, R. A. (2013). Keeping management effects separate from environmental effects in terrestrial carbon accounting. *Global Change Biology*, 19, 2609–2612.

Hovi, A., Lindberg, E., Lang, M., et al. (2019). Seasonal dynamics of albedo across European boreal forests: Analysis of MODIS albedo and structural metrics from airborne LiDAR. *Remote Sensing of Environment*, 224, 365–381.

Howard, R. A., and Howard, E. S. (1983). *Alexander Anderson's Geography and History of St. Vincent, West Indies*. Cambridge, MA: Arnold Arboretum.

Howe, L. C., MacInnis, B., Krosnick, J. A., Markowitz, E. M., and Socolow, R. (2019). Acknowledging uncertainty impacts public acceptance of climate scientists' predictions. *Nature Climate Change*, 9, 863–867.

Hu, X., Shi, L., Lin, L., and Magliulo, V. (2020). Improving surface roughness lengths estimation using machine learning algorithms. *Agricultural and Forest Meteorology*, 287, 107956, DOI: https://doi.org/10.1016/j.agrformet.2020.107956.

Hubau, W., Lewis, S. L., Phillips, O. L., et al. (2020). Asynchronous carbon sink saturation in African and Amazonian tropical forests. *Nature*, 579, 80–87.

Hudiburg, T. W., Law, B. E., Moomaw, W. R., Harmon, M. E., and Stenzel, J. E. (2019). Meeting GHG reduction targets requires accounting for all forest sector emissions. *Environmental Research Letters*, 14, 095005, DOI: https://doi.org/10.1088/1748-9326/ab28bb.

Hulme, M. (2009). On the origin of "the greenhouse effect": John Tyndall's 1859 interrogation of nature. *Weather*, 64, 121–123.

Humboldt, A. von (1808). *Ansichten der Natur mit wissenschaftlichen Erläuterungen*. Tübingen: J. G. Cotta.

Humboldt, A. von (1817). Des lignes isothermes et de la distribution de la chaleur sur le globe.

Mémoires de physique et de chimie, de la société d'Arcueil, 3, 462–602.

Humboldt, A. von (1820–21). On isothermal lines, and the distribution of heat over the globe. *Edinburgh Philosophical Journal*, 3, 1–20, 256–274; 4, 23–37, 262–281; 5, 28–39.

Humboldt, A. von (1826). *Ansichten der Natur, mit wissenschaftlichen Erläuterungen*, 2 vols. Stuttgart & Tübingen: J. G. Cotta.

Humboldt, A. von (1831). *Fragmens de géologie et de climatologie asiatiques*, 2 vols. Paris: Gide; Pihan Delaforest; Delaunay.

Humboldt, A. von (1843). *Asie centrale: recherches sur les chaines de montagnes et la climatologie comparée*, 3 vols. Paris: Gide.

Humboldt, A. von (1846–58). *Cosmos: Sketch of a Physical Description of the Universe*, translated by E. Sabine, 4 vols. London: Longman, Brown, Green & Longmans; John Murray.

Humboldt, A. von (1849). *Ansichten der Natur, mit wissenschaftlichen Erläuterungen*, 2 vols. Stuttgart & Tübingen: J. G. Cotta.

Humboldt, A. von (1850). *Views of Nature: Or Contemplations on the Sublime Phenomena of Creation; with Scientific Illustrations*, translated by E. C. Otté and H. G. Bohn. London: Henry G. Bohn.

Humboldt, A. von (2009). *Briefe aus Russland 1829*, edited by E. Knobloch, I. Schwarz and C. Suckow, introductory essay by O. Ette. Berlin: Akademie Verlag.

Humboldt, A. von (2014). *Views of Nature*, translated by M. W. Person, edited by S. T. Jackson and L. D. Walls. Chicago: University of Chicago Press.

Humboldt, A. von, and Bonpland, A. (1814–29). *Personal Narrative of Travels to the Equinoctial Regions of the New Continent, during the Years 1799–1804*, translated by H. M. Williams, 7 vols. London: Longman, Hurst, Rees, Orme & Brown.

Humboldt, A. von, and Bonpland, A. (2009). *Essay on the Geography of Plants*, edited with an introduction by S. T. Jackson, translated by S. Romanowski. Chicago: University of Chicago Press.

Hume, D. (1752). Of the populousness of antient nations. In *Political Discourses*. Edinburgh: R. Fleming, pp. 155–261.

Humphrey, V., Berg, A., Ciais, P., et al. (2021). Soil moisture–atmosphere feedback dominates land carbon uptake variability. *Nature*, 592, 65–69.

Humphrey, V., Zscheischler, J., Ciais, P., et al. (2018). Sensitivity of atmospheric CO_2 growth rate to observed changes in terrestrial water storage. *Nature*, 560, 628–631.

Hunt, A. (2016). *Reviving Roman Religion: Sacred Trees in the Roman World*. Cambridge, UK: Cambridge University Press.

Huntley, B., and Prentice, I. C. (1993). Holocene vegetation and climates of Europe. In *Global Climates since the Last Glacial Maximum*, edited by H. E. Wright, Jr., J. E. Kutzbach, T. Webb, III, et al. Minneapolis: University of Minnesota Press, pp. 136–168.

Hurmekoski, E., Smyth, C. E., Stern, T., Verkerk, P. J., and Asada, R. (2021). Substitution impacts of wood use at the market level: A systematic review. *Environmental Research Letters*, 16, 123004, DOI: https://doi.org/10.1088/1748-9326/ac386f.

Hursh, C. R. (1948). *Local Climate in the Copper Basin of Tennessee as Modified by the Removal of Vegetation*, Circular Number 774. Washington, DC: U.S. Department of Agriculture.

Hurtt, G. C., Chini, L., Sahajpal, R., et al. (2020). Harmonization of global land use change and management for the period 850–2100 (LUH2) for CMIP6. *Geoscientific Model Development*, 13, 5425–5464.

Hutchison, B. A., Matt, D. R., McMillen, R. T., et al. (1986). The architecture of a deciduous forest canopy in eastern Tennessee, U.S.A. *Journal of Ecology*, 74, 635–646.

Huuskonen, S., Domisch, T., Finér, L., et al. (2021). What is the potential for replacing monocultures with mixed-species stands to enhance ecosystem services in boreal forests in Fennoscandia? *Forest Ecology and Management*, 479, 118558, DOI: https://doi.org/10.1016/j.foreco.2020.118558.

Iglesias, V., Balch, J. K., and Travis, W. R. (2022). U.S. fires became larger, more frequent, and more widespread in the 2000s. *Science Advances*, 8, eabc0020, DOI: https://doi.org/10.1126/sciadv.abc0020.

Imlay, G. (1793). *A Description of the Western Territory of North America*. Dublin: William Jones.

Irvine, P. J., Ridgwell, A., and Lunt, D. J. (2011). Climatic effects of surface albedo geoengineering. *Journal of Geophysical Research*, 116, D24112, DOI: https://doi.org/10.1029/2011JD016281.

Irving, W. (1848). *The Sketchbook of Geoffrey Crayon, Gent. – The Author's Revised Edition*. New York: George P. Putnam.

Jackson, R. (2020). Eunice Foote, John Tyndall and a question of priority. *Notes and Records*, 74, 105–118.

Jackson, R. B., Jobbágy, E. G., Avissar, R., et al. (2005). Trading water for carbon with biological carbon sequestration. *Science*, 310, 1944–1947.

Jackson, R. B., Randerson, J. T., Canadell, J. G., et al. (2008). Protecting climate with forests. *Environmental Research Letters*, 3, 044006, DOI: https://doi.org/10.1088/1748–9326/3/4/044006.

Jahren, A. H. (2007). The Arctic forest of the middle Eocene. *Annual Review of Earth and Planetary Sciences*, 35, 509–540.

James, N. D. G. (1981). *A History of English Forestry*. Oxford: Basil Blackwell.

James, N. D. G. (1996). A history of forestry and monographic forestry literature in Germany, France, and the United Kingdom. In *The Literature of Forestry and Agroforestry*, edited by P. McDonald and J. Lassoie. Ithaca, NY: Cornell University Press, pp. 15–44.

Jamieson, T. F. (1860). *The Tweeddale Prize Essay on the Rainfall*. Edinburgh: William Blackwood.

Janisch, H. R. (1908). *Extracts from the St. Helena Records (Second Edition) and Chronicles of Cape Commanders*. Jamestown, St. Helena: Benjamin Grant.

Jefferson, T. (1787). *Notes on the State of Virginia*. London: John Stockdale.

Jefferson, T. (1903). To Lewis E. Beck, July 16, 1824. In *The Writings of Thomas Jefferson*, vol. 16, edited by A. E. Bergh. Washington, DC: Thomas Jefferson Memorial Association of the United States, pp. 71–72.

Jefferson, T. (1954). To Jean Baptiste Le Roy, November 13, 1786. In *The Papers of Thomas Jefferson*, vol. 10. *22 June to 31 December 1786*, edited by J. P. Boyd. Princeton, NJ: Princeton University Press, pp. 524–530.

Jefferson, T. (1955). To Buffon, October 1, 1787. In *The Papers of Thomas Jefferson*, vol. 12. *7 August 1787 to 31 March 1788*, edited by J. P. Boyd. Princeton, NJ: Princeton University Press, pp. 194–195.

Jefferson, T. (2006). To Nathaniel Chapman, December 11, 1809. In *The Papers of Thomas Jefferson, Retirement Series*, vol. 2. *16 November 1809 to 11 August 1810*, edited by J. J. Looney. Princeton, NJ: Princeton University Press, pp. 70–72.

Jia, G., Shevliakova, E., Artaxo, P., et al. (2019). Land–climate interactions. In *Climate Change and Land: An IPCC Special Report on Climate Change, Desertification, Land Degradation, Sustainable Land Management, Food Security, and Greenhouse Gas Fluxes in Terrestrial Ecosystems*, edited by P. R. Shukla, J. Skea, E. Calvo Buendia, et al. Geneva: World Meteorological Organization, pp. 131–247.

Jiang, M., Medlyn, B. E., Drake, J. E., et al. (2020). The fate of carbon in a mature forest under carbon dioxide enrichment. *Nature*, 580, 227–231.

Jiang, Y., Wang, G., Liu, W., et al. (2021). Modeled response of South American climate to three decades of deforestation. *Journal of Climate*, 34, 2189–2203.

Johnson, E. (1910). *Johnson's Wonder-Working Providence, 1628–1651*, edited by J. F. Jameson. New York: Charles Scribner's Sons.

Joussaume, S., Taylor, K. E., Braconnot, P., et al. (1999). Monsoon changes for 6000 years ago: Results of 18 simulations from the Paleoclimate Modeling Intercomparison Project (PMIP). *Geophysical Research Letters*, 26, 859–862.

Jouzel, J., Masson-Delmotte, V., Cattani, O., et al. (2007). Orbital and millennial Antarctic climate variability over the past 800,000 years. *Science*, 317, 793–796.

Juang, J.-Y., Katul, G., Siqueira, M., Stoy, P., and Novick, K. (2007). Separating the effects of albedo from eco-physiological changes on surface temperature along a successional chronosequence in the southeastern United States. *Geophysical Research Letters*, 34, L21408, DOI: https://doi.org/10.1029/2007GL031296.

Kalliokoski, T., Bäck, J., Boy, M., et al. (2020). Mitigation impact of different harvest scenarios of Finnish forests that account for albedo, aerosols, and trade-offs of carbon sequestration and avoided emissions. *Frontiers in Forests and Global Change*, 3, 562044, DOI: https://doi.org/10.3389/ffgc.2020.562044.

Kalm, P. (1770–71). *Travels into North America*, translated by J. R. Forster, 3 vols. Warrington: William Eyres; London: T. Lowndes.

Kalm, P. (1937). *The America of 1750: Peter Kalm's Travels in North America; The English Version of 1770*, edited and translated by A. B. Benson, 2 vols. New York: Wilson-Erickson.

Kasahara, A., and Washington, W. M. (1967). NCAR global general circulation model of the atmosphere. *Monthly Weather Review*, 95, 389–402.

Kasahara, A., and Washington, W. M. (1971). General circulation experiments with a six-layer NCAR model, including orography, cloudiness and surface temperature calculations. *Journal of the Atmospheric Sciences*, 28, 657–701.

Kedzie, R. C. (1867). The influence of forest trees on agriculture. In *Sixth Annual Report of the Secretary of the State Board of Agriculture of the State of Michigan, for the Year 1867*. Lansing: John A. Kerr, pp. 465–483.

Kedzie, R. C., Woodman, J. J., and Fellows, O. H. (1866). Report of the committee. In *Fifth Annual Report of the Secretary of the State Board of Agriculture of the State of Michigan, for the Year 1866*. Lansing: John A. Kerr, Appendix pp. 3–31.

Keeling, R. F., Piper, S. C. and Heimann, M. (1996). Global and hemispheric CO_2 sinks deduced from changes in atmospheric O_2 concentration. *Nature*, 381, 218–221.

Keenan, T. F., and Williams, C. A. (2018). The terrestrial carbon sink. *Annual Review of Environment and Resources*, 43, 219 –243.

Keenan, T. F., Luo, X., De Kauwe, M. G., et al. (2021). A constraint on historic growth in global photosynthesis due to increasing CO_2. *Nature*, 600, 253–258.

Keever, C. (1950). Causes of succession on old fields of the Piedmont, North Carolina. *Ecological Monographs*, 20, 229–250.

Keever, C. (1983). A retrospective view of old-field succession after 35 years. *American Midland Naturalist*, 110, 397–404.

Keith, H., Vardon, M., Obst, C., et al. (2021). Evaluating nature-based solutions for climate mitigation and conservation requires comprehensive carbon accounting. *Science of the Total Environment*, 769, 144341, DOI: https://doi.org/10.1016/j.scitotenv.2020.144341.

Kellomäki, S., Väisänen, H., Kirschbaum, M. U. F., Kirsikka-Aho, S., and Peltola, H. (2021). Effects of different management options of Norway spruce on radiative forcing through changes in carbon stocks and albedo. *Forestry*, 94, 588–597.

Kemena, T. P., Matthes, K., Martin, T., Wahl, S., and Oschlies, A. (2018). Atmospheric feedbacks in North Africa from an irrigated, afforested Sahara. *Climate Dynamics*, 50, 4561–4581.

Kemp, M. (2008). Looking at the face of the Earth. *Nature*, 456, 876.

Kerner von Marilaun, A. (1888). *Pflanzenleben*, vol. 1. *Gestalt und Leben der Pflanze*. Leipzig: Bibliographischen Instituts.

Kerner von Marilaun, A. (1894). *The Natural History of Plants: Their Forms, Growth, Reproduction, and Distribution*, vol. 1. *Biology and Configuration of Plants*, translated by F. W. Oliver. London: Blackie.

Khanna, J., Medvigy, D., Fueglistaler, S., and Walko, R. (2017). Regional dry-season climate changes due to three decades of Amazonian deforestation. *Nature Climate Change*, 7, 200–204.

Kittredge, J. (1948). *Forest Influences: The Effects of Woody Vegetation on Climate, Water, and Soil, with Applications to the Conservation of Water and the Control of Floods and Erosion*. New York: McGraw-Hill.

Klages, J. P., Salzmann, U., Bickert, T., et al. (2020). Temperate rainforests near the South Pole during peak Cretaceous warmth. *Nature*, 580, 81–86.

Kleidon, A., Fraedrich, K., and Heimann, M. (2000). A green planet versus a desert world: Estimating the maximum effect of vegetation on the land surface climate. *Climatic Change*, 44, 471–493.

Klein, R. G. (1995). Anatomy, behavior, and modern human origins. *Journal of World Prehistory*, 9, 167–198.

Kooperman, G. J., Chen, Y., Hoffman, F. M., et al. (2018). Forest response to rising CO_2 drives zonally asymmetric rainfall change over tropical land. *Nature Climate Change*, 8, 434–440.

Kopenawa, D., and Albert, B. (2013). *The Falling Sky: Words of a Yanomami Shaman*, translated by N. Elliott and A. Dundy. Cambridge, MA: Harvard University Press.

Koren, I., Kaufman, Y. J., Remer, L. A., and Martins, J. V. (2004). Measurement of the effect of Amazon smoke on inhibition of cloud formation. *Science*, 303, 1342–1345.

Koren, I., Martins, J. V., Remer, L. A., and Afargan, H. (2008). Smoke invigoration versus inhibition of clouds over the Amazon. *Science*, 321, 946–949.

Koster, R. D., Dirmeyer, P. A., Guo, Z., et al. (2004). Regions of strong coupling between soil moisture and precipitation. *Science*, 305, 1138–1140.

Kotok, E. I. (1940). The forester's dependence on the science of meteorology. *Bulletin of the American Meteorological Society*, 21, 383–384, 397–406.

Kreidenweis, U., Humpenöder, F., Stevanović, M., et al. (2016). Afforestation to mitigate climate change: Impacts on food prices under consideration of albedo effects. *Environmental Research Letters*, 11, 085001, DOI: https://doi.org/10.1088/1748-9326/11/8/085001.

Kröniger, K., De Roo, F., Brugger, P., et al. (2018). Effect of secondary circulations on the surface–atmosphere exchange of energy at an isolated semi-arid forest. *Boundary-Layer Meteorology*, 169, 209–232.

Kucharski, F., Zeng, N., and Kalnay, E. (2013). A further assessment of vegetation feedback on decadal Sahel rainfall variability. *Climate Dynamics*, 40, 1453–1466.

Kulmala, M., Ezhova, E., Kalliokoski, T., et al. (2020). CarbonSink+: Accounting for multiple climate feedbacks from forests. *Boreal Environmental Research*, 25, 145–159.

Kulmala, M., Nieminen, T., Nikandrova, A., et al. (2014). CO_2-induced terrestrial climate feedback mechanism: From carbon sink to aerosol source and back. *Boreal Environment Research*, 19 (suppl. B), 122–131.

Kulmala, M., Suni, T., Lehtinen, K. E. J., et al. (2004). A new feedback mechanism linking forests, aerosols, and climate. *Atmospheric Chemistry and Physics*, 4, 557–562.

Kupperman, K. O. (1982). The puzzle of the American climate in the early colonial period. *American Historical Review*, 87, 1262–1289.

Kurtén, T., Kulmala, M., Dal Maso, M., et al. (2003). Estimation of different forest-related contributions to the radiative balance using observations in southern Finland. *Boreal Environment Research*, 8, 275–285.

Kurz, W. A., Dymond, C. C., Stinson, G., et al. (2008). Mountain pine beetle and forest carbon feedback to climate change. *Nature*, 452, 987–990.

Kutzbach, J., Bonan, G., Foley, J., and Harrison, S. P. (1996). Vegetation and soil feedbacks on the response of the African monsoon to orbital forcing in the early to middle Holocene. *Nature*, 384, 623–626.

Kutzleb, C. R. (1971). Can forests bring rain to the Plains? *Forest History*, 15(3), 14–21.

Kuusinen, N., Tomppo, E., Shuai, Y., and Berninger, F. (2014). Effects of forest age on albedo in boreal forests estimated from MODIS and Landsat albedo retrievals. *Remote Sensing of Environment*, 145, 145–153.

Laguë, M. M., Bonan, G. B., and Swann, A. L. S. (2019). Separating the impact of individual land surface properties on the terrestrial surface energy budget in both the coupled and uncoupled land-atmosphere system. *Journal of Climate*, 32, 5725–5744.

Laguë, M. M., Pietschnig, M., Ragen, S., Smith, T. A., and Battisti, D. S. (2021a). Terrestrial evaporation and global climate: Lessons from northland, a planet with a hemispheric continent. *Journal of Climate*, 34, 2253–2276.

Laguë, M. M., and Swann, A. L. S. (2016). Progressive midlatitude afforestation: Impacts on clouds, global energy transport, and precipitation. *Journal of Climate*, 29, 5561–5573.

Laguë, M. M., Swann, A. L. S., and Boos, W. R. (2021b). Radiative feedbacks on land surface change and associated tropical precipitation shifts. *Journal of Climate*, 34, 6651–6672.

Lamarck, J.-B. (1794). *Recherches sur les causes des principaux faits physiques*, 2 vols. Paris: Maradan.

Lamarck, J.-B. (1801/02). *Annuaire météorologique, pour l'an X de l'ère de la République française*, vol. 3. Paris: Maillard.

Lamarck, J.-B. (1802). *Hydrogéologie*. Paris: Agasse; Maillard.

Lamarck, J.-B. (1820). *Système analytique des connaissances positives de l'homme*. Paris: A. Belin.

Lamarck, J.-B. (1964). *Hydrogeology*, translated by A. V. Carozzi. Urbana, IL: University of Illinois Press.

Lamb, H. H. (1977). *Climate: Present, Past and Future*, vol. 2. *Climatic History and the Future*. London: Methuen.

Lamb, H. H. (1995). *Climate, History and the Modern World*, 2nd ed. London: Routledge.

Lange, M. (2005). Ecological laws: What would they be and why would they matter? *Oikos*, 110, 394–403.

Lapham, I. A., Knapp J. G., and Crocker, H. (1867). *Report on the Disastrous Effects of the Destruction of Forest Trees, Now Going on So Rapidly in the State of Wisconsin*. Madison: Atwood & Rublee.

Larsen, J. A. (1980). *The Boreal Ecosystem*. New York: Academic Press.

Lathière, J., Hewitt, C. N., and Beerling D. J. (2010). Sensitivity of isoprene emissions from the terrestrial biosphere to 20th century changes in atmospheric CO_2 concentration, climate, and land use. *Global Biogeochemical Cycles*, 24, GB1004, DOI: https://doi.org/10.1029/2009GB003548.

Lawrence, D., and Vandecar, K. (2015). Effects of tropical deforestation on climate and agriculture. *Nature Climate Change*, 5, 27–36.

Lawton, J. H. (1999). Are there general laws in ecology? *Oikos*, 84, 177–192.

Lazarus, M. H., and Pardoe, H. S. (2003). *Catalogue of Botanical Prints and Drawings at the National Museums & Galleries of Wales*. Cardiff: National Museums & Galleries of Wales.

Lecoy, A. (1879). The forest question in New Zealand. *Transactions and Proceedings of the New Zealand Institute*, 12, 3–23.

Lee, D. (1850). Agricultural meteorology. In *Report of the Commissioner of Patents, for the Year 1849*. Part. II. *Agriculture*. Washington, DC: Office of Printers to the Senate, pp. 38–48.

Lee, X., Goulden, M. L., Hollinger, D. Y., et al. (2011). Observed increase in local cooling effect of deforestation at higher latitudes. *Nature*, 479, 384–387.

Legg, S. (2014). Debating the climatological role of forests in Australia, 1827–1949: A survey of the popular press. In *Climate, Science, and Colonization: Histories from Australia and New Zealand*, edited by J. Beattie, E. O'Gorman, and M. Henry. New York: Palgrave Macmillan, pp. 119–136.

Legg, S. M. (2018). Views from the Antipodes: The "forest influence" debate in the Australian and New Zealand press, 1827–1956. *Australian Geographer*, 49, 41–60.

Le Jeune, P. (1634). *Relation de ce qui s'est passe en la Nouvelle France en l'annee 1633*. Paris: Sebastien Cramoisy.

Lejeune, Q., Davin, E. L., Gudmundsson, L., Winckler, J., and Seneviratne, S. I. (2018). Historical deforestation locally increased the intensity of hot days in northern mid-latitudes. *Nature Climate Change*, 8, 386–390.

Lejeune, Q., Seneviratne, S. I., and Davin, E. L. (2017). Historical land-cover change impacts on climate: Comparative assessment of LUCID and CMIP5 multimodel experiments. *Journal of Climate*, 30, 1439–1459.

Lembke, J. (2005). *Virgil's Georgics: A New Verse Translation*. New Haven, CT: Yale University Press.

Lemordant, L., Gentine, P., Swann, A. S., Cook, B. I., and Scheff, J. (2018). Critical impact of vegetation physiology on the continental hydrologic cycle in response to increasing CO_2. *Proceedings of the National Academy of Sciences USA*, 115, 4093–4098.

Lenton, T. M., Held, H., Kriegler, E., et al. (2008). Tipping elements in the Earth's climate system. *Proceedings of the National Academy of Sciences USA*, 105, 1786–1793.

Lenton, T. M., Rockström, J., Gaffney, O., et al. (2019). Climate tipping points – too risky to bet against. *Nature*, 575, 592–595.

Leonardi, C., and Stagi, F. (2019). *The Architecture of Trees*. Hudson, NY: Princeton Architectural Press.

Le Roy, C.-G. (1757). Forêt. In *Encyclopédie, ou dictionnaire raisonné des sciences, des arts et des métiers*, vol. 7, edited by D. Diderot and J. Le Rond d'Alembert. Paris: Briasson; David; Le Breton; Durand, pp. 129–132.

Le Roy, J. B. (1954). From Jean Baptiste Le Roy, September 28, 1786. In *The Papers of Thomas Jefferson*, vol. 10. *22 June to 31 December 1786*, edited by J. P. Boyd. Princeton: Princeton University Press, pp. 410–411.

Lescarbot, M. (1609a). *Histoire de la Nouvelle France*. Paris: Jean Milot.

Lescarbot, M. (1609b). *Nova Francia: Or the Description of That Part of New France, Which is One Continent with Virginia*, translated by P. E. London: George Bishop.

Leslie, A. B., Beaulieu, J., Holman, G., et al. (2018). An overview of extant conifer evolution from the perspective of the fossil record. *American Journal of Botany*, 105, 1531–1544.

Leslie, J. (1819). On heat and climate. *Annals of Philosophy*, 14, 5–27.

Leslie, J. (1804). *An Experimental Inquiry into the Nature, and Propagation, of Heat*. London: J. Mawman.

Leuzinger, S., and Körner, C. (2007). Tree species diversity affects canopy leaf temperatures in a mature temperate forest. *Agricultural and Forest Meteorology*, 146, 29–37.

Levis, S., and Bonan, G. B. (2004). Simulating springtime temperature patterns in the Community Atmosphere Model coupled to the Community Land Model using prognostic leaf area. *Journal of Climate*, 17, 4531–4540.

Levis, S., Bonan, G. B., and Bonfils, C. (2004). Soil feedback drives the mid-Holocene North African monsoon northward in fully coupled CCSM2 simulations with a dynamic vegetation model. *Climate Dynamics*, 23, 791–802.

Levis, S., Foley, J. A., and Pollard, D. (1999). CO_2, climate, and vegetation feedbacks at the Last Glacial Maximum. *Journal of Geophysical Research*, 104D, 31191–31198.

Lewis, S. L., Brando, P. M., Phillips, O. L., van der Heijden, G. M. F., and Nepstad, D. (2011). The 2010 Amazon drought. *Science*, 331, 554.

Lewis, S. L., Wheeler, C. E., Mitchard, E. T. A., and Koch, A. (2019). Regenerate natural forests to store carbon. *Nature*, 568, 25–28.

Li, Q., Wei, X., Zhang, M., et al. (2017). Forest cover change and water yield in large forested watersheds: A global synthetic assessment.
Ecohydrology, 10, e1838, DOI: https://doi.org/10.1002/eco.1838.

Li, Y., Brando, P. M., Morton, D. C., et al. (2022a). Deforestation-induced climate change reduces carbon storage in remaining tropical forests. *Nature Communications*, 13, 1964, DOI: https://doi.org/10.1038/s41467-022-29601-0.

Li, Y., de Noblet-Ducoudré, N., Davin, E. L., et al. (2016). The role of spatial scale and background climate in the latitudinal temperature response to deforestation. *Earth System Dynamics*, 7, 167–181.

Li, Y., Kalnay, E., Motesharrei, S., et al. (2018a). Climate model shows large-scale wind and solar farms in the Sahara increase rain and vegetation. *Science*, 361, 1019–1022.

Li, Y., Liu, Y., Bohrer, G., et al. (2022b). Impacts of forest loss on local climate across the conterminous United States: Evidence from satellite time-series observations. *Science of The Total Environment*, 802, 149651, DOI: https://doi.org/10.1016/j.scitotenv.2021.149651.

Li, Y., Piao, S., Li, L. Z. X., et al. (2018b). Divergent hydrological response to large-scale afforestation and vegetation greening in China. *Science Advances*, 4, eaar4182, DOI: https://doi.org/10.1126/sciadv.aar4182.

Li, Y., Piao, S., Chen, A., Ciais, P., and Li, L. Z. X. (2020). Local and teleconnected temperature effects of afforestation and vegetation greening in China. *National Science Review*, 7, 897–912.

Li, Y., Randerson, J. T., Mahowald, N. M., and Lawrence, P. J. (2021). Deforestation strengthens atmospheric transport of mineral dust and phosphorus from North Africa to the Amazon. *Journal of Climate*, 34, 6087–6096.

Li, Y., Zhao, M., Motesharrei, S., et al. (2015). Local cooling and warming effects of forests based on satellite observations. *Nature Communications*, 6, 6603, DOI: https://doi.org/10.1038/ncomms7603.

Liebig, J. von (1862). *Die Chemie in ihrer Anwendung auf Agricultur und Physiologie*, 7th ed., 2 vols. Braunschweig: Friedrich Vieweg.

Lihavainen, H., Kerminen, V.-M., Tunved, P., et al. (2009). Observational signature of the direct radiative effect by natural boreal forest aerosols and its relation to the corresponding first indirect effect. *Journal of Geophysical Research*, 114, D20206, DOI: https://doi.org/10.1029/2009JD012078.

Lihavainen, H., Asmi, E., Aaltonen, V., Makkonen, U., and Kerminen, V.-M. (2015). Direct radiative feedback due to biogenic secondary organic aerosol estimated from boreal forest site observations. *Environmental Research Letters*, 10, 104005, DOI: https://doi.org/10.1088/1748-9326/10/10/104005.

Likens, G. E., Bormann, F. H., Pierce, R. S., Eaton, J. S., and Johnson, N. M. (1977). *Biogeochemistry of a Forested Ecosystem*. New York: Springer-Verlag.

Linquist, S., Gregory, T. R., Elliott, T. A., et al. (2016). Yes! There are resilient generalizations (or "laws") in ecology. *Quarterly Review of Biology*, 91, 119–131.

Lintunen, J., Rautiainen, A., and Uusivuori, J. (2022). Which is more important, carbon or albedo? Optimizing harvest rotations for timber and climate benefits in a changing climate. *American Journal of Agricultural Economics*, 104, 134–160.

Liu, H., and Randerson, J. T. (2008). Interannual variability of surface energy exchange depends on stand age in a boreal forest fire chronosequence. *Journal of Geophysical Research*, 113, G01006, DOI: https://doi.org/10.1029/2007JG000483.

Liu, H., Randerson, J. T., Lindfors, J., and Chapin, F. S., III (2005). Changes in the surface energy budget after fire in boreal ecosystems of interior Alaska: An annual perspective. *Journal of Geophysical Research*, 110, D13101, DOI: https://doi.org/10.1029/2004JD005158.

Liu, J., Li, S., Ouyang, Z., Tam, C., and Chen, X. (2008). Ecological and socioeconomic effects of China's policies for ecosystem services. *Proceedings of the National Academy of Sciences USA*, 105, 9477–9482.

Liu, L., Cheng, Y., Wang, S., et al. (2020). Impact of biomass burning aerosols on radiation, clouds, and precipitation over the Amazon: Relative importance of aerosol–cloud and aerosol–radiation interactions. *Atmospheric Chemistry and Physics*, 20, 13283–13301.

Locher, F., and Fressoz, J.-B. (2012). Modernity's frail climate: A climate history of environmental reflexivity. *Critical Inquiry*, 38, 579–598.

Löffelholz-Colberg, F. F. von (1872). *Die Bedeutung und Wichtigkeit des Waldes, Ursachen und Folgen der Entwaldung, die Wiederbewaldung mit Rücksicht auf Pflanzenphysiologie, Klimatologie, Meteorologie, Forststatistik, Forstgeographie und die forstlichen Verhältnisse aller Länder. . ..* Leipzig: Heinrich Schmidt.

Lombardozzi, D., Levis, S., Bonan, G., Hess, P. G., and Sparks, J. P. (2015). The influence of chronic ozone exposure on global carbon and water cycles. *Journal of Climate*, 28, 292–305.

Long, E. (1774). *The History of Jamaica: or, General Survey of the Antient and Modern State of That Island*, 3 vols. London: T. Lowndes.

Longfellow, H. W. (1863). *Tales of a Wayside Inn*. Boston: Ticknor & Fields.

Loomis, E. (1868). *A Treatise on Meteorology, with a Collection of Meteorological Tables*. New York: Harper.

Lorentz, B., and Parade, A. (1837). *Cours élémentaire de culture des bois, créé a l'école royale forestière de Nancy*. Paris: Huzard; Nancy: George-Grimblot.

Lorentz, B., and Parade, A. (1883). *Cours élémentaire de culture des bois, créé a l'école forestière de Nancy*, 6th ed. Paris: Octave Doin.

Lorenz, E. N. (1970). Climatic change as a mathematical problem. *Journal of Applied Meteorology*, 9, 325–329.

Lorenz, J. R. (1877). *Über Bedeutung und Vertretung der land- und forstwirthschaftlichen Meteorologie*. Vienna: Faesy & Frick.

Lorenz, J. R., and Rothe, C. (1874). *Lehrbuch der Klimatologie mit besonderer Rücksicht auf Land- und Forstwirthschaft*. Vienna: Wilhelm Braumüller.

Lorenz, R., Pitman, A. J., and Sisson, S. A. (2016). Does Amazonian deforestation cause global effects; can we be sure? *Journal of Geophysical Research: Atmospheres*, 121, 5567–5584.

Lorenz von Liburnau, J. R. (1878). *Wald, Klima und Wasser*. Munich: R. Oldenbourg.

Lorenz von Liburnau, J. R. (1879). *Bericht für den zweiten internationalen Meteorologen-Congress über die Frage: Wie können die meteorologischen Institute sich der Land- und Forstwirtschaft förderlich erweisen?* Vienna: Carl Fromme.

Lorenz von Liburnau, J. R. (1890). *Resultate Forstlich-Meteorologischer Beobachtungen insbesondere in den Jahren 1885–1887*, part 1: *Untersuchungen über die Temperatur und die Feuchtigkeit der Luft unter, in und über den Baumkronen des Waldes, sowie im Freilande*, Mittheilungen vom forstlichen Versuchswesen in Österreich. XII Heft. Vienna: K. & K. Hof-Buchhandlung W. Frick.

Lorey, T. (1888). *Handbuch der Forstwissenschaft: forstliche Produktionslehre. I.* Tübingen: H. Laupp'schen.

Loskutova, M. (2020). Quantifying scarcity: Deforestation in the Upper Volga region and early debates over climate change in nineteenth-century Russia. *European Review of History: Revue européenne d'histoire*, 27, 253–272.

Loudon, J. C. (1838). *Arboretum et Fruticetum Britannicum; or, The Trees and Shrubs of Britain*, 8 vols. London: Longman, Orme, Brown, Green, & Longmans.

Lovejoy, T. E., and Nobre, C. (2018). Amazon tipping point. *Science Advances*, 4, eaat2340, DOI: https://doi.org/10.1126/sciadv.aat2340.

Lovell, J. (1826). *Meteorological Register for the Years 1822, 1823, 1824 & 1825, from Observations Made by the Surgeons of the Army, at the Military Posts of the United States*. Washington, DC: Edward de Krafft.

Lovelock, J. E. (1979). *Gaia: A New Look at Life on Earth*. Oxford: Oxford University Press.

Lovelock, J. E., and Margulis, L. (1974). Atmospheric homeostasis by and for the biosphere: The gaia hypothesis. *Tellus*, 26, 2–10.

Lowell, P. (1906). *Mars and its Canals*. New York: Macmillan.

Lowell, P. (1908). *Mars as the Abode of Life*. New York: Macmillan.

Lu, Z., Zhang, Q., Miller, P. A., et al. (2021). Impacts of large-scale Sahara solar farms on global climate and vegetation cover. *Geophysical Research Letters*, 48, e2020GL090789, DOI: https://doi.org/10.1029/2020GL090789.

Lusardi, J. P. (1973). Appendix A, Barnes' supplications of 1531 and 1534. In *The Yale Edition of the Complete Works of St. Thomas More*, vol. 8. *The Confutation of Tyndale's Answer, Part II The Text, Books V–IX, Appendices*, edited by L. A. Schuster, R. C. Marius, J. P. Lusardi, and R. J. Schoeck. New Haven, CT: Yale University Press, pp. 1035–1061.

Lüthi, D., Le Floch, M., Bereiter, B., et al. (2008). High-resolution carbon dioxide concentration record 650,000–800,000 years before present. *Nature*, 453, 379–382.

Lutz, D. A., and Howarth, R. B. (2014). Valuing albedo as an ecosystem service: Implications for forest management. *Climatic Change*, 124, 53–63.

Lutz, D. A., Burakowski, E. A., Murphy, M. B., et al. (2016). Trade-offs between three forest ecosystem services across the state of New Hampshire, USA: Timber, carbon, and albedo. *Ecological Applications*, 26, 146–161.

Luyssaert, S., Jammet, M., Stoy, P. C., et al. (2014). Land management and land-cover change have impacts of similar magnitude on surface temperature. *Nature Climate Change*, 4, 389–393.

Luyssaert, S., Marie, G., Valade, A., et al. (2018). Trade-offs in using European forests to meet climate objectives. *Nature*, 562, 259–262.

Lyell, C. (1834). *Principles of Geology*, 3rd ed., 4 vols. London: John Murray.

Lynch, P. (2008). The origins of computer weather prediction and climate modeling. *Journal of Computational Physics*, 227, 3431–3444.

Mabey, R. (2016). *The Cabaret of Plants: Forty Thousand Years of Plant Life and the Human Imagination*. New York: W. W. Norton.

Mahmood, R., Pielke, R. A., Sr., Hubbard, K. G., et al. (2014). Land cover changes and their biogeophysical effects on climate. *International Journal of Climatology*, 34, 929–953.

Mahrt, L., and Ek, M. (1993). Spatial variability of turbulent fluxes and roughness lengths in HAPEX-MOBILHY. *Boundary-Layer Meteorology*, 65, 381–400.

Makar, P. A., Akingunola, A., Chen, J., et al. (2021). Forest-fire aerosol–weather feedbacks over western North America using a high-resolution, online coupled air-quality model. *Atmospheric Chemistry and Physics*, 21, 10557–10587.

Malhi, Y. (2017). The concept of the Anthropocene. *Annual Review of Environment and Resources*, 42, 77–104.

Malmer, A., Murdiyarso, D., Bruijnzeel, L. A., and Ilstedt, U. (2010). Carbon sequestration in tropical forests and water: A critical look at the basis for commonly used generalizations. *Global Change Biology*, 15, 599–604.

Malone, J. J. (1964). *Pine Trees and Politics: The Naval Stores and Forest Policy in Colonial New England, 1691–1775*. Seattle: University of Washington Press.

Malte-Brun, C. (1834). *A System of Universal Geography, or a Description of all the Parts of the World*, with additions and corrections by J. G. Percival, 3 vols. Boston: Samuel Walker.

Manabe, S. (1969). Climate and the ocean circulation. I. The atmospheric circulation and the hydrology of the Earth's surface. *Monthly Weather Review*, 97, 739–774.

Manabe, S., Smagorinsky, J., and Strickler, R. F. (1965). Simulated climatology of a general circulation model with a hydrologic cycle. *Monthly Weather Review*, 93, 769–798.

Maness, H., Kushner, P. J., and Fung, I. (2013). Summertime climate response to mountain pine beetle disturbance in British Columbia. *Nature Geoscience*, 6, 65–70.

Mann, C. R., and Twiss, G. R. (1910). *Physics*, rev. ed. Chicago: Scott, Foresman & Company.

Manning, A. C., and Keeling, R. F. (2006). Global oceanic and land biotic carbon sinks from the Scripps atmospheric oxygen flask sampling network. *Tellus*, 58B, 95–116.

Manwood, J. (1598). *A Treatise and Discourse of the Lawes of the Forrest*. London: Thomas Wight & Bonham Norton.

Marengo, J. A., Souza, C. M., Jr., Thonicke, K., et al. (2018). Changes in climate and land use over the Amazon region: Current and future variability and trends. *Frontiers in Earth Science*, 6, 228, DOI: https://doi.org/10.3389/feart.2018.00228.

Marks, P. L. (1974). The role of pin cherry (*Prunus pensylvanica* L.) in the maintenance of stability in northern hardwood ecosystems. *Ecological Monographs*, 44, 73–88.

Marsh, G. P. (1864). *Man and Nature; or, Physical Geography as Modified by Human Action*. New York: Charles Scribner.

Martin, S. T., Andreae, M. O., Artaxo, P., et al. (2010). Sources and properties of Amazonian aerosol particles. *Reviews of Geophysics*, 48,

RG2002, DOI: https://doi.org/10.1029/2008RG000280.

Massman, W. J. (1997). An analytical one-dimensional model of momentum transfer by vegetation of arbitrary structure. *Boundary-Layer Meteorology*, 83, 407–421.

Mather, C. (1721). *The Christian Philosopher: A Collection of the Best Discoveries in Nature, with Religious Improvements*. London: Eman. Matthews.

Mathieu, A. (1855). *Description des bois des essences forestières les plus importantes*. Nancy: Grimbolt & Veuve Raybois.

Mathieu, A. (1877). *Flore forestière: description et histoire des végétaux ligneux qui croissant spontanément en France et des essences importantes de l'Algérie*, 3rd ed. Paris: Berger-Levrault & Cie; Vueve Bouchard-Huzard.

Mathieu, A. (1878). *Météorologie comparée agricole et forestière: Rapport à M. le sous-secrétaire d'État, président du conseil d'administration des forêts*. Exposition universelle de 1878. Ministère de l'Agriculture et du commerce. Administration des forêts. Paris: Imprimerie Nationale.

Matsuda, M., Tadaki, Y., Izuhara, S., Takumi, A., and Ohshima, Y. (1987). Seasonal variations of the physical environment of larch forest. *Journal of Agricultural Meteorology*, 43, 3–13.

Matteson, K. (2015). *Forests in Revolutionary France: Conservation, Community, and Conflict, 1669–1848*. New York: Cambridge University Press.

Matthews, H. D., Weaver, A. J., Meissner, K. J., Gillett, N. P., and Eby, M. (2004). Natural and anthropogenic climate change: Incorporating historical land cover change, vegetation dynamics and the global carbon cycle. *Climate Dynamics*, 22, 461–479.

Matthews, W. H., Kellogg, W. W., and Robinson, G. D. (1971). *Man's Impact on the Climate*. Cambridge, MA: Massachusetts Institute of Technology Press.

Matthies, B. D., and Valsta, L. T. (2016). Optimal forest species mixture with carbon storage and albedo effect for climate change mitigation. *Ecological Economics*, 123, 95–105.

Mayr, E. (1982). *The Growth of Biological Thought: Diversity, Evolution, and Inheritance*. Cambridge, MA: Harvard University Press.

McCurry, M. R., Cantrill, D. J., Smith, P. M., et al. (2022). A Lagerstätte from Australia provides insight into the nature of Miocene mesic ecosystems. *Science Advances*, 8, eabm1406, DOI: https://doi.org/10.1126/sciadv.abm1406.

McDowell, N., Allen, C. D., Anderson-Teixeira, K., et al. (2018). Drivers and mechanisms of tree mortality in moist tropical forests. *New Phytologist*, 219, 851–869.

McDowell, N., Pockman, W. T., Allen, C. D., et al. (2008). Mechanisms of plant survival and mortality during drought: Why do some plants survive while others succumb to drought? *New Phytologist*, 178, 719–739.

McDowell, N. G., Allen, C. D., Anderson-Teixeira, K., et al. (2020). Pervasive shifts in forest dynamics in a changing world. *Science*, 368, eaaz9463, DOI: https://doi.org/10.1126/science .aaz9463.

McDowell, N. G., Fisher, R. A., Xu, C., et al. (2013). Evaluating theories of drought-induced vegetation mortality using a multimodel–experiment framework. *New Phytologist*, 200, 304–321.

McGuire, K. J., and Likens, G. E. (2011). Historical roots of forest hydrology and biogeochemistry. In *Forest Hydrology and Biogeochemistry: Synthesis of Past Research and Future Directions*, edited by D. F. Levia, D. Carlyle-Moses, and T. Tanaka. Dordrecht: Springer, pp. 3–26.

McIntosh, R. P. (1985). *The Background of Ecology: Concept and Theory*. Cambridge, UK: Cambridge University Press.

McLachlan, J. S., Clark, J. S., and Manos, P. S. (2005). Molecular indicators of tree migration capacity under rapid climate change. *Ecology*, 86, 2088–2098.

McLeod, A. R., and Long, S. P. (1999). Free-air carbon dioxide enrichment (FACE) in global change research: A review. *Advances in Ecological Research*, 28, 1–56.

Meacham, J. (2012). *Thomas Jefferson: The Art of Power*. New York: Random House.

Medhurst, J., Parsby, J., Linder, S., et al. (2006). A whole-tree chamber system for examining tree-level physiological responses of field-grown trees to environmental variation and climate change. *Plant, Cell and Environment*, 29, 1853–1869.

Medlyn, B. E., Zaehle, S., De Kauwe, M. G., et al. (2015). Using ecosystem experiments to improve vegetation models. *Nature Climate Change*, 5, 528–534.

Medrano, H., Gulías, J., Chaves, M. M., Galmés, J., and Flexas, J. (2012). Photosynthetic water-use efficiency. In *Terrestrial Photosynthesis in a Changing Environment: A Molecular, Physiological and Ecological Approach*, edited by J. Flexas, F. Loreto, and H. Medrano. Cambridge, UK: Cambridge University Press, pp. 523–536.

Medvigy, D., Walko, R. L., Otte, M. J., and Avissar, R. (2013). Simulated changes in Northwest U.S. climate in response to Amazon deforestation. *Journal of Climate*, 26, 9115–9136.

Meier, R., Schwaab, J., Seneviratne, S. I., et al. (2021). Empirical estimate of forestation-induced precipitation changes in Europe. *Nature Geoscience*, 14, 473–478.

Meinshausen, M., Nicholls, Z. R. J., Lewis, J., et al. (2020). The shared socio-economic pathway (SSP) greenhouse gas concentrations and their extensions to 2500. *Geoscientific Model Development*, 13, 3571–3605.

Meir, P., Wood, T. E., Galbraith, D. R., et al. (2015). Threshold responses to soil moisture deficit by trees and soil in tropical rain forests: Insights from field experiments. *BioScience*, 65, 882–892.

Meissner, K. J., Weaver, A. J., Matthews, H. D., and Cox, P. M. (2003). The role of land surface dynamics in glacial inception: A study with the UVic Earth System Model. *Climate Dynamics*, 21, 515–537.

Melville, A. D. (1987). *Ovid Metamorphoses*, translated by A. D. Melville with introduction and notes by E. J. Kenney. Oxford: Oxford University Press.

Menzies, N. K. (1994). *Forest and Land Management in Imperial China*. New York: St. Martin's Press.

Mercado, L. M., Bellouin, N., Sitch, S., et al. (2009). Impact of changes in diffuse radiation on the global land carbon sink. *Nature*, 458, 1014–1017.

Meusel, J. G. (1781). *Historische Litteratur für das Jahr 1781*, vol. 2. Erlangen: Palm.

Michaux, F. A. (1819). *The North American Sylva*, 2 vols. Paris: C. D'Hautel.

Mikkelson, K. M., Bearup, L. A., Maxwell, R. M., et al. (2013). Bark beetle infestation impacts on nutrient cycling, water quality and interdependent hydrological effects. *Biogeochemistry*, 115, 1–21.

Miller, S. D., Goulden, M. L., Hutyra, L. R., et al. (2011). Reduced impact logging minimally alters tropical rainforest carbon and energy exchange. *Proceedings of the National Academy of Sciences USA*, 108, 19431–19435.

Miralles, D. G., Gash, J. H., Holmes, T. R. H., de Jeu, R. A. M., and Dolman, A. J. (2010). Global canopy interception from satellite observations. *Journal of Geophysical Research*, 115, D16122, DOI: https://doi.org/10.1029/2009JD013530.

Miralles, D. G., Gentine, P., Seneviratne, S. I., and Teuling, A. J. (2019). Land-atmospheric feedbacks during droughts and heatwaves: State of the science and current challenges. *Annals of the New York Academy of Sciences*, 1436, 19–35.

Miralles, D. G., Teuling, A. J., van Heerwaarden, C. C., and Vilà-Guerau de Arellano, J. (2014). Mega-heatwave temperatures due to combined soil desiccation and atmospheric heat accumulation. *Nature Geoscience*, 7, 345–349.

Mirbel, C.-F. Brisseau de (1815). *Élémens de physiologie végétale et de botanique*, 3 vols. Paris: Magimel.

Mitchell, J. (1767). *The Present State of Great Britain and North America, with Regard to Agriculture, Population, Trade, and Manufactures, Impartially Considered*. London: T. Becket & P. A. de Hondt.

Möller, D. (2020). *Chemistry of the Climate System*, vol. 2: *History, Change and Sustainability*, 3rd ed. Berlin: De Gruyter.

Moomaw, W. R., Masino, S. A., and Faison, E. K. (2019). Intact forests in the United States: Proforestation mitigates climate change and serves the greatest good. *Frontiers in Forests and Global Change*, 2, 27, DOI: https://doi.org/10.3389/ffgc.2019.00027.

Moon, D. (2010). The debate over climate change in the steppe region in nineteenth-century Russia. *The Russian Review*, 69, 251–275.

Moon, D. (2013). *The Plough that Broke the Steppes: Agriculture and Environment on Russia's Grasslands, 1700–1914*. Oxford: Oxford University Press.

Moore, W. L. (1910). *A Report on the Influence of Forests on Climate and on Floods*. Washington, DC: Government Printing Office.

More, T. (1973). The Confutation of Tyndale's Answer, Books V–IX. In *The Yale Edition of the Complete Works of St. Thomas More*, vol. 8. *The Confutation of Tyndale's Answer, Part II The Text, Books V–IX, Appendices*, edited by L. A. Schuster, R. C. Marius, J. P. Lusardi, and R. J. Schoeck. New Haven, CT: Yale University Press, pp. 575–1034.

Moreau de Jonnès, A. (1825). *Premier mémoire en réponse a la question proposée par l'Académie Royale de Bruxelles: Quels sont les changemens que peut occasioner le déboisement de forêts considérables sur les contrées et communes adjacentes. . ..* Brussels: P. J. de Mat.

Morecroft, M. D., Taylor, M. E., and Oliver, H. R. (1998). Air and soil microclimates of deciduous woodland compared to an open site. *Agricultural and Forest Meteorology*, 90, 141–156.

Moret, P., Muriel, P., Jaramillo, R., and Dangles, O. (2019). Humboldt's *Tableau Physique* revisited. *Proceedings of the National Academy of Sciences USA*, 116, 12889–12894.

Morice, C. P., Kennedy, J. J., Rayner, N. A., et al. (2021). An updated assessment of near-surface temperature change from 1850: The HadCRUT5 data set. *Journal of Geophysical Research: Atmospheres*, 126, e2019JD032361, DOI: https://doi.org/10.1029/2019JD032361.

Morueta-Holme, N., Engemann, K., Sandoval-Acuña, P., et al. (2015). Strong upslope shifts in Chimborazo's vegetation over two centuries since Humboldt. *Proceedings of the National Academy of Sciences USA*, 112, 12741–12745.

Mueller, F. (1867). *Australian Vegetation, Indigenous or Introduced, Considered Especially in its Bearings on the Occupation of the Territory, and with a View of Unfolding its Resources*. Melbourne: Blundell.

Mueller, N. D., Butler, E. E., McKinnon, K. A., et al. (2016). Cooling of US Midwest summer temperature extremes from cropland intensification. *Nature Climate Change*, 6, 317–322.

Muir, J. (1894). *The Mountains of California.* New York: Century.

Munns, E. N. (1930). An East African estimate of forest influences on climate and water supply. *Forest Worker*, 6(1), 24–25.

Murchison, R. I., Verneuil, E. de., and Keyserling, A. von (1845). *The Geology of Russia in Europe and the Ural Mountains*, 2 vols. London: John Murray.

Murray, J. (1831). On raining trees. *Magazine of Natural History, and Journal of Zoology, Botany, Mineralogy, Geology, and Meteorology*, 4, 32–34.

Müttrich, A. (1890). Ueber den Einfluß des Waldes auf die periodischen Veränderungen der Lufttemperatur. *Zeitschrift für Forst- und Jagdwesen*, 22, 385–400, 449–458, 513–526.

Müttrich, A. (1892). Ueber den Einfluß des Waldes auf die Größe der atmosphärischen Niederschläge. *Zeitschrift für Forst- und Jagdwesen*, 24, 27–42.

Nabuurs, G.-J., Delacote, P., Ellison, D., et al. (2017). By 2050 the mitigation effects of EU forests could nearly double through climate smart forestry. *Forests*, 8, 484, DOI: https://doi.org/10.3390/f8120484.

Nakai, T., Sumida, A., Daikoku, K., et al. (2008). Parameterisation of aerodynamic roughness over boreal, cool- and warm-temperate forests. *Agricultural and Forest Meteorology*, 148, 1916–1925.

Napier, M. (1842). *The Encyclopædia Britannica, or Dictionary of Arts, Sciences, and General Literature*, 7th ed., 21 vols. Edinburgh: Adam & Charles Black.

Nash, L. K. (1952). *Plants and the Atmosphere* (Harvard Case Histories in Experimental Science, Case 5). Cambridge, MA: Harvard University Press.

National Research Council (1986). *Earth System Science – Overview: A Program for Global Change.* Washington, DC: National Academies Press.

Naudts, K., Chen, Y., McGrath, M. J., et al. (2016). Europe's forest management did not mitigate climate warming. *Science*, 351, 597–600.

Negrón, J. F., and Cain, B. (2019). Mountain pine beetle in Colorado: A story of changing forests. *Journal of Forestry*, 117, 144–151.

Negrón-Juárez, R., Baker, D. B., Zeng, H., Henkel, T. K., and Chambers, J. Q. (2010). Assessing hurricane-induced tree mortality in U.S. Gulf Coast forest ecosystems. *Journal of Geophysical Research*, 115, G04030, DOI: https://doi.org/10.1029/2009JG001221.

Negrón-Juárez, R. I., Holm, J. A., Marra, D. M., et al. (2018). Vulnerability of Amazon forests to storm-driven tree mortality. *Environmental Research Letters*, 13, 054021, DOI: https://doi.org/10.1088/1748-9326/aabe9f.

Nepstad, D. C., Stickler, C. M., Soares-Filho, B., and Merry, F. (2008). Interactions among Amazon land use, forests and climate: Prospects for a near-term forest tipping point. *Philosophical Transactions of the Royal Society B*, 363, 1737–1746.

Newbold, T. J. (1839). Notice of river dunes on the banks of the Hogri and Pennaur. *Madras Journal of Literature and Science*, 9(23), 309–310.

Newell, R. E. (1971). The Amazon forest and atmospheric general circulation. In *Man's Impact on the Climate*, edited by W. H. Matthews, W. W. Kellogg, and G. D. Robinson. Cambridge, MA: Massachusetts Institute of Technology Press, pp. 457–459.

Nicholson, H. (1676). An extract of a letter &c. from Dublin May the 10th, 1676. *Philosophical Transactions*, 11(127), 647–653.

Nicholson, J. W. (1929). *The Influence of Forests on Climate and Water Supply in Kenya,* Forestry Department Pamphlet Number 2. Nairobi: East African Standard.

Nicholson, S. E. (2013). The West African Sahel: A review of recent studies on the rainfall regime and its interannual variability. *International Scholarly Research Notices: Meteorology*, 453521, DOI: https://doi.org/10.1155/2013/453521.

Niinemets, Ü., and Anten, N. P. R. (2009). Packing the photosynthetic machinery: From leaf to canopy. In *Photosynthesis in silico: Understanding Complexity from Molecules to Ecosystems*, edited by A. Laisk, L. Nedbal, and Govindjee. Dordrecht: Springer, pp. 363–399.

Niinemets, Ü., Keenan, T. F., and Hallik, L. (2015). A worldwide analysis of within-canopy variations in leaf structural, chemical and physiological traits across plant functional types. *New Phytologist*, 205, 973–993.

Nisbet, J. (1893). *The Climatic and National-Economic Influence of Forests*. London: Eyre & Spottiswoode.

Nisbet, J. (1894). The climatic and national-economic influence of forests. *Nature*, 49, 302–305.

Nisbet, J. (1905). *The Forester: A Practical Treatise on British Forestry and Arboriculture for Landowners, Land Agents, and Foresters*, 2 vols. Edinburgh: William Blackwood.

Nobre, C. A., and Borma, L. S. (2009). "Tipping points" for the Amazon forest. *Current Opinion in Environmental Sustainability*, 1, 28–36.

Nobre, C. A., Silva Dias, M. A., Culf, A. D., et al. (2004). The Amazonian climate. In *Vegetation, Water, Humans and the Climate: A New Perspective on an Interactive System*, edited by P. Kabat, M. Claussen, P. A. Dirmeyer, et al. Berlin: Springer, pp. 79–92.

Norby, R. J., and Zak, D. R. (2011). Ecological lessons from free-air CO_2 enrichment (FACE) experiments. *Annual Review of Ecology, Evolution, and Systematics*, 42, 181–203.

Nördlinger, T. (1885). *Der Einfluss des Waldes auf die Luft- und Bodenwärme*. Berlin: Paul Parey.

Novick, K. A., and Katul, G. G. (2020). The duality of reforestation impacts on surface and air temperature. *Journal of Geophysical Research: Biogeosciences*, 125, DOI: https://doi.org/10.1029/2019JG005543.

Oberthaler, E., Pénot, S., Sellink, M., Spronk, R., and Hoppe-Harnoncourt, A. (2018). *Bruegel: The Hand of the Master – Exhibition Catalogue of the Kunsthistorisches Museum Vienna*. Vienna: KHM-Museumsverband.

O'Halloran, T. L., Law, B. E., Goulden, M. L., et al. (2012). Radiative forcing of natural forest disturbances. *Global Change Biology*, 18, 555–565.

Olearius, A. (1662). *The Voyages & Travels of the Ambassadors from the Duke of Holstein, to the Great Duke of Muscovy, and the King of Persia...*, rendered into English by John Davies. London: Thomas Dring; John Starkey.

Oleson, K. W., Bonan, G. B., Levis, S., and Vertenstein, M. (2004). Effects of land use change on North American climate: Impact of surface datasets and model biogeophysics. *Climate Dynamics*, 23, 117–132.

O'Neill, B. C., Tebaldi, C., van Vuuren, D. P., et al. (2016). The Scenario Model Intercomparison Project (ScenarioMIP) for CMIP6. *Geoscientific Model Development*, 9, 3461–3482.

O'Neill, B. C., Kriegler, E., Ebi, K. L., et al. (2017). The roads ahead: Narratives for shared socioeconomic pathways describing world futures in the 21st century. *Global Environmental Change*, 42, 169–180.

Oosting, H. J. (1942). An ecological analysis of the plant communities of Piedmont, North Carolina. *American Midland Naturalist*, 28, 1–126.

Ornstein, L., Aleinov, I., and Rind, D. (2009). Irrigated afforestation of the Sahara and Australian Outback to end global warming. *Climatic Change*, 97, 409–437.

Oswald, F. L. (1877). The climatic influence of vegetation. – A plea for our forests. *Popular Science Monthly*, 11(August), 385–390.

Otto-Bliesner, B. L., Brady, E. C., Fasullo, F., et al. (2016). Climate variability and change since 850 CE: An ensemble approach with the Community Earth System Model (CESM). *Bulletin of the American Meteorological Society*, 97, 735–754.

Oudin, L., Andréassian, V., Lerat, L., and Michel, C. (2008). Has land cover a significant impact on mean annual streamflow? An international assessment using 1508 catchments. *Journal of Hydrology*, 357, 303–316.

Overpeck, J. T., Webb, R. S., and Webb, T., III (1992). Mapping eastern North American vegetation change of the past 18 ka: No-analogs and the future. *Geology*, 20, 1071–1074.

Oyama, M. D., and Nobre, C. A. (2003). A new climate-vegetation equilibrium state for Tropical South America. *Geophysical Research Letters*, 30, 2199, DOI: https://doi.org/10.1029/2003GL018600.

Paasonen, P., Asmi, A., Petäjä, T., et al. (2013). Warming-induced increase in aerosol number concentration likely to moderate climate change. *Nature Geoscience*, 6, 438–442.

Pakenham, T. (1996). *Meetings with Remarkable Trees*. London: Weidenfeld & Nicolson.

Pakenham, T. (2002). *Remarkable Trees of the World*. London: Weidenfeld & Nicolson.

Pan, Y., Birdsey, R. A., Phillips, O. L., and Jackson, R. B. (2013). The structure, distribution, and biomass of the world's forests. *Annual Review of Ecology, Evolution, and Systematics*, 44, 593–622.

Park, J.-H., Goldstein, A. H., Timkovsky, J., et al. (2013). Active atmosphere-ecosystem exchange of the vast majority of detected volatile organic compounds. *Science*, 341, 643–647.

Parker, D. H. (2008). *A Critical Edition of Robert Barnes' A Supplication Unto the Most Gracyous Prince Kynge Henry The VIII, 1534*. Toronto: University of Toronto Press.

Parks, S. A., and Abatzoglou, J. T. (2020). Warmer and drier fire seasons contribute to increases in area burned at high severity in western US forests from 1985 to 2017. *Geophysical Research Letters*, 47, e2020GL089858, DOI: https://doi.org/10.1029/2020GL089858.

Parsons, L. A., Jung, J., Masuda, Y. J., et al. (2021). Tropical deforestation accelerates local warming and loss of safe outdoor working hours. *One Earth*, 4, 1730–1740.

Paso y Troncoso, F. del (1905). *Papeles de Nueva España, segunda serie: Geografía y estadística*, vol. 4. Madrid: Establicimiento Tipográfico "Sucesores de Rivadeneyra."

Pastorello, G., Trotta, C., Canfora, E., et al. (2020). The FLUXNET2015 dataset and the ONEFlux processing pipeline for eddy covariance data. *Scientific Data*, 7, 225, DOI: https://doi.org/10.1038/s41597-020-0534-3.

Pausata, F. S. R., Gaetani, M., Messori, G., et al. (2020). The greening of the Sahara: Past changes and future implications. *One Earth*, 2, 235–250.

Payne, C. (2017). *Silent Witnesses: Trees in British Art, 1760–1870*. Bristol: Sansom.

Pelloutier, S. (1740). *Histoire des Celtes, et particulierement des Gaulois et des Germains, depuis les tems fabuleux, jusqu'à la prise de Rome par les Gaulois*. La Haye: Isaac Beauregard.

Peng, S.-S., Piao, S., Zeng, Z., et al. (2014). Afforestation in China cools local land surface temperature. *Proceedings of the National Academy of Sciences USA*, 111, 2915–2919.

Peppercorne, F. S. (1879). Influence of forests on climate and rainfall. *Transactions and Proceedings of the New Zealand Institute*, 12, 24–32.

Perugini, L., Caporaso, L., Marconi, S., et al. (2017). Biophysical effects on temperature and precipitation due to land cover change. *Environmental Research Letters*, 12, 053002, DOI: https://doi.org/10.1088/1748-9326/aa6b3f.

Petäjä, T., Tabakova, K., Manninen, A., et al. (2022). Influence of biogenic emissions from boreal forests on aerosol–cloud interactions. *Nature Geoscience*, 15, 42–47.

Phillips, O. L., Aragão, L. E. O. C., Lewis, S. L., et al. (2009). Drought sensitivity of the Amazon rainforest. *Science*, 323, 1344–1347.

Piao, S., Wang, X., Park, T., et al. (2020). Characteristics, drivers and feedbacks of global greening. *Nature Reviews Earth and Environment*, 1, 14–27.

Pielke, R. A., Lee, T. J., Copeland, J. H., et al. (1997). Use of USGS-provided data to improve weather and climate simulations. *Ecological Applications*, 7, 3–21.

Pielke, R. A., Rodriguez, J. H., Eastman, J. L., Walko, R. L., and Stocker, R. A. (1993). Influence of albedo variability in complex terrain on mesoscale systems. *Journal of Climate*, 6, 1798–1806.

Pielke, R. A., Sr., Pitman, A., Niyogi, D., et al. (2011). Land use/land cover changes and climate: Modeling analysis and observational evidence. *WIREs Climate Change*, 2, 828–850.

Pigafetta, A. (1906). *Magellan's Voyage around the World*, translated by J. A. Robertson, 2 vols. Cleveland, OH: Arthur H. Clark.

Pigafetta, A. (1985). *Primer viaje alrededor del mundo*, edición de Leoncio Cabrero. Madrid: Historia 16.

Pinchot, G. (1905a). *A Primer of Forestry. Part II. – Practical Forestry*, Bulletin Number 24, Part II. U.S. Department of Agriculture, Bureau of Forestry. Washington, DC: Government Printing Office.

Pinchot, G. (1905b). *The Use of the National Forest Reserves: Regulations and Instructions*. Washington, DC: U.S. Department of Agriculture.

Pinchot, G. (2017). Dear Forester, February 1, 1905. In *Gifford Pinchot: Selected Writings*, edited by C. Miller. University Park, PA: Pennsylvania State University Press, pp. 39-42.

Piper, R. U. (1855). *The Trees of America*. Boston: William White.

Pitman, A. J., and Lorenz, R. (2016). Scale dependence of the simulated impact of Amazonian deforestation on regional climate. *Environmental Research Letters*, 11, 094025, DOI: https://doi.org/10.1088/1748-9326/11/9/094025.

Pliny (1963). *Natural History*, vol. 8. *Books 28–32*, translated by W. H. S. Jones (Loeb Classical Library 418). Cambridge, MA: Harvard University Press.

Poivre, P. (1768). *Voyages d'un philosophe: ou, observations sur les moeurs et les arts des peuples de l'Afrique, de l'Asie et de l'Amerique*. Yverdon: n.p.

Poivre, P. (1797). *Oeuvres complettes de P. Poivre, intendant des Isles de France et de Bourbon, correspondant de l'académie des sciences, etc.* Paris: Fuchs.

Pongratz, J., Reick, C. H., Raddatz, T., and Claussen, M. (2010). Biogeophysical versus biogeochemical climate response to historical anthropogenic land cover change. *Geophysical Research Letters*, 37, L08702, DOI: https://doi.org/10.1029/2010GL043010.

Pongratz, J., Reick, C. H., Houghton, R. A., and House, J. I. (2014). Terminology as a key uncertainty in net land use and land cover change carbon flux estimates. *Earth System Dynamics*, 5, 177–195.

Pongratz, J., Schwingshackl, C., Bultan, S., et al. (2021). Land use effects on climate: Current state, recent progress, and emerging topics. *Current Climate Change Reports*, 7, 99–120.

Popp, A., Calvin, K., Fujimori, S., et al. (2017). Land-use futures in the shared socio-economic pathways. *Global Environmental Change*, 42, 331–345.

Pöschl, U., Martin, S. T., Sinha, B., et al. (2010). Rainforest aerosols as biogenic nuclei of clouds and precipitation in the Amazon. *Science*, 329, 1513–1516.

Postel, S. L., and Thompson, B. H., Jr. (2005). Watershed protection: Capturing the benefits of nature's water supply services. *Natural Resources Forum*, 29, 98–108.

Potter, S., Solvik, K., Erb, A., et al. (2020). Climate change decreases the cooling effect from postfire albedo in boreal North America. *Global Change Biology*, 26, 1592–1607.

Poyatos, R., Granda, V., Molowny-Horas, R., et al. (2016). SAPFLUXNET: Towards a global database of sap flow measurements. *Tree Physiology*, 36, 1449–1455.

Prentice, I. C., Jolly, D., and BIOME6000 (2000). Mid-Holocene and glacial-maximum vegetation geography of the northern continents and Africa. *Journal of Biogeography*, 27, 507–519.

Prévost, A.-F., l'abbé (1744). *Voyages du capitaine Robert Lade en differentes parties de l'Afrique, de l'Asie et de l'Amerique*, 2 vols. Paris: Didot.

Priestley, J. (1772). Observations on different kinds of air. *Philosophical Transactions*, 62, 147–264.

Pross, J., Contreras, L., Bijl, P. K., et al. (2012). Persistent near-tropical warmth on the Antarctic continent during the early Eocene epoch. *Nature*, 488, 73–77.

Pukkala, T. (2018). Effect of species composition on ecosystem services in European boreal forest. *Journal of Forestry Research*, 29, 261–272.

Purkyně, E. (1875). Etwas über die Waldfrage, Wasserfrage und Sumpffrage. *Oesterreichische Monatsschrift für Forstwesen*, 25, 479–525.

Purkyně, E. (1876). Ueber die Wald- und Wasserfrage. *Oesterreichische Monatsschrift für Forstwesen*, 26, 136–151, 161–178, 179–204, 209–251, 267–291, 327–349, 405–426, 473–498.

Purkyně, E. (1877). Ueber die Wald- und Wasserfrage. *Oesterreichische Monatsschrift für Forstwesen*, 27, 102–143.

Quesada, B., Arneth, A., and de Noblet-Ducoudré, N. (2017). Atmospheric, radiative, and hydrologic effects of future land use and land cover changes: A global and multimodel climate picture. *Journal of Geophysical Research: Atmospheres*, 122, 5113–5131.

R. C. (1912). Forests and rainfall. *Nature*, 89, 662–664.

R. D. (1908). Les forêts de la planète Mars. *Revue des eaux et forêts*, 47, 404–406.

Rackham, O. (1986). *The History of the Countryside*. London: J. M. Dent.

Rae, J. W. B., Zhang, Y. G., Liu, X., et al. (2021). Atmospheric CO_2 over the past 66 million years from marine archives. *Annual Review of Earth and Planetary Sciences*, 49, 609–641.

Rajan, S. R. (2006). *Modernizing Nature: Forestry and Imperial Eco-Development 1800–1950*. Oxford: Oxford University Press.

Ramsay, D. (1809). *The History of South-Carolina, from Its First Settlement in 1670, to the Year 1808*, 2 vols. Charleston, SC: David Longworth.

Randerson, J. T., Liu, H., Flanner, M. G., et al. (2006). The impact of boreal forest fire on climate warming. *Science*, 314, 1130–1132.

Rap, A., Scott, C. E., Reddington, C. L., et al. (2018). Enhanced global primary production by biogenic aerosol via diffuse radiation fertilization. *Nature Geoscience*, 11, 640–644.

Rauch, F.-A. (1792). *Plan nourricier, ou recherches sur les moyens à mettre en usage pour assurer à jamais le pain au peuple français. . . .* Didot jeune: Paris.

Rauch, F. A. (1802). *Harmonie hydro-végétale et météorologique, ou recherches sur les moyens de recréer avec nos forêts la force des températures et la régularité des saisons, par des plantations raisonnées*, 2 vols. Paris: Levrault.

Rauch, F.-A. (1818). *Régénération de la nature végétale, ou recherches sur les moyens de recréer, dans tous les climats, les anciennes températures et l'ordre primitif des saisons, par des plantations raisonnées*, 2 vols. Paris: Didot l'Aîné.

Raupach, M. R. (1988). Canopy transport processes. In *Flow and Transport in the Natural Environment: Advances and Applications*, edited by W. L. Steffen and O. T. Denmead. Berlin: Springer-Verlag, pp. 95–127.

Raupach, M. R. (1994). Simplified expressions for vegetation roughness length and zero-plane displacement as functions of canopy height and area index. *Boundary-Layer Meteorology*, 71, 211–216.

Raynor, G. S. (1971). Wind and temperature structure in a coniferous forest and a contiguous field. *Forest Science*, 17, 351–363.

Reifsnyder, W. E. (1973). Forest meteorology in the seventies. *Bulletin of the American Meteorological Society*, 54, 326–330.

Ren, J., Adam, J. C., Hicke, J. A., et al. (2021). How does water yield respond to mountain pine beetle infestation in a semiarid forest? *Hydrology and Earth System Sciences*, 25, 4681–4699.

Renou, E. (1866). Théorie de la pluie. *Annuaire de la société météorologique de France*, 14, 89–106.

Restrepo-Coupe, N., Albert, L. P., Longo, M., et al. (2021). Understanding water and energy fluxes in the Amazonia: Lessons from an observation-model intercomparison. *Global Change Biology*, 27, 1802–1819.

Riahi, K., van Vuuren, D. P., Kriegler, E., et al. (2017). The shared socioeconomic pathways and their energy, land use, and greenhouse gas emissions implications: An overview. *Global Environmental Change*, 42, 153–168.

Ribbentrop, B. (1900). *Forestry in British India*. Calcutta: Office of the Superintendent of Government Printing.

Richardson, L. F. (1922). *Weather Prediction by Numerical Processes*. Cambridge, UK: Cambridge University Press.

Ridgway, R. (1872). Notes on the vegetation of the lower Wabash Valley. *American Naturalist*, 6, 658–665.

Ridgwell, A., Singarayer, J. S., Hetherington, A. M., and Valdes, P. J. (2009). Tackling regional climate change by leaf albedo bio-geoengineering. *Current Biology*, 19, 146–150.

Roberts, J., Cabral, O. M. R., and De Aguiar, L. F. (1990). Stomatal and boundary-layer conductances in an Amazonian terra firme rain forest. *Journal of Applied Ecology*, 27, 336–353.

Roberts, P., Boivin, N., Lee-Thorp, J., Petraglia, M., and Stock, J. (2016). Tropical forests and the genus *Homo*. *Evolutionary Anthropology*, 25, 306–317.

Robertson, W. (1777). *The History of America*, 2 vols. London: W. Strahan.

Robinson, N. A. (2014). The Charter of the Forest: Evolving human rights in nature. In *Magna Carta and the Rule of Law*, edited by D. B. Magraw, A. Martinez, and R. E. Brownell II. Chicago: American Bar Association, pp. 311–377.

Rodman, K. C., Veblen, T. T., Battaglia, M. A., et al. (2020). A changing climate is snuffing out post-fire recovery in montane forests. *Global Ecology and Biogeography*, 29, 2039–2051.

Rodrigues, L. (2007). Dr. Alexander Gibson and the emergence of conservationism and desiccationism in Bombay: 1838 to 1860. In *Proceedings of the Indian History Congress, 67th Session, Calicut*

2006–07. Delhi: Indian History Congress, pp. 655–665.

Rodríguez-Fonseca, B., Mohino, E., Mechoso, C. R., et al. (2015). Variability and predictability of West African droughts: A review on the role of sea surface temperature anomalies. *Journal of Climate*, 28, 4034–4060.

Roe, S., Streck, C., Obersteiner, M., et al. (2019). Contribution of the land sector to a 1.5 °C world. *Nature Climate Change*, 9, 817–828.

Roe, S., Streck, C., Beach, R., et al. (2021). Land-based measures to mitigate climate change: Potential and feasibility by country. *Global Change Biology*, 27, 6025–6058.

Rogelj, J., Popp, A., Calvin, K. V., et al. (2018a). Scenarios towards limiting global mean temperature increase below 1.5 °C. *Nature Climate Change*, 8, 325–332.

Rogelj, J., Shindell, D., Jiang, K., et al. (2018b). Mitigation pathways compatible with 1.5°C in the context of sustainable development. In *Global Warming of 1.5°C: An IPCC Special Report on the Impacts of Global Warming of 1.5°C Above Pre-Industrial Levels and Related Global Greenhouse Gas Emission Pathways, in the Context of Strengthening the Global Response to the Threat of Climate Change, Sustainable Development, and Efforts to Eradicate Poverty*, edited by V. Masson-Delmotte, P. Zhai, H.-O. Pörtner, et al. Geneva: World Meteorological Organization, pp. 93–174.

Rogers, B. M., Randerson, J. T., and Bonan, G. B. (2013). High-latitude cooling associated with landscape changes from North American boreal forest fires. *Biogeosciences*, 10, 699–718.

Rogers, B. M., Soja, A. J., Goulden, M. L., and Randerson, J. T. (2015). Influence of tree species on continental differences in boreal fires and climate feedbacks. *Nature Geoscience*, 8, 228–234.

Rogers, H. (1873). Report on the effects of the cutting down of forests on the climate and health of the Mauritius. *Transactions of the Botanical Society of Edinburgh*, 11, 115–118.

Rohatyn, S., Rotenberg, E., Ramati, E., et al. (2018). Differential impacts of land use and precipitation on "ecosystem water yield." *Water Resources Research*, 54, 5457–5470.

Rooke, H. (1790). *Descriptions and Sketches of Some Remarkable Oaks, in the Park at Welbeck, in the County of Nottingham, a Seat of His Grace The Duke of Portland*. London: J. Nichols.

Rotenberg, E., and Yakir, D. (2010). Contribution of semi-arid forests to the climate system. *Science*, 327, 451–454.

Rotenberg, E., and Yakir, D. (2011). Distinct patterns of changes in surface energy budget associated with forestation in the semiarid region. *Global Change Biology*, 17, 1536–1548.

Rothwell, G. W., Mapes, G., and Mapes, R. H. (1997). Late Paleozoic conifers of North America: Structure, diversity and occurrences. *Review of Palaeobotany and Palynology*, 95, 95–113.

Rothwell, G. W., Mapes, G., Stockey, R. A., and Hilton, J. (2012). The seed cone *Eathiestrobus* gen. nov.: Fossil evidence for a Jurassic origin of Pinaceae. *American Journal of Botany*, 99, 708–720.

Royer, D. L., Osborne, C. P., and Beerling, D. J. (2003). Carbon loss by deciduous trees in a CO_2-rich ancient polar environment. *Nature*, 424, 60–62.

Rugendas, J. M. (1827). *Voyage pittoresque dans le Brésil*. Paris: Engelmann & Cie.

Rush, B. (1786). An enquiry into the cause of the increase of bilious and intermitting fevers in Pennsylvania, with hints for preventing them. *Transactions of the American Philosophical Society*, 2, 206–212.

Rutkow, E. (2012). *American Canopy: Trees, Forests, and the Making of a Nation*. New York: Scribner.

Ryan, M. G. (2013). Three decades of research at Flakaliden advancing whole-tree physiology, forest ecosystem and global change research. *Tree Physiology*, 33, 1123–1131.

Saberwal, V. K. (1998). Science and the desiccationist discourse of the 20th century. *Environment and History*, 4, 309–343.

Sagan, C., Toon, O. B., and Pollack, J. B. (1979). Anthropogenic albedo changes and the Earth's climate. *Science*, 206, 1363–1368.

Saint, J. (2021). *Ariadne*. New York: Flatiron Books.

Sallon, S., Solowey, E., Cohen, Y., et al. (2008). Germination, genetics, and growth of an ancient date seed. *Science*, 320, 1464.

Sallon, S., Cherif, E., Chabrillange, N., et al. (2020). Origins and insights into the historic Judean date palm based on genetic analysis of germinated ancient seeds and morphometric studies. *Science Advances*, 6, eaax0384, DOI: https://doi.org/10.1126/sciadv.aax0384.

Santopuoli, G., Temperli, C., Alberdi, I., et al. (2021). Pan-European sustainable forest management indicators for assessing Climate-Smart Forestry in Europe. *Canadian Journal of Forest Research*, 51, 1741–1750.

Sargent, C.S. (1882). The protection of forests. *North American Review*, 135, 386–401.

Sargent, C. S., Abbot, H. L., Agassiz, A., et al. (1897). *Report of the Committee Appointed by the National Academy of Sciences upon the Inauguration of a Forest Policy For the Forested Lands of the United States to the Secretary of the Interior, May 1, 1897*. Washington, DC: Government Printing Office.

Saunders, C. J. (1993). *The Forest of Medieval Romance: Avernus, Broceliande, Arden.* Cambridge, UK: D. S. Brewer.

SCEP (1970). *Man's Impact on the Global Environment – Assessment and Recommendations for Actions: Report of the Study of Critical Environmental Problems (SCEP)*. Cambridge, MA: Massachusetts Institute of Technology Press.

Schacht, H. (1853). *Der Baum: Studien über Bau und Leben der höheren Gewächse*. Berlin: G. W. F. Müller.

Scheffers, B. R., Phillips, B. L., Laurance, W. F., et al. (2013). Increasing arboreality with altitude: A novel biogeographic dimension. *Proceedings of the Royal Society B*, 280, 20131581, DOI: https://doi.org/10.1098/rspb.2013.1581.

Schiff, A. L. (1962). *Fire and Water: Scientific Heresy in the Forest Service.* Cambridge, MA: Harvard University Press.

Schimel, D., Schneider, F. D., Bloom, A., et al. (2019). Flux towers in the sky: Global ecology from space. *New Phytologist*, 224, 570–584.

Schleiden, M. J. (1848a). *Die Pflanze und ihr Leben*. Leipzig: Wilhelm Engelmann.

Schleiden, M. J. (1848b). *The Plant; A Biography*, translated by A. Henfrey. London: Hippolyte Bailliere.

Schlich, W. (1889). *A Manual of Forestry*, vol. 1. *The Utility of Forests, and Fundamental Principles of Sylviculture*. London: Bradbury, Agnew & Co.

Schlich, W. (1910). Forests and forestry. In *The Encyclopædia Britannica: A Dictionary of Arts, Sciences, Literature and General Information*, 11th ed., vol. 10. New York: Encyclopædia Britannica Company, pp. 645–651.

Schmidt, M. (2019). *Gilgamesh: The Life of a Poem*. Princeton: Princeton University Press.

Schneider, S. H., and Dickinson, R. E. (1974). Climate modeling. *Reviews of Geophysics and Space Physics*, 12, 447–493.

Schofield, P. F. (1875). Forests and rainfall. *Popular Science Monthly*, 8(November), 111–112.

Schöpf, J. D. (1875). *The Climate and Diseases of America*, translated by J. R. Chadwick. Boston: H. O. Houghton.

Schott, C. A. (1872). *Tables and Results of the Precipitation*, in *Rain and Snow, in the United States: And at Some Stations in Adjacent Parts of North America, and in Central and South America*. Washington, DC: Smithsonian Institution.

Schott, C. A. (1876). *Tables, Distribution, and Variations of the Atmospheric Temperature in the United States, and Some Adjacent Parts of America*. Washington, DC: Smithsonian Institution.

Schubert, S. D., Stewart, R. E., Wang, H., et al. (2016). Global meteorological drought: A synthesis of current understanding with a focus on SST drivers of precipitation deficits. *Journal of Climate*, 29, 3989–4019.

Schubert, S. D., Suarez, M. J., Pegion, P. J., Koster, R. D., and Bacmeister, J. T. (2004). On the cause of the 1930s Dust Bowl. *Science*, 303, 1855–1859.

Schultz, N. M., Lawrence, P. J., and Lee, X. (2017). Global satellite data highlights the diurnal asymmetry of the surface temperature response to deforestation. *Journal of Geophysical Research: Biogeosciences*, 122, 903–917.

Schulze, E. D., Sierra, C. A., Egenolf, V., et al. (2020). The climate change mitigation effect of bioenergy from sustainably managed forests in Central Europe. *GCB Bioenergy*, 12, 186–197.

Schuster, L. A. (1973). Thomas More's polemical career, 1523–1533. In *The Yale Edition of the Complete Works of St. Thomas More*, vol. 8. *The Confutation of Tyndale's Answer, Part III Introduction, Commentary, Glossary, Index*, edited by L. A. Schuster, R. C. Marius, J. P. Lusardi, and R. J. Schoeck. New Haven, CT: Yale University Press, pp. 1135–1268.

Schwaab, J., Davin, E. L., Bebi, P., et al. (2020). Increasing the broad-leaved tree fraction in European forests mitigates hot temperature extremes. *Scientific Reports*, 10, 14153, DOI: https://doi.org/10.1038/s41598-020-71055-1.

Schwartz, M. D., and Karl, T. R. (1990). Spring phenology: Nature's experiment to detect the effect of "green-up" on surface maximum temperatures. *Monthly Weather Review*, 118, 883–890.

Schwingshackl, C., Davin, E. L., Hirschi, M., et al. (2019). Regional climate model projections underestimate future warming due to missing plant physiological CO_2 response. *Environmental Research Letters*, 14, 114019, DOI: https://doi.org/10.1088/1748-9326/ab4949.

Scott, C. E., Monks, S. A., Spracklen, D. V., et al. (2018). Impact on short-lived climate forcers increases projected warming due to deforestation. *Nature Communications*, 9, 157, DOI: https://doi.org/10.1038/s41467-017-02412-4.

Scott, C. E., Rap, A., Spracklen, D. V., et al. (2014). The direct and indirect radiative effects of biogenic secondary organic aerosol. *Atmospheric Chemistry and Physics*, 14, 447–470.

Seager, R., and Hoerling, M. (2014). Atmosphere and ocean origins of North American droughts. *Journal of Climate*, 27, 4581–4606.

Searchinger, T. D., Hamburg, S. P., Melillo, J., et al. (2009). Fixing a critical climate accounting error. *Science*, 326, 527–528.

Segal, M., Avissar, R., McCumber, M. C., and Pielke, R. A. (1988). Evaluation of vegetation effects on the generation and modification of mesoscale circulations. *Journal of the Atmospheric Sciences*, 45, 2268–2292.

Sellers, P. J., Bounoua, L., Collatz, G. J., et al. (1996a). Comparison of radiative and physiological effects of doubled atmospheric CO_2 on climate. *Science*, 271, 1402–1406.

Sellers, P. J., Mintz, Y., Sud, Y. C., and Dalcher, A. (1986). A simple biosphere model (SiB) for use within general circulation models. *Journal of the Atmospheric Sciences*, 43, 505–531.

Sellers, P. J., Randall, D. A., Collatz, G. J., et al. (1996b). A revised land surface parameterization (SiB2) for atmospheric GCMs. Part I: Model formulation. *Journal of Climate*, 9, 676–705.

Sellink, M. (2019). Leading the eye and staging the composition: Some remarks on Pieter Bruegel the Elder's compositional techniques. In *Bruegel: The Hand of the Master. The 450th Anniversary Edition – Essays in Context*, edited by A. Hoppe-Harnoncourt, E. Oberthaler, S. Pénot, M. Sellink, and R. Spronk. Vienna: KHM-Museumsverband, pp. 336–352.

Seneviratne, S. I., Corti, T., Davin, E. L., et al. (2010). Investigating soil moisture-climate interactions in a changing climate: A review. *Earth-Science Reviews*, 99, 125–161.

Seneviratne, S. I., Phipps, S. J., Pitman, A. J., et al. (2018). Land radiative management as contributor to regional-scale climate adaptation and mitigation. *Nature Geoscience*, 11, 88–96.

Senf, C., Buras, A., Zang, C. S., Rammig, A., and Seidl, R. (2020). Excess forest mortality is consistently linked to drought across Europe. *Nature Communications*, 11, 6200, DOI: https://doi.org/10.1038/s41467-020-19924-1.

Senf, C., Pflugmacher, D., Zhiqiang, Y., et al. (2018). Canopy mortality has doubled in Europe's temperate forests over the last three decades. *Nature Communications*, 9, 4978, DOI: https://doi.org/10.1038/s41467-018-07539-6.

Seth, A., and Giorgi, F. (1996). Three-dimensional model study of organized mesoscale circulations induced by vegetation. *Journal of Geophysical Research*, 101D, 7371–7391.

Shapiro, A. (2014). A grand experiment: USDA Forest Service experimental forests and ranges. In *USDA Forest Service Experimental Forests and Ranges: Research for the Long Term*, edited by D. C. Hayes, S. L. Stout, R. H. Crawford, and A. P. Hoover. New York: Springer, pp. 3–23.

Shaw, R. H., and Pereira, A. R. (1982). Aerodynamic roughness of a plant canopy: A numerical experiment. *Agricultural Meteorology*, 26, 51–65.

Sheil, D. (2014). How plants water our planet: Advances and imperatives. *Trends in Plant Science*, 19, 209–211.

Sheil, D., and Murdiyarso, D. (2009). How forests attract rain: An examination of a new hypothesis. *Bioscience*, 59, 341–347.

Shugart, H. H. (1984). *A Theory of Forest Dynamics: The Ecological Implications of Forest Succession Models*. New York: Springer-Verlag.

Shugart, H. H. (1987). Dynamic ecosystem consequences of tree birth and death patterns. *BioScience*, 37, 596–602.

Shugart, H. H. (1998). *Terrestrial Ecosystems in Changing Environments*. Cambridge, UK: Cambridge University Press.

Shugart, H. H., Leemans, R., and Bonan, G. B. (1992). *A Systems Analysis of the Global Boreal Forest*. Cambridge, UK: Cambridge University Press.

Shukla, J., and Mintz, Y. (1982). Influence of land-surface evapotranspiration on the Earth's climate. *Science*, 215, 1498–1501.

Shukla, P. R., Skea, J., Slade, R., et al. (2019). Technical summary. In *Climate Change and Land: An IPCC Special Report on Climate Change, Desertification, Land Degradation, Sustainable Land Management, Food Security, and Greenhouse Gas Fluxes in Terrestrial Ecosystems* edited by P. R. Shukla, J. Skea, E. Calvo Buendia, et al. Geneva: World Meteorological Organization, pp. 37–74.

Simmons, C. T., and Matthews, H. D. (2016). Assessing the implications of human land-use change for the transient climate response to cumulative carbon emissions. *Environmental Research Letters*, 11, 035001, DOI: https://doi.org/10.1088/1748-9326/11/3/035001.

Sitch, S., Cox, P. M., Collins, W. J., and Huntingford, C. (2007). Indirect radiative forcing of climate change through ozone effects on the land-carbon sink. *Nature*, 448, 791–794.

Skea, R. (2013). *Vincent's Trees: Paintings and Drawings by Van Gogh*. New York: Thames & Hudson.

Skea, R. (2015). *Monet's Trees: Paintings and Drawings by Claude Monet*. New York: Thames & Hudson.

Skinner, C. B., Poulsen, C. J., and Mankin, J. S. (2018). Amplification of heat extremes by plant CO$_2$ physiological forcing. *Nature Communications*, 9, 1094, DOI: https://doi.org/10.1038/s41467-018-03472-w.

Slinski, K. M., Hogue, T. S., Porter, A. T., and McCray, J. E. (2016). Recent bark beetle outbreaks have little impact on streamflow in the Western United States. *Environmental Research Letters*, 11, 074010, DOI: https://doi.org/10.1088/1748-9326/11/7/074010.

Sloan, P. R. (1981). Buffon's preface to the *Vegetable Staticks* of Stephen Hales (1735). In *From Natural History to the History of Nature: Readings from Buffon and His Critics*, edited by J. Lyon and P. R. Sloan. Notre Dame, IN: University of Notre Dame Press, pp. 35–40.

SMIC (1971). *Inadvertent Climate Modification: Report of the Study of Man's Impact on Climate (SMIC)*. Cambridge, MA: Massachusetts Institute of Technology Press.

Smith, M. B. (1906). *The First Forty Years of Washington Society*, edited by G. Hunt. New York: Charles Scribner's Sons.

Smith, M. N., Stark, S. C., Taylor, T. C., et al. (2019a). Seasonal and drought-related changes in leaf area profiles depend on height and light environment in an Amazon forest. *New Phytologist*, 222, 1284–1297.

Smith, P., Adams, J., Beerling, D. J., et al. (2019b). Impacts of land-based greenhouse gas removal options on ecosystem services and the United Nations sustainable development goals. *Annual Review of Environment and Resources*, 44, 255–286.

Smith, P., Arneth, A., Barnes, D. K. A., et al. (2022). How do we best synergize climate mitigation actions to co-benefit biodiversity? *Global Change Biology*, 28, 2555–2577.

Smith, P., Davis, S. J., Creutzig, F., et al. (2016). Biophysical and economic limits to negative CO$_2$ emissions. *Nature Climate Change*, 6, 42–50.

Snyder, P. K. (2010). The influence of tropical deforestation on the Northern Hemisphere climate by atmospheric teleconnections. *Earth Interactions*, 14(4), 1–34.

Snyder, P. K., Delire, C., and Foley, J. A. (2004). Evaluating the influence of different vegetation biomes on the global climate. *Climate Dynamics*, 23, 279–302.

Solomon, A. M., West, D. C., and Solomon, J. A. (1981). Simulating the role of climate change and species immigration in forest succession. In *Forest Succession: Concepts and Application*, edited by D. C. West, H. H. Shugart, and D. B. Botkin. New York: Springer-Verlag, pp. 154–177.

Soltis, D., Soltis, P., Endress, P., et al. (2018). *Phylogeny and Evolution of the Angiosperms*, revised and updated edition. Chicago: University of Chicago Press.

Song, X.-P., Hansen, M. C., Stehman, S. V., et al. (2018). Global land change from 1982 to 2016. *Nature*, 560, 639–643.

Sonntag, S., Pongratz, J., Reick, C. H., and Schmidt, H. (2016). Reforestation in a high-CO_2 world: Higher mitigation potential than expected, lower adaptation potential than hoped for. *Geophysical Research Letters*, 43, 6546–6553, DOI: https://doi.org/10.1002/2016GL068824.

Spencer, A. R. T., Mapes, G., Bateman, R. M., Hilton, J., and Rothwell, G. W. (2015). Middle Jurassic evidence for the origin of Cupressaceae: A paleobotanical context for the roles of regulatory genetics and development in the evolution of conifer seed cones. *American Journal of Botany*, 102, 942–961.

Spracklen, D. V., and Garcia-Carreras, L. (2015). The impact of Amazonian deforestation on Amazon basin rainfall. *Geophysical Research Letters*, 42, 9546–9552.

Spracklen, D. V., Arnold, S. R., and Taylor, C. M. (2012). Observations of increased tropical rainfall preceded by air passage over forests. *Nature*, 489, 282–285.

Spracklen, D. V., Baker, J. C. A., Garcia-Carreras, L., and Marsham, J. H. (2018). The effects of tropical vegetation on rainfall. *Annual Review of Environment and Resources*, 43, 193–218.

Spracklen, D. V., Bonn, B., and Carslaw, K. S. (2008). Boreal forests, aerosols and the impacts on clouds and climate. *Philosophical Transactions of the Royal Society A*, 366, 4613–4626.

Staal, A., Fetzer, I., Wang-Erlandsson, L., et al. (2020). Hysteresis of tropical forests in the 21st century. *Nature Communications*, 11, 4978, DOI: https://doi.org/10.1038/s41467-020-18728-7.

Staal, A., Tuinenburg, O. A., Bosmans, J. H. C., et al. (2018). Forest-rainfall cascades buffer against drought across the Amazon. *Nature Climate Change*, 8, 539–543.

Stafford, F. (2016). *The Long, Long Life of Trees*. New Haven, CT: Yale University Press.

Starr, F., Jr. (1866). American forests; their destruction and preservation. In *Report of the Commissioner of Agriculture for the Year 1865*. Washington, DC: Government Printing Office, pp. 210–234.

Staudt, K., Serafimovich, A., Siebicke, L., Pyles, R. D., and Falge, E. (2011). Vertical structure of evapotranspiration at a forest site (a case study). *Agricultural and Forest Meteorology*, 151, 709–729.

Steffen, W., Persson, A., Deutsch, L., et al. (2011). The Anthropocene: From global change to planetary stewardship. *Ambio*, 40, 739–761.

Steffen, W., Richardson, K., Rockström, J., et al. (2020). The emergence and evolution of Earth System Science. *Nature Reviews: Earth and Environment*, 1, 54–63.

Steffen, W., Rockström, J., Richardson, K., et al. (2018). Trajectories of the Earth system in the Anthropocene. *Proceedings of the National Academy of Sciences USA*, 115, 8252–8259.

Stegner, W. (1990). It all began with conservation. *Smithsonian*, 21(1), 35–43.

Stehr, N., and von Storch, H. (2000). *Eduard Brückner: The Sources and Consequences of Climate Change and Climate Variability in Historical Times*. Dordrecht: Kluwer Academic Publishers.

Stein, W. E., Berry, C. M., Hernick, L. V., and Mannolini, F. (2012). Surprisingly complex community discovered in the mid-Devonian fossil forest at Gilboa. *Nature*, 483, 78–81.

Stein, W. E., Berry, C. M., Morris, J. L., et al. (2020). Mid-Devonian *Archaeopteris* roots signal revolutionary change in earliest fossil forests. *Current Biology*, 30, 421–431.e2.

Stephens, L., Fuller, D., Boivin, N., et al. (2019). Archaeological assessment reveals Earth's early transformation through land use. *Science*, 365, 897–902.

Still, C., Powell, R., Aubrecht, D., et al. (2019). Thermal imaging in plant and ecosystem ecology: Applications and challenges. *Ecosphere*, 10, e02768, DOI: https://doi.org/10.1002/ecs2.2768.

Still, C. J., Rastogi, B., Page, G. F. M., et al. (2021). Imaging canopy temperature: Shedding (thermal) light on ecosystem processes. *New Phytologist*, 230, 1746–1753.

Stoy, P. C., Katul, G. G., Siqueira, M. B., et al. (2006). Separating the effects of climate and vegetation on evapotranspiration along a successional chronosequence in the southeastern US. *Global Change Biology*, 12, 2115–2135.

Strassberg, R. E. (1994). *Inscribed Landscapes: Travel Writing from Imperial China*. Berkeley, University of California Press.

Strassburg, B. B. N., Iribarrem, A., Beyer, H. L., et al. (2020). Global priority areas for ecosystem restoration. *Nature*, 586, 724–729.

Stringer, C. (2016). The origin and evolution of *Homo sapiens*. *Philosophical Transactions of the Royal Society B*, 371, 20150237, DOI: https://doi.org/10.1098/rstb.2015.0237.

Stringer, C., and Galway-Witham, J. (2017). On the origin of our species. *Nature*, 546, 212–214.

Strutt, J. G. (1822). *Sylva Britannica; or, Portraits of Forest Trees, Distinguished for Their Antiquity, Magnitude, or Beauty*, folio ed. London: Colnaghi.

Strutt, J. G. (1830). *Sylva Britannica; or, Portraits of Forest Trees, Distinguished for Their Antiquity, Magnitude, or Beauty*. London: J. G. Strutt.

Stubbe, H. (1667). Observations made by a curious and learned person, sailing from England, to the Caribe-Islands. *Philosophical Transactions*, 2 (27), 494–502.

Stuenzi, S. M., and Schaepman-Strub, G. (2020). Vegetation trajectories and shortwave radiative forcing following boreal forest disturbance in eastern Siberia. *Journal of Geophysical Research: Biogeosciences*, 125, e2019JG005395, DOI: https://doi.org/10.1029/2019JG005395.

Suess, E. (1875). *Die Entstehung der Alpen*. Vienna: Wilhelm Braumüller.

Surell, A. (1841). *Étude sur les torrents des hautes-alpes*. Paris: Carilian-Goeury & V. Dalmont.

Swank, W. T., and Douglass, J. E. (1974). Streamflow greatly reduced by converting deciduous hardwood stands to pine. *Science*, 185, 857–859.

Swank, W. T., and Miner, N. H. (1968). Conversion of hardwood-covered watersheds to white pine reduces water yield. *Water Resources Research*, 4, 947–954.

Swank, W. T., Swift, L. W., Jr., and Douglass, J. E. (1988). Streamflow changes associated with forest cutting, species conversions, and natural disturbances. In *Forest Hydrology and Ecology at Coweeta*, edited by W. T. Swank and D. A. Crossley, Jr. New York: Springer-Verlag, pp. 297–312.

Swann, A. L., Fung, I. Y., Levis, S., Bonan, G. B., and Doney, S. C. (2010). Changes in Arctic vegetation amplify high-latitude warming through the greenhouse effect. *Proceedings of the National Academy of Sciences USA*, 107, 1295–1300.

Swann, A. L. S., Fung, I. Y., and Chiang, J. C. H. (2012). Mid-latitude afforestation shifts general circulation and tropical precipitation. *Proceedings of the National Academy of Sciences USA*, 109, 712–716.

Swann, A. L. S., Fung, I. Y., Liu, Y., and Chiang, J. C. H. (2014). Remote vegetation feedbacks and the mid-Holocene green Sahara. *Journal of Climate*, 27, 4857–4870.

Swann, A. L. S., Laguë, M. M., Garcia, E. S., et al. (2018). Continental-scale consequences of tree die-offs in North America: Identifying where forest loss matters most. *Environmental Research Letters*, 13, 055014, DOI: https://doi.org/10.1088/1748-9326/aaba0f.

Swann, A. L. S., Longo, M., Knox, R. G., Lee, E., and Moorcroft, P. R. (2015). Future deforestation in the Amazon and consequences for South American climate. *Agricultural and Forest Meteorology*, 214-215, 12–24.

Tagesson, T., Schurgers, G., Horion, S., et al. (2020). Recent divergence in the contributions of tropical and boreal forests to the terrestrial carbon sink. *Nature Ecology and Evolution*, 4, 202–209.

Tansley, A. G. (1935). The use and abuse of vegetational concepts and terms. *Ecology*, 16, 284–307.

Taylor, C. M., Parker, D. J., and Harris, P. P. (2007). An observational case study of mesoscale atmospheric circulations induced by soil moisture. *Geophysical Research Letters*, 34, L15801, DOI: https://doi.org/10.1029/2007GL030572.

Ter-Mikaelian, M. T., Colombo, S. J., and Chen, J. (2015). The burning question: Does forest bioenergy reduce carbon emissions? A review of common misconceptions about forest carbon accounting. *Journal of Forestry*, 113, 57–68.

Teuling, A. J. (2018). A forest evapotranspiration paradox investigated using lysimeter data. *Vadose Zone Journal*, 17, 170031, DOI: https://doi.org/10.2136/vzj2017.01.0031.

Teuling, A. J., Seneviratne, S. I., Stöckli, R., et al. (2010). Contrasting response of European forest and grassland energy exchange to heatwaves. *Nature Geoscience*, 3, 722–727.

Teuling, A. J., Taylor, C. M., Meirink, J. F., et al. (2017). Observational evidence for cloud cover enhancement over western European forests. *Nature Communications*, 8, 14065, DOI: https://doi.org/10.1038/ncomms14065.

Theophrastus (1990). *De causis plantarum*, vol. 3. *Books 5–6*, edited and translated by B. Einarson and G. K. K. Link (Loeb Classical Library 475). Cambridge, MA: Harvard University Press.

Thomas, C. K., Law, B. E., Irvine, J., et al. (2009). Seasonal hydrology explains interannual and seasonal variation in carbon and water exchange in a semiarid mature ponderosa pine forest in central Oregon. *Journal of Geophysical Research*, 114, G04006, DOI: https://doi.org/10.1029/2009JG001010.

Thomas, W. L., Jr. (1956). Introductory: About the symposium, about the people, about the theme. In *Man's Role in Changing the Face of the Earth*, edited by W. L. Thomas, Jr. Chicago: University of Chicago Press, pp. xxi–xxxviii.

Thompson, J. R., Carpenter, D. N., Cogbill, C. V., and Foster, D. R. (2013). Four centuries of change in northeastern United States forests. *PLoS ONE*, 8(9), e72540, DOI: https://doi.org/10.1371/journal.pone.0072540.

Thompson, K. (1980). Forests and climate change in America: Some early views. *Climatic Change*, 3, 47–64.

Thompson, K. (1981). The question of climatic stability in America before 1900. *Climatic Change*, 3, 227–241.

Thompson, M. P., Adams, D., and Sessions, J. (2009). Radiative forcing and the optimal rotation age. *Ecological Economics*, 68, 2713–2720.

Thompson, O. E., and Pinker, R. T. (1975). Wind and temperature profile characteristics in a tropical evergreen forest in Thailand. *Tellus*, 27, 562–573.

Thoreau, H. D. (1863a). The succession of forest trees. In *Excursions*. Boston: Ticknor & Fields, pp. 135–160.

Thoreau, H. D. (1863b). Autumnal tints. In *Excursions*. Boston: Ticknor & Fields, pp. 215–265.

Thoreau, H. D. (1906). *The Writings of Henry David Thoreau: Journal*, edited by B. Torrey, 14 vols. Boston: Houghton Mifflin.

Thoreau, H. D. (2009). *The Maine Woods: A Fully Annotated Edition*, edited by J. S. Cramer. New Haven, CT: Yale University Press.

Thornhill, G. D., Ryder, C. L., Highwood, E. J., Shaffrey, L. C., and Johnson, B. T. (2018). The effect of South American biomass burning aerosol emissions on the regional climate. *Atmospheric Chemistry and Physics*, 18, 5321–5342.

Thornthwaite, C. W. (1956). Modification of rural microclimates. In *Man's Role in Changing the Face of the Earth*, edited by W. L. Thomas, Jr. Chicago: University of Chicago Press, pp. 567–583.

Thornton, P. E., Doney, S. C., Lindsay, K., et al. (2009). Carbon–nitrogen interactions regulate climate–carbon cycle feedbacks: Results from an atmosphere–ocean general circulation model. *Biogeosciences*, 6, 2099–2120.

Thwaites, R. G. (1897a). *The Jesuit Relations and Allied Documents: Travels and Explorations of the Jesuit Missionaries in New France, 1610–1791*, vol. 3. *Acadia: 1611–1616*. Cleveland: Burrows.

Thwaites, R. G. (1897b). *The Jesuit Relations and Allied Documents: Travels and Explorations of the Jesuit Missionaries in New France, 1610–1791*, vol. 5. *Quebec: 1632–1633*. Cleveland: Burrows.

Tierney, J. E., Pausata, F. S. R., deMenocal, P. B. (2017). Rainfall regimes of the Green Sahara. *Science Advances*, 3, e1601503, DOI: https://doi.org/10.1126/sciadv.1601503.

Tilley, M. P. (1950). *A Dictionary of the Proverbs in England in the Sixteenth and Seventeenth Centuries*. Ann Arbor: University of Michigan Press.

Tilmann, J. P. (1971) *An Appraisal of the Geographical Works of Albertus Magnus and His Contributions to Geographical Thought* (Michigan Geographical Publication No. 4). Ann Arbor: Department of Geography, University of Michigan.

Todd, D., and Abbe, C. (1891). Additional results of the United States scientific expedition to West Africa. *Nature*, 43, 563–565.

Totman, C. (1989). *The Green Archipelago: Forestry in Preindustrial Japan*. Berkeley: University of California Press.

Travassos-Britto, B., Pardini, R., El-Hani, C. N., and Prado, P. I. (2021). Towards a pragmatic view of theories in ecology. *Oikos*, 130, 821–830.

Travers, W. T. L. (1870). On the changes effected in the natural features of a new country by the introduction of civilized races. Part III. *Transactions and Proceedings of the New Zealand Institute*, 3, 326–336.

Tristram, H. B. (1868). *The Natural History of the Bible: Being a Review of the Physical Geography, Geology, and Meteorology of the Holy Land*, 2nd ed. London: Society for Promoting Christian Knowledge.

Tunved, P., Hansson, H.-C., Kerminen, V.-M., et al. (2006). High natural aerosol loading over boreal forests. *Science*, 312, 261–263.

Turner, O. (1849). *Pioneer History of the Holland Purchase of Western New York*. Buffalo, NY: Jewett, Thomas; George H. Derby.

Twohy, C. H., Toohey, D. W., Levin, E. J. T., et al. (2021). Biomass burning smoke and its influence on clouds over the western U.S. *Geophysical Research Letters*, 48, e2021GL094224, DOI: https://doi.org/10.1029/2021GL094224.

Tyndall, J. (1863). On radiation through the earth's atmosphere. *The London, Edinburgh, and Dublin Philosophical Magazine and Journal of Science*, series 4, 25, 200–206.

Unger, N. (2013). Isoprene emission variability through the twentieth century. *Journal of Geophysical Research: Atmospheres*, 118, 13606–13613.

Unger, N. (2014). Human land-use-driven reduction of forest volatiles cools global climate. *Nature Climate Change*, 4, 907–910.

Uno, K. T., Polissar, P. J., Jackson, K. E., and deMenocal, P. B. (2016). Neogene biomarker record of vegetation change in eastern Africa. *Proceedings of the National Academy of Sciences USA*, 113, 6355–6363.

Urbanski, S., Barford, C., Wofsy, S., et al. (2007). Factors controlling CO_2 exchange on timescales from hourly to decadal at Harvard Forest. *Journal of Geophysical Research*, 112, G02020, DOI: https://doi.org/10.1029/2006JG000293.

Ure, A. (1821). *A Dictionary of Chemistry*. London: Thomas & George Underwood; J. Highley & Son; and others.

Van Bueren, G. (2015). More Magna than Magna Carta: Magna Carta's sister – the Charter of the Forest. In *Magna Carta and Its Modern Legacy*, edited by R. Hazell and J. Melton. New York: Cambridge University Press, pp. 194–211.

Van Cleve, K., and Viereck, L. A. (1981). Forest succession in relation to nutrient cycling in the boreal forest of Alaska. In *Forest Succession: Concepts and Application*, edited by D. C. West, H. H. Shugart, and D. B. Botkin. New York: Springer-Verlag, pp. 185–211.

Van Cleve, K., Chapin, F. S., III, Flanagan, P. W., Viereck, L. A., and Dyrness, C. T. (1986). *Forest Ecosystems in the Alaskan Taiga: A Synthesis of Structure and Function*. New York: Springer-Verlag.

Van Cleve, K., Dyrness, C. T., Viereck, L. A., et al. (1983). Taiga ecosystems in interior Alaska. *BioScience*, 33, 39–44.

van der Bles, A. M., van der Linden, S., Freeman, A. L. J., and Spiegelhalter, D. J. (2020). The effects of communicating uncertainty on public trust in facts and numbers. *Proceedings of the National Academy of Sciences USA*, 117, 7672–7683.

Vanderhoof, M. K., and Williams, C. A. (2015). Persistence of MODIS evapotranspiration impacts from mountain pine beetle outbreaks in lodgepole pine forests, south-central Rocky Mountains. *Agricultural and Forest Meteorology*, 200, 78–91.

Vanderhoof, M., Williams, C. A., Ghimire, B., and Rogan, J. (2013). Impact of mountain pine beetle outbreaks on forest albedo and radiative forcing, as derived from Moderate Resolution Imaging Spectroradiometer, Rocky Mountains, USA. *Journal of Geophysical Research: Biogeosciences*, 118, 1461–1471.

Vanderhoof, M., Williams, C. A., Shuai, Y., et al. (2014). Albedo-induced radiative forcing from mountain pine beetle outbreaks in forests, south-central Rocky Mountains: Magnitude, persistence, and relation to outbreak severity. *Biogeosciences*, 11, 563–575.

van der Werf, G. R., Randerson, J. T., Giglio, L., et al. (2017). Global fire emissions estimates during 1997–2016. *Earth System Science Data*, 9, 697–720.

Vargas Zeppetello, L. R., Cook-Patton, S. C., Parsons, L. A., et al. (2022). Consistent cooling benefits of silvopasture in the tropics. *Nature Communications*, 13, 708, DOI: https://doi.org/10.1038/s41467-022-28388-4.

Vargas Zeppetello, L. R., Parsons, L. A., Spector, J. T., et al. (2020). Large scale tropical deforestation drives extreme warming. *Environmental Research Letters*, 15, 084012, DOI: https://doi.org/10.1088/1748-9326/ab96d2.

Venäläinen, A., Lehtonen, I., Laapas, M., et al. (2020). Climate change induces multiple risks to boreal forests and forestry in Finland: A literature review. *Global Change Biology*, 26, 4178–4196.

Verkerk, P. J., Costanza, R., Hetemäki, L., et al. (2020). Climate-Smart Forestry: The missing link. *Forest Policy and Economics*, 115, 102164, DOI: https://doi.org/10.1016/j.forpol.2020.102164.

Vernadsky, V. I. (1998). *The Biosphere*, forward by L. Margulis and colleagues; introduction by J. Grinevald; translated by D. B. Langmuir; revised and annotated by M. A. S. McMenamin. New York: Springer.

Vincent, P. (1637). *A True Relation of the Late Battell Fought in New England, between the English, and the Salvages: With the Present State of Things There*. London: Nathanael Butter & John Bellamie (Text Creation Partnership, Ann Arbor, Michigan; http://name.umdl.umich.edu/A14439.0001.001).

Vitruvius (1914). *The Ten Books on Architecture*, translated by M. H. Morgan. Cambridge, MA: Harvard University Press.

Voeikov, A. (1885a). Der Einfluss der Wälder auf das Klima. *Petermanns geographische Mitteilungen*, 31, 81–87.

Voeikov, A. (1885b). Die Regenverhältnisse des malayischen Archipels. *Zeitschrift der österreichischen Gesellschaft für Meteorologie*, 20, 113–138, 201–211, 250–263.

Voeikov, A. (1886). On the influence of forests upon climate. *Quarterly Journal of the Royal Meteorological Society*, 12, 26–38.

Voeikov, A. (1887). *Die Klimate der Erde*, 2 vols. Jena, Germany: Hermann Costenoble.

Voeikov, A. (1878). Einfluss der Wälder und der Irrigation auf das Klima. *Zeitschrift der Österreichischen Gesellschaft für Meteorologie*, 13, 47–48.

Voeikov, A. (1888). Klimatologische Zeit- und Streitfragen. *Meteorologische Zeitschrift*, 5, 17–21, 191–195.

Voeikov, A. (1901). De l'influence de l'homme sur la terre. *Annales de géographie*, 10, 97–114, 193–215.

Vogel, B. (2011). The letter from Dublin: Climate change, colonialism, and the Royal Society in the seventeenth century. *Osiris*, 26, 111–128.

Volney, C.-F. (1804). *View of the Climate and Soil of the United States of America*. London: J. Johnson.

von Arx, G., Dobbertin, M., and Rebetez, M. (2012). Spatio-temporal effects of forest canopy on understory microclimate in a long-term experiment in Switzerland. *Agricultural and Forest Meteorology*, 166–167, 144–55.

von Arx, G., Pannatier, E. G., Thimonier, A., and Rebetez, M. (2013). Microclimate in forests with varying leaf area index and soil moisture: Potential implications for seedling establishment in a changing climate. *Journal of Ecology*, 101, 1201–1213.

von Randow, C., Manzi, A. O., Kruijt, B., et al. (2004). Comparative measurements and seasonal variations in energy and carbon exchange over forest and pasture in South West Amazonia. *Theoretical and Applied Climatology*, 78, 5–26.

Walker, A. P., De Kauwe, M. G., Bastos, A., et al. (2021). Integrating the evidence for a terrestrial carbon sink caused by increasing atmospheric CO_2. *New Phytologist*, 229, 2413–2445.

Walker, X. J., Baltzer, J. L., Cumming, S. G., et al. (2019). Increasing wildfires threaten historic carbon sink of boreal forest soils. *Nature*, 572, 520–523.

Wang, G., and Eltahir, E. A. B. (2000a). Biosphere–atmosphere interactions over West Africa. II: Multiple climate equilibria. *Quarterly Journal of the Royal Meteorological Society*, 126, 1261–1280.

Wang, G., and Eltahir, E. A. B. (2000b). Ecosystem dynamics and the Sahel drought. *Geophysical Research Letters*, 27, 795–798.

Wang, G., and Eltahir, E. A. B. (2000c). Role of vegetation dynamics in enhancing the low-frequency variability of the Sahel rainfall. *Water Resources Research*, 36, 1013–1021.

Wang, G., Eltahir, E. A. B., Foley, J. A., Pollard, D., and Levis, S. (2004). Decadal variability of rainfall in the Sahel: Results from the coupled GENESIS–IBIS atmosphere–biosphere model. *Climate Dynamics*, 22, 625–637.

Wang, G., Yu, M., and Xue, Y. (2016). Modeling the potential contribution of land cover changes to the late twentieth century Sahel drought using a regional climate model: Impact of lateral boundary conditions. *Climate Dynamics*, 47, 3457–3477.

Wang, J., Chagnon, F. J. F., Williams, E. R., et al. (2009). Impact of deforestation in the Amazon basin on cloud climatology. *Proceedings of the National Academy of Sciences USA*, 106, 3670–3674.

Wang, J. A., Baccini, A., Farina, M., Randerson, J. T., and Friedl, M. A. (2021). Disturbance suppresses the aboveground carbon sink in North American boreal forests. *Nature Climate Change*, 11, 435–441.

Warde, P. (2006). Fear of wood shortage and the reality of the woodland in Europe, c.1450–1850. *History Workshop Journal*, 62 (Autumn), 28–57.

Warde, P. (2018). *The Invention of Sustainability: Nature and Destiny, c. 1500–1870*. Cambridge, UK: Cambridge University Press.

Watson, J. E. M., Evans, T., Venter, O., et al. (2018). The exceptional value of intact forest ecosystems. *Nature Ecology and Evolution*, 2, 599–610.

Watson, W. C. (1866). Forests – their influence, uses and reproduction. In *Transactions of the New York State Agricultural Society for the Year 1865*, vol. 25. Albany, NY: Cornelius Wendell, pp. 288–303.

Watts, F. (1872). *Report of the Commissioner of Agriculture for the Year 1871*. Washington, DC: Government Printing Office.

Watts, F. (1874). *Report of the Commissioner of Agriculture for the Year 1872*. Washington, DC: Government Printing Office.

Webb, T., III, Bartlein, P. J., Harrison, S. P., and Anderson, K. H. (1993). Vegetation, lake levels, and climate in eastern North America for the past 18,000 years. In *Global Climates since the Last Glacial Maximum*, edited by H. E. Wright, Jr.,

J. E. Kutzbach, T. Webb, III, et al. Minneapolis: University of Minnesota Press, pp. 415–467.

Webster, N. (1790). *A Collection of Essays and Fugitiv Writings: On Moral, Historical, Political and Literary Subjects*. Boston: I. Thomas & E. T. Andrews.

Webster, N. (1799). On the effects of evergreens on climate. *Transactions of the Society for the Promotion of Agriculture, Arts, and Manufactures, Instituted in the State of New York*, 1(4), 51–52.

Webster, N. (1809). Experiments respecting dew, intended to ascertain whether dew is the descent of vapour during the night, or the perspiration of the earth, or of plants; or whether it is not the effect of condensation. *Memoirs of the American Academy of Arts and Sciences*, 3(1), 95–103.

Webster, N. (1810). A dissertation on the supposed change in the temperature of winter. *Memoirs of the Connecticut Academy of Arts and Sciences*, 1 (1), 1–68.

Webster, W. H. B. (1834). *Narrative of a Voyage to the Southern Atlantic Ocean, in the Years 1828, 29, 30, Performed in H. M. Sloop Chanticleer*, 2 vols. London: Richard Bentley.

Weld, I. (1799). *Travels through the States of North America, and the Provinces of Upper and Lower Canada, during the Years 1795, 1796, and 1797*. London: John Stockdale.

Weppelmann, S. (2019). Introduction: Bruegel between 2019 and 2069. In *Bruegel: The Hand of the Master. The 450th Anniversary Edition – Essays in Context*, edited by A. Hoppe-Harnoncourt, E. Oberthaler, S. Pénot, M. Sellink, and R. Spronk. Vienna: KHM-Museumsverband, pp. 8–9.

Werth, D., and Avissar, R. (2002). The local and global effects of Amazon deforestation. *Journal of Geophysical Research*, 107, 8087, DOI: https://doi.org/10.1029/2001JD000717.

West, D. C., Shugart, H. H., and Botkin, D. B. (1981). *Forest Succession: Concepts and Application*. New York: Springer-Verlag.

Wex, G. (1873). Ueber die Wasserabnahme in den Quellen, Flüssen und Strömen. *Zeitschrift des oesterreichischen Ingenieur- und Architekten-Vereins*, 25, 23–30, 63–76, 101–119.

Wex, G. (1880). *Second Treatise on the Decrease of Water in Springs, Creeks, and Rivers,*

Contemporaneously with an Increase in Height of Floods in Cultivated Countries, translated by G. Weitzel. Washington, DC: Government Printing Office.

Wex, G. (1881). *First Treatise on the Decrease of Water in Springs, Creeks, and Rivers, Contemporaneously with an Increase in Height of Floods in Cultivated Countries*, translated by G. Weitzel. Washington, DC: Government Printing Office.

Whitbourne, R. (1620). *A Discourse and Discovery of New-Found-Land*. London: William Barret.

White, G. (1789). *The Natural History and Antiquities of Selborne, in the County of Southampton*. London: T. Bensley.

Whitney, G. G. (1994). *From Coastal Wilderness to Fruited Plain: A History of Environmental Change in Temperate North America, 1500 to the Present*. Cambridge, UK: Cambridge University Press.

Whittaker, R. H. (1956). Vegetation of the Great Smoky Mountains. *Ecological Monographs*, 26, 1–80.

Whittaker, R. H. (1975). *Communities and Ecosystems*, 2nd ed. New York: MacMillan.

Whittaker, R. H., Bormann, F. H., Likens, G. E., and Siccama, T. G. (1974). The Hubbard Brook Ecosystem Study: Forest biomass and production. *Ecological Monographs*, 44, 233–252.

Wieder, W. R., Cleveland, C. C., Smith, W. K., and Todd-Brown, K. (2015). Future productivity and carbon storage limited by terrestrial nutrient availability. *Nature Geoscience*, 8, 441–444.

Wilber, C. D. (1881). *The Great Valleys and Prairies of Nebraska and the Northwest*. Omaha: Daily Republican Print.

Williams, A. P., Abatzoglou, J. T., Gershunov, A., et al. (2019). Observed impacts of anthropogenic climate change on wildfire in California. *Earth's Future*, 7, 892–910.

Williams, C. A., Gu, H., and Jiao, T. (2021). Climate impacts of US forest loss span net warming to net cooling. *Science Advances*, 7, eaax8859, DOI: https://doi.org/10.1126/sciadv.aax8859.

Williams, C. A., Reichstein, M., Buchmann, N., et al. (2012). Climate and vegetation controls on the surface water balance: Synthesis of evapotranspiration measured across a global network of flux towers. *Water Resources Research*, 48, W06523, DOI: https://doi.org/10.1029/2011WR011586.

Williams, C. J., Johnson, A. H., LePage, B. A., Vann, D. R., and Sweda, T. (2003). Reconstruction of Tertiary *Metasequoia* forests. II. Structure, biomass, and productivity of Eocene floodplain forests in the Canadian Arctic. *Paleobiology*, 29, 271–292.

Williams, E., Rosenfeld, D., Madden, N., et al. (2002). Contrasting convective regimes over the Amazon: Implications for cloud electrification. *Journal of Geophysical Research*, 107D, 8082, DOI: https://doi.org/10.1029/2001JD000380.

Williams, J. W., Shuman, B. N., Webb, T., III, Bartlein, P. J., and Leduc, P. L. (2004). Late-Quaternary vegetation dynamics in North America: Scaling from taxa to biomes. *Ecological Monographs*, 74, 309–334.

Williams, M. (1989). *Americans and Their Forests: A Historical Geography*. Cambridge, UK: Cambridge University Press.

Williams, M. (2003). *Deforesting the Earth: From Prehistory to Global Crisis*. Chicago: University of Chicago Press.

Williams, S. (1786). Experiments on evaporation, and meteorological observations made at Bradford in New-England, in 1772. *Transactions of the American Philosophical Society*, 2, 118–141.

Williams, S. (1794). *The Natural and Civil History of Vermont*. Walpole, NH: Isaiah Thomas & David Carlisle.

Williamson, H. (1771). An attempt to account for the change of climate, which has been observed in the middle colonies in North-America. *Transactions of the American Philosophical Society*, 1, 272–280.

Williamson, H. (1773). Dans lequel on tâche de rendre raison du changement de climat qu'on a observé dans les colonies situées dans l'intérieur des terres de l'Amérique septentrionale. *Observations sur la physique, sur l'histoire naturelle et sur les arts*, 1, 430–436.

Williamson, H. (1811). *Observations on the Climate in Different Parts of America, Compared with the Climate in Corresponding Parts of the Other Continent*. New York: T. & J. Swords.

Willis, E. P., and Hooke, W. H. (2006). Cleveland Abbe and American meteorology, 1871–1901.

Bulletin of the American Meteorological Society,
87, 315–326.

Wilson, H. M. (1898). The relation of forestation to
water-supply. *The Engineering Magazine* 14(5),
807–816.

Wilson, J. F. (1865a). Water supply in the basin of the
River Orange, or 'Gariep, South Africa. *Journal of
the Royal Geographical Society,* 35, 106–129.

Wilson, J. F. (1865b). On the progressing desiccation
of the basin of the Orange River in Southern
Africa. *Proceedings of the Royal Geographical
Society,* 9, 106–109.

Wilson, J. P., Montañez, I. P., White, J. D., et al.
(2017). Dynamic Carboniferous tropical forests:
New views of plant function and potential for
physiological forcing of climate. *New Phytologist,*
215, 1333–1353.

Wilson, J. S. (1867). *Report of the Commissioner of
General Land Office, for the Year 1867.*
Washington, DC: Government Printing Office.

Wilson, J. S. (1868). *Report of the Commissioner of
General Land Office for the Year 1868.*
Washington, DC: Government Printing Office.

Winckler, J., Lejeune, Q., Reick, C. H., and
Pongratz, J. (2019a). Nonlocal effects dominate
the global mean surface temperature response to
the biogeophysical effects of deforestation.
Geophysical Research Letters, 46, 745–755.

Winckler, J., Reick, C. H., Bright, R. M., and
Pongratz, J. (2019b). Importance of surface
roughness for the local biogeophysical effects of
deforestation. *Journal of Geophysical Research:
Atmospheres,* 124, 8605–8618.

Winckler, J., Reick, C. H., Luyssaert, S., et al.
(2019c). Different response of surface temperature
and air temperature to deforestation in climate
models. *Earth System Dynamics,* 10, 473–484.

Windisch, M. G., Davin, E. L., and Seneviratne, S. I.
(2021). Prioritizing forestation based on
biogeochemical and local biogeophysical impacts.
Nature Climate Change, 11, 867–871.

Windisch, M. G., Humpenöder, F., Lejeune, Q., et al.
(2022). Accounting for local temperature effect
substantially alters afforestation patterns.
Environmental Research Letters, 17, 024030,
DOI: https://doi.org/10.1088/1748-9326/ac4f0e.

Wood, T. (1875). Should the farmers of America
oppose further the destruction of our forests? In

*Report of the Transactions of the Pennsylvania
State Agricultural Society, for the Years 1874–75,*
vol. 10. Harrisburg: B. F. Meyers, pp. 153–155.

Woodward, J. (1699). Some thoughts and
experiments concerning vegetation. *Philosophical
Transactions,* 21(253), 193–227.

Worden, S., Fu, R., Chakraborty, S., Liu, J., and
Worden, J. (2021). Where does moisture come
from over the Congo Basin? *Journal of
Geophysical Research: Biogeosciences,* 126,
e2020JG006024, DOI: https://doi.org/10.1029/
2020JG006024.

Wright, I. J., Reich, P. B., Westoby, M., et al. (2004).
The worldwide leaf economics spectrum. *Nature,*
428, 821–827.

Wright, J. S., Fu, R., Worden, J. R., et al. (2017).
Rainforest-initiated wet season onset over the
southern Amazon. *Proceedings of the National
Academy of Sciences USA,* 114, 8481–8486.

Wulf, A. (2015). *The Invention of Nature: Alexander
von Humboldt's New World.* New York: Alfred
A. Knopf.

Xu, L., Saatchi, S. S., Yang, Y., et al. (2021). Changes
in global terrestrial live biomass over the 21st
century. *Science Advances,* 7, eabe9829, DOI:
https://doi.org/10.1126/sciadv.abe9829.

Xu, X., Riley, W. J., Koven, C. D., Jia, G., and
Zhang, X. (2020). Earlier leaf-out warms air in the
north. *Nature Climate Change,* 10, 370–375.

Xue, Y. (2006). Interactions and feedbacks between
climate and dryland vegetations. In *Dryland
Ecohydrology,* edited by P. D'Odorico and
A. Porporato. Dordrecht: Springer, pp. 85–105.

Xue, Y., and Shukla, J. (1993). The influence of
land surface properties on Sahel climate. Part I:
Desertification. *Journal of Climate,* 6,
2232–2245.

Xue, Y., and Shukla, J. (1996). The influence of land
surface properties on Sahel climate. Part II:
Afforestation. *Journal of Climate,* 9, 3260–3275.

Xue, Y., Hutjes, R. W. A., Harding, R. J., et al.
(2004). The Sahelian climate. In *Vegetation,
Water, Humans and the Climate: A New
Perspective on an Interactive System,* edited by
P. Kabat, M. Claussen, P. A. Dirmeyer, et al.
Berlin: Springer, pp. 59–77.

Yang, Y., Saatchi, S. S., Xu, L., et al. (2018). Post-
drought decline of the Amazon carbon sink.

Nature Communications, 9, 3172, DOI: https://doi.org/10.1038/s41467-018-05668-6.

Yli-Juuti, T., Mielonen, T., Heikkinen, L., et al. (2021). Significance of the organic aerosol driven climate feedback in the boreal area. *Nature Communications*, 12, 5637, DOI: https://doi.org/10.1038/s41467-021-25850-7.

Yosef, G., Walko, R., Avisar, R., et al. (2018). Large-scale semi-arid afforestation can enhance precipitation and carbon sequestration potential. *Scientific Reports*, 8, 996, DOI: https://doi.org/10.1038/s41598-018-19265-6.

Young, C. R. (1979). *The Royal Forests of Medieval England*. Philadelphia: University of Pennsylvania Press.

Yousefpour, R., Augustynczik, A. L. D., Reyer, C. P. O., et al. (2018). Realizing mitigation efficiency of European commercial forests by climate smart forestry. *Scientific Reports*, 8, 345, DOI: https://doi.org/10.1038/s41598-017-18778-w.

Yu, L., Liu, Y., Liu, T., and Yan, F. (2020). Impact of recent vegetation greening on temperature and precipitation over China. *Agricultural and Forest Meteorology*, 295, 108197, DOI: https://doi.org/10.1016/j.agrformet.2020.108197.

Zaehle, S., Friend, A. D., Friedlingstein, P., et al. (2010). Carbon and nitrogen cycle dynamics in the O-CN land surface model: 2. Role of the nitrogen cycle in the historical terrestrial carbon balance. *Global Biogeochemical Cycles*, 24, GB1006, DOI: https://doi.org/10.1029/2009GB003522.

Zaehle, S., Medlyn, B. E., De Kauwe, M. G., et al. (2014). Evaluation of 11 terrestrial carbon–nitrogen cycle models against observations from two temperate Free-Air CO_2 Enrichment studies. *New Phytologist*, 202, 803–822.

Zaitchik, B. F., Macalady, A. K., Bonneau, L. R., and Smith, R. B. (2006). Europe's 2003 heat wave: A satellite view of impacts and land-atmosphere feedbacks. *International Journal of Climatology*, 26, 743–769.

Zarakas, C. M., Swann, A. L. S., Laguë, M. M., Armour, K. C., and Randerson, J. T. (2020). Plant physiology increases the magnitude and spread of the transient climate response to CO_2 in CMIP6 earth system models. *Journal of Climate*, 33, 8561–8577.

Zemp, D. C., Schleussner, C.-F., Barbosa, H. M. J., et al. (2017). Self-amplified Amazon forest loss due to vegetation-atmosphere feedbacks. *Nature Communications*, 8, 14681, DOI: https://doi.org/10.1038/ncomms14681.

Zeng, N., and Yoon, J. (2009). Expansion of the world's deserts due to vegetation-albedo feedback under global warming. *Geophysical Research Letters*, 36, L17401, DOI: https://doi.org/10.1029/2009GL039699.

Zeng, N., Neelin, J. D., Lau, K.-M., and Tucker, C. J. (1999). Enhancement of interdecadal climate variability in the Sahel by vegetation interaction. *Science*, 286, 1537–1540.

Zeng, Z., Piao, S., Li, L. Z. X., et al. (2017). Climate mitigation from vegetation biophysical feedbacks during the past three decades. *Nature Climate Change*, 7, 432–436.

Zhang, L., Dawes, W. R., and Walker, G. R. (2001). Response of mean annual evapotranspiration to vegetation changes at catchment scale. *Water Resources Research*, 37, 701–708.

Zhang, M., Lee, X., Yu, G., et al. (2014). Response of surface air temperature to small-scale land clearing across latitudes. *Environmental Research Letters*, 9, 034002, DOI: https://doi.org/10.1088/1748-9326/9/3/034002.

Zhang, M., Liu, N., Harper, R., et al. (2017). A global review on hydrological responses to forest change across multiple spatial scales: Importance of scale, climate, forest type and hydrological regime. *Journal of Hydrology*, 546, 44–59.

Zhang, Q., Barnes, M., Benson, M., et al. (2020a). Reforestation and surface cooling in temperate zones: Mechanisms and implications. *Global Change Biology*, 26, 3384–3401.

Zhang, X., Du, J., Zhang, L., et al. (2020b). Impact of afforestation on surface ozone in the North China Plain during the three-decade period. *Agricultural and Forest Meteorology*, 287, 107979, DOI: https://doi.org/10.1016/j.agrformet.2020.107979.

Zhang, X., Huang, T., Zhang, L., et al. (2016a). Three-North Shelter Forest Program contribution to long-term increasing trends of biogenic isoprene emissions in northern China. *Atmospheric Chemistry and Physics*, 16, 6949–6960.

Zhang, Y., Peng, C., Li, W., et al. (2016b). Multiple afforestation programs accelerate the greenness in the "Three North" region of China from 1982 to 2013. *Ecological Indicators*, 61, 404–412.

Zhang, Z., Li, X., and Liu, H. (2022). Biophysical feedback of forest canopy height on land surface temperature over contiguous United States. *Environmental Research Letters*, 17, 034002, DOI: https://doi.org/10.1088/1748-9326/ac4657.

Zhang, Z., Zhang, F., Wang, L., Lin, A., and Zhao, L. (2021). Biophysical climate impact of forests with different age classes in mid- and high-latitude North America. *Forest Ecology and Management*, 494, 119327, DOI: https://doi.org/10.1016/j.foreco.2021.119327.

Zhao, K., and Jackson, R. B. (2014). Biophysical forcings of land-use changes from potential forestry activities in North America. *Ecological Monographs*, 84, 329–353.

Zhu, Z., Piao, S., Myneni, R. B., et al. (2016). Greening of the Earth and its drivers. *Nature Climate Change*, 6, 791–795.

Zhuang, Y., Fu, R., Santer, B. D., Dickinson, R. E., and Hall, A. (2021). Quantifying contributions of natural variability and anthropogenic forcings on increased fire weather risk over the western United States. *Proceedings of the National Academy of Sciences USA*, 118, e2111875118, DOI: https://doi.org/10.1073/pnas.2111875118.

Zilberstein, A. (2016). *A Temperate Empire: Making Climate Change in Early America*. Oxford: Oxford University Press.

Zipes, J. (1988). *The Brothers Grimm: From Enchanted Forests to the Modern World*. New York: Routledge.

Zon, R. (1912). Forests and water in the light of scientific investigation. In *Final Report of the National Waterways Commission*. Washington, DC: Government Printing Office, pp. 203–302.

Zon, R. (1913). The relation of forests in the Atlantic Plain to the humidity of the central states and prairie region. *Proceedings of the Society of American Foresters*, 8, 139–153.

Zötl, G. von (1831). *Handbuch der Forstwirthschaft im Hochgebirge*. Vienna: Carl Gerold.

Index

atmospheric measurements, 110, 159
Boussingault, 64
climate feedback, 164–165, 172, 180
CO_2 fertilization, 163–164
Ébelmen, 65
preindustrial, 161
terrestrial carbon sink, 163–165
cedar of Lebanon, 75, 95
Cézanne, Ernest, 57
Charney, Jule, 197
Chekhov, Anton, 45
China
afforestation, 179, 196
historical deforestation, 85
Chrétien de Troyes, 93
Church, Frederic Edwin, 86, 89, 215
Clarke, William Branwhite, 41
Clavé, Jules, 47, 57, 65, 83, 184
Clayton, John, 27
Cleghorn, Hugh, 40
climate change
continental drift, 109
glacial cycles, 108
greenhouse gases, 110–112
solar radiation, 109
thermohaline circulation, 109
climate model, 78, 111, 136–138, 168,
170, 177, 183, 187
deforestation, 174, 185, 186, 190, 194
desert world, 169
dryland degradation, 198
grassland world, 173–174
climate service index, 181, 205–206
climate zones, 103–104, 105–106
climate-smart forestry, 207–210
rotation cycle, 208, 209, 211–212
cloud forest, 151
Cole, Thomas, 85, 88
Columbus, Christopher, 26, 27, 33
conduction, 116
convection, 116
Cooper, Ellwood, 48
Cotta, Heinrich, 57
Coweeta Hydrologic Laboratory, 154
Cowthorpe oak, 86
Curtis, George, 71
Curtis, John, 77

Dalzell, Nicol, 40
Darwin, Charles, 91
Denes, Agnes, 90
Denys, Nicolas, 14

Disney, Walt, 91
Douglas, William O., 95
Dove, Heinrich, 58, 188
Drori, Jonathan, 92
Dubos, Jean-Baptiste, 16
Duhamel du Monceau, Henri-Louis, 14,
85
Dunbar, James, 16
Dunbar, William, 20
Dust Bowl, 179
Dwight, Timothy, 18, 21, 34

Earth system model, 80, 111, 136, 137,
168
Earth system science, 6, 8, 80, 225
Lamarck, 54
Ébelmen, Jacques-Joseph, 64
Ebermayer, Ernst, 59–60, 61, 63, 67, 139,
145
mesoscale circulations, 177
oxygen, 65
roots, 59
US government reports, 47, 48, 71
water vapor, 65
ecosystem, definition, 11
eddy covariance, 56, 128–131
evapotranspiration, 152, 155–156
forest air temperature, 148, 176
surface fluxes, 121, 179
surface temperature, 176, 194
El Hierro, rainmaking tree, 26, 36
El Niño, 108, 223
Elliott, Richard, 49
emerald planet, 6, 224
Emerson, Ralph Waldo, 65
energy balance
forest, 120–126
land surface, 114–117
leaf, 117–120
planetary, 101–102
eucalyptus, 41, 48
Europe
forest evapotranspiration, 155, 156,
179
forest influences on climate, 194, 210,
211
forest microclimate, 145
heatwaves, 179
paleovegetation, 220
evaporation, 16, 27, 105, 116, 117, 151.
See also evapotranspiration
Aughey, 49

Boussingault, 35
Brown, 43
Clavé, 57
Curtis, 72
Ebermayer, 59
Harrington, 71
Humboldt, 34
leaf area, 124, 152, 172, 180, 195
Lorenz, 61
Mathieu, 56
Richardson, 77
Sargent, 72
Williams, 34
Zon, 74
evapotranspiration, 27, 105, 120, 124,
151, 152, 172, 175, 205
atmospheric boundary layer, 121, 125,
170, 175
biogeography, 214
deforestation, 154, 170, 173, 185, 188
drought, 50, 155, 179, 198, 199, 210
eddy covariance, 130, 152, 155–156
effect on climate, 78, 169, 170, 172,
173, 174, 194
forest, 152, 154, 155–156, 186, 191
Gannett, 149
Harrington, 71
heatwaves, 179, 210
Hubbard Brook, 134, 154
lysimeter, 156
mesoscale circulations, 177
mountain pine beetle, 195
precipitation recycling, 185
remote sensing, 136
roots, 152, 155, 172, 173, 179, 186, 210
surface temperature, 148, 194, 196,
206, 210
Voeikov, 45
water balance, 133, 151
Zon, 73
Evelyn, John, 27, 29, 31, 82, 84, 87, 92,
118, 229

Fabre, Jean-Antoine, 34, 44
fairy tales, 92, 93
Fernández de Oviedo y Valdés, Gonzalo,
26
Fernow, Bernhard, 70, 73, 76
Forest Influences, 70–72, 76
Philosophical Society of Washington, 3
Ferrel cell, 103